CAMBRIDGE STUDIES IN
ADVANCED MATHEMATICS 78

Completely Bounded Maps and Operator Algebras

In this book the reader is provided with a tour of the principal results and ideas in the theories of completely positive maps, completely bounded maps, dilation theory, operator spaces, and operator algebras, together with some of their main applications.

The author assumes only that the reader has a basic background in functional analysis and C^*-algebras, and the presentation is self-contained and paced appropriately for graduate students new to the subject. The book could be used as a text for a course or for independent reading; with this in mind, many exercises are included. Experts will also want this book for their library, since the author presents new and simpler proofs of some of the major results in the area, and many applications are also included.

This will be an indispensable introduction to the theory of operator spaces for all who want to know more.

Already published

2 K. Petersen *Ergodic theory*
3 P.T. Johnstone *Stone spaces*
5 J.-P. Kahane *Some random series of functions, 2nd edition*
7 J. Lambek & P.J. Scott *Introduction to higher-order categorical logic*
8 H. Matsumura *Commutative ring theory*
9 C.B. Thomas *Characteristic classes and the cohomology of finite groups*
10 M. Aschbacher *Finite group theory*
11 J.L. Alperin *Local representation theory*
12 P. Koosis *The logarithmic integral I*
14 S.J. Patterson *An introduction to the theory of the Riemann zeta-function*
15 H.J. Baues *Algebraic homotopy*
16 V.S. Varadarajan *Introduction to harmonic analysis on semisimple Lie groups*
17 W. Dicks & M. Dunwoody *Groups acting on graphs*
18 L.J. Corwin & F.P. Greenleaf *Representations of nilpotent Lie groups and their applications*
19 R. Fritsch & R. Piccinini *Cellular structures in topology*
20 H. Klingen *Introductory lectures on Siegel modular forms*
21 P. Koosis *The logarithmic integral II*
22 M.J. Collins *Representations and characters of finite groups*
24 H. Kunita *Stochastic flows and stochastic differential equations*
25 P. Wojtaszczyk *Banach spaces for analysts*
26 J.E. Gilbert & M.A.M. Murray *Clifford algebras and Dirac operators in harmonic analysis*
27 A. Frohlich & M.J. Taylor *Algebraic number theory*
28 K. Goebel & W.A. Kirk *Topics in metric fixed point theory*
29 J.F. Humphreys *Reflection groups and Coxeter groups*
30 D.J. Benson *Representations and cohomology I*
31 D.J. Benson *Representations and cohomology II*
32 C. Allday & V. Puppe *Cohomological methods in transformation groups*
33 C. Soule et al. *Lectures on Arakelov geometry*
34 A. Ambrosetti & G. Prodi *A primer of nonlinear analysis*
35 J. Palis & F. Takens *Hyperbolicity, stability and chaos at homoclinic bifurcations*
37 Y. Meyer *Wavelets and operators 1*
38 C. Weibel *An introduction to homological algebra*
39 W. Bruns & J. Herzog *Cohen-Macaulay rings*
40 V. Snaith *Explicit Brauer induction*
41 G. Laumon *Cohomology of Drinfeld modular varieties I*
42 E.B. Davies *Spectral theory and differential operators*
43 J. Diestel, H. Jarchow, & A. Tonge *Absolutely summing operators*
44 P. Mattila *Geometry of sets and measures in Euclidean spaces*
45 R. Pinsky *Positive harmonic functions and diffusion*
46 G. Tenenbaum *Introduction to analytic and probabilistic number theory*
47 C. Peskine *An algebraic introduction to complex projective geometry*
48 Y. Meyer & R. Coifman *Wavelets*
49 R. Stanley *Enumerative combinatorics I*
50 I. Porteous *Clifford algebras and the classical groups*
51 M. Audin *Spinning tops*
52 V. Jurdjevic *Geometric control theory*
53 H. Volklein *Groups as Galois groups*
54 J. Le Potier *Lectures on vector bundles*
55 D. Bump *Automorphic forms and representations*
56 G. Laumon *Cohomology of Drinfeld modular varieties II*
57 D. M. Clark & B. A. Davey *Natural dualities for the working algebraist*
58 J. McCleary *A user's guide to spectral sequences II*
59 P. Taylor *Practical foundations of mathematics*
60 M.P. Brodmann & R.Y. Sharp *Local cohomology*
61 J.D. Dixon et al. *Analytic pro-P groups*
62 R. Stanley *Enumerative combinatorics II*
63 R. M. Dudley *Uniform central limit theorems*
64 J. Jost & X. Li-Jost *Calculus of variations*
65 A.J. Berrick & M.E. Keating *An introduction to rings and modules*
66 S. Morosawa *Holomorphic dynamics*
67 A.J. Berrick & M.E. Keating *Categories and modules with K-theory in view*
68 K. Sato *Levy processes and infinitely divisible distributions*
69 H. Hida *Modular forms and Galois cohomology*
70 R. Iorio & V. Iorio *Fourier analysis and partial differential equations*
71 R. Blei *Analysis in integer and fractional dimensions*
72 F. Borceaux & G. Janelidze *Galois theories*
73 B. Bollobas *Random graphs*

COMPLETELY BOUNDED MAPS AND OPERATOR ALGEBRAS

VERN PAULSEN
University of Houston

PUBLISHED BY THE PRESS SYNDICATE OF THE UNIVERSITY OF CAMBRIDGE
The Pitt Building, Trumpington Street, Cambridge, United Kingdom

CAMBRIDGE UNIVERSITY PRESS
The Edinburgh Building, Cambridge CB2 2RU, UK
40 West 20th Street, New York, NY 10011-4211, USA
477 Williamstown Road, Port Melbourne, VIC 3207, Australia
Ruiz de Alarcón 13, 28014 Madrid, Spain
Dock House, The Waterfront, Cape Town 8001, South Africa

http://www.cambridge.org

First published 2002

Printed in the United Kingdom at the University Press, Cambridge

Typeface Times 10/13 pt. *System* LaTeX 2ε [TB]

A catalog record for this book is available from the British Library.

Library of Congress Cataloging in Publication Data

Paulsen, Vern I., 1951–

Completely bounded maps and operator algebras / Vern Paulsen.

p. cm. – (Cambridge studies in advanced mathematics; 78)

Includes bibliographical references and index.

ISBN 0-521-81669-6 (alk. paper)

1. Operator algebras. 2. Dilation theory (Operator theory) 3. Mappings (Mathematics)
I. Title. II. Series.

QA326 .P37 2002

512′.55 – dc21 2002024624

ISBN 0 521 81669 6 hardback

To John, Ival, Effie, Susan, Stephen and Lisa.
My past, present and future.

Contents

Preface

This book is intended to give the reader an introduction to the principal results and ideas in the theories of completely positive maps, completely bounded maps, dilation theory, operator spaces, and operator algebras, together with some of their main applications. It is intended to be self-contained and accessible to any reader who has had a first course in functional analysis that included an introduction to C^*-algebras. It could be used as a text for a course or for independent reading. With this in mind, we have included plenty of exercises.

We have made no attempt at giving a full state-of-the-art exposition of any of these fields. Instead, we have tried to give the reader an introduction to many of the important techniques and results of these fields, together with a feel for their connections and some of the important applications of the ideas. However, we present new proofs and approaches to some of the well-known results in this area, which should make this book of interest to the expert in this area as well as to the beginner.

The quickest route to a result is often not the most illuminating. Consequently, we occasionally present more than one proof of some results. For example, scattered throughout the text and exercises are five different proofs of a key inequality of von Neumann. We feel that such redundancy can lead to a deeper understanding of the material.

In an effort to establish a common core of knowledge that we can assume the reader is familiar with, we have adopted R.G. Douglas's *Banach Algebra Techniques in Operator Theory* as a basic text. Results that appear in that text we have assumed are known, and we have attempted to give a full accounting of all other facts by either presenting them, leaving them as an exercise, or giving a reference. Consequently, parts of the text may seem unnecessarily elementary to some readers. For example, readers with a background in Banach spaces or C^*-algebras may find our discussions of the tensor theory a bit naïve.

We now turn our attention to a description of the contents of this book.

The first seven chapters develop the theory of positive and completely positive maps together with their connections with dilation theory. Dilations are a technique for studying operators on a Hilbert space by representing a given operator as the restriction of a (hopefully) better-understood operator, acting on a larger Hilbert space, to the original space. The operator on the larger space is referred to as a *dilation* of the original operator. Thus, dilation theory involves essentially geometric constructions. We shall see that many of the classic theorems that characterize which sequences of complex numbers are the moments of a measure are really dilation theorems.

One of the better-known dilation theorems is due to Sz.-Nagy and asserts that every contraction operator can be dilated to a unitary operator. Thus to prove some results about contraction operators it is enough to show that they are true for unitary operators. The most famous application of this idea is Sz.-Nagy's elegant proof of an inequality of von Neumann to the effect that the norm of a polynomial in a contraction operator is at most the supremum of the absolute value of the polynomial over the unit disk.

Ando generalized Sz.-Nagy's and von Neumann's results to pairs of commuting contractions, but various analogues of these theorems are known to fail for three or more commuting contractions. Work of Sarason and of Sz.-Nagy and Foias showed that many classical results about analytic functions, including the Nevanlinna–Pick theory, Nehari's theorem, and Caratheodory's completion theorem are consequences of these results about contraction operators. Thus, one finds that there is an operator-theoretic obstruction to generalizing many of these classic results. These results are the focus of Chapter 5.

W.F. Stinespring introduced the theory of completely positive maps as a means of giving abstract necessary and sufficient conditions for the existence of dilations. In many ways completely positive maps play the same role as positive measures when commutative C^*-algebras are replaced by noncommutative C^*-algebras. The connections between completely positive maps and dilation theory were broadened further by Arveson, who developed a deep structure theory for these maps, including an operator-valued Hahn–Banach extension theorem.

Completely positive maps also play a central role in the theory of tensor products of C^*-algebras. Characterizations of nuclear C^*-algebras and injectivity are given in terms of these maps. In noncommutative harmonic analysis they arise in the guise of positive definite operator-valued functions on groups.

In spite of the broad range of applications of completely positive maps, this text is one of the few places where one can find a full introduction to their theory.

In the early 1980s, motivated largely by the work of Wittstock and Haagerup, researchers began extending much of the theory of completely positive maps to the family of completely bounded maps. To the extent that completely positive maps are the analogue of positive measures, completely bounded maps are the analogue of bounded measures.

This newer family of maps also allows for the development of a theory that ties together many questions about the existence or nonexistence of similarities or what are sometimes referred to as skew dilations. Two famous problems of this type are Kadison's and Halmos's similarity conjectures. The theory of completely bounded maps has had an enormous impact on both of these conjectures. Kadison conjectured that every bounded homomorphism of a C^*-algebra into the algebra of operators on a Hilbert space is similar to a $*$-homomorphism. Halmos conjectured that every polynomially bounded operator is similar to a contraction.

In Chapters 8 and 9, we develop the basic theory of completely bounded maps and their connections with similarity questions.

In Chapter 10 we study polynomially bounded operators and present Pisier's counterexample to the Halmos conjecture. The Kadison conjecture still remains unresolved at the time of this writing, but in Chapter 19 we present Pisier's theory of similarity and factorization degrees, which we believe is the most hopeful route towards a solution of the Kadison conjecture.

Attempts to generalize von Neumann's results and the theory of polynomially bounded operators to domains other than the unit disk led to the concepts of spectral and K-spectral sets. We study the applications of the theory of completely bounded maps to these ideas in Chapter 11.

In Chapter 12 we get our first introduction to tensor theory in order to further develop some of the multivariable analogues of von Neumann's inequality.

In order to discuss completely positive or completely bounded maps between two spaces, the domains and ranges of these maps need to be what is known as an *operator system* or *operator space*, respectively. Such spaces arise naturally as subspaces of the space of bounded operators on a Hilbert space, and this is how operator systems and operator spaces were originally defined. However, results of Choi and Effros and of Ruan gave abstract characterizations of operator systems and operator spaces that enabled researchers to treat their theory and the corresponding theories of completely positive and completely bounded maps in a way that was free of dependence on this underlying Hilbert space.

These abstract characterizations have had an impact on this field similar to the impact that the Gelfand–Naimark–Segal theorem has had on the study of C^*-algebras. These characterizations have also allowed for the development of many parallels with ideas from the theory of Banach spaces and bounded linear maps, which have in turn led to a deeper understanding of many results in the theory of C^*-algebras and von Neumann algebras.

We present these characterizations in Chapter 13 and give a quick introduction to the rapidly growing field of operator spaces in Chapter 14, where we carefully examine some of the more important examples of operator spaces.

The abstract characterization of operator spaces led to the Blecher–Ruan–Sinclair abstract characterization of operator algebras. The last chapters of this book are devoted to a development of this theory and some of its applications.

We give two separate developments of the Blecher–Ruan–Sinclair theory. First, we present a new proof based on Hamana's theory of injective envelopes, which we develop in Chapters 15 and 16. We then develop the theory of the Haagerup tensor product and the representation theorems for multilinear maps that are completely bounded in the sense of Christensen and Sinclair, and give a proof of the Blecher–Ruan–Sinclair theorem based on this theory. Our development of the Haagerup tensor theory also uses the theory of the injective envelope in a novel fashion.

The remaining two chapters of the book develop some applications of the Blecher–Ruan–Sinclair theorem. First, we develop the theory of the universal operator algebra of a unital algebra and its applications, including new proofs of Nevanlinna's factorization theorem for analytic functions on the disk and Agler's generalization of Nevanlinna's theorem to analytic functions on the bidisk. Finally, in the last chapter, we present Pisier's theory of the universal operator algebra of an operator space, his results on similarity and factorization degree, and their applications to Kadison's similarity conjecture.

This book grew out of my earlier lecture notes on completely positive and completely bounded maps [161], that have been out-of-print for over a decade.

I would like to acknowledge all the friends and colleagues who have helped to make this book possible. David Blecher, Ken Davidson, Gilles Pisier, and Roger Smith have been tremendous sources of information and ideas. The reader should be grateful to Aristides Katavolos, whose proofreading of selected chapters led to a further polishing of the entire manuscript. Ron Douglas has been a constant source of support throughout my academic career and provided the original impetus to write this book. The author would also like to thank the Mathematics Department at Rice University where portions of this book were written, Roger Astley of Cambridge University Press for his support and advice, the proofreaders at TechBooks for making many of my thoughts flow smoother without altering their mathematical content, and Robin Campbell who typed nearly the entire manuscript and contributed much to its overall look. Finally, without my family's patience and endurance this project would have not been possible.

While writing this book I was partially supported by a grant from the National Science Foundation.

Chapter 1
Introduction

It is assumed throughout this book that the reader is familiar with operator theory and the basic properties of C^*-algebras (see for example [76] and [8, Chapter 1]). We concentrate primarily on giving a self-contained exposition of the theory of completely positive and completely bounded maps between C^*-algebras and the applications of these maps to the study of operator algebras, similarity questions, and dilation theory. In particular, we assume that the reader is familiar with the material necessary for the Gelfand–Naimark–Segal theorem, which states that every C^*-algebra has a one-to-one, $*$-preserving, norm-preserving representation as a norm-closed, $*$-closed algebra of operators on a Hilbert space.

In this chapter we introduce some of the key concepts that will be studied in this book.

As well as having a norm, a C^*-algebra also has an order structure, induced by the cone of positive elements. Recall that an element of a C^*-algebra is *positive* if and only if it is self-adjoint and its spectrum is contained in the nonnegative reals, or equivalently, if it is of the form a^*a for some element a. Since the property of being positive is preserved by $*$-isomorphism, if a C^*-algebra is represented as an algebra of operators on a Hilbert space, then the positive elements of the C^*-algebra coincide with the positive operators that are contained in the representation of the algebra. An equivalent characterization of positivity for an operator on a Hilbert space is that A is a *positive operator* provided that the inner product $\langle Ax, x\rangle$ is nonnegative for every vector x in the space. We shall write $a \geq 0$ to denote that a is positive.

The positive elements in a C^*-algebra $\mathcal{A}$ are a norm-closed, convex cone in the C^*-algebra, denoted by $\mathcal{A}^+$. If h is a self-adjoint element, then it is easy to see, via the functional calculus, that h is the difference of two positive elements.

Indeed, if we let

$$f^{+}(x) = \begin{cases} x, & x \geq 0, \\ 0, & x < 0, \end{cases} \qquad f^{-}(x) = \begin{cases} 0, & x \geq 0, \\ -x, & x < 0, \end{cases}$$

then using the functional calculus we have that $h = f^{+}(h) - f^{-}(h)$, with $f^{+}(h)$ and $f^{-}(h)$ both positive. In particular, we see that the real linear span of the positive elements is the set of self-adjoint elements, which is also norm-closed.

Using the *Cartesian decomposition* of an arbitrary element a of $\mathcal{A}$, namely, $a = h + ik$ with $h = h^*$, $k = k^*$, we see that

$$a = (p_1 - p_2) + i(p_3 - p_4),$$

with p_i positive, $i = 1, 2, 3, 4$. Thus, the complex linear span of $\mathcal{A}^{+}$ is $\mathcal{A}$.

In addition to having its own norm and order structure, a C^*-algebra is also equipped with a whole sequence of norms and order structures on a set of C^*-algebras naturally associated with the original algebra, and this additional structure will play a central role in this book.

To see how to obtain this additional structure, let $\mathcal{A}$ be our C^*-algebra, let M_n denote the $n \times n$ complex matrices, and let $M_n(\mathcal{A})$ denote the set of $n \times n$ matrices with entries from $\mathcal{A}$. We'll denote a typical element of $M_n(\mathcal{A})$ by $(a_{i,j})$.

There is a natural way to make $M_n(\mathcal{A})$ into a $*$-algebra. Namely, for $(a_{i,j})$ and $(b_{i,j})$ in $M_n(\mathcal{A})$, set

$$(a_{i,j}) \cdot (b_{i,j}) = \left(\sum_{k=1}^{n} a_{i,k} b_{k,j} \right)$$

and

$$(a_{i,j})^* = (a_{j,i}^*).$$

What is not so obvious is that there is a unique way to introduce a norm such that $M_n(\mathcal{A})$ becomes a C^*-algebra.

To see how this is done, we begin with the most basic of all C^*-algebras, $B(\mathcal{H})$, the bounded linear operators on a Hilbert space $\mathcal{H}$.

If we let $\mathcal{H}^{(n)}$ denote the direct sum of n copies of $\mathcal{H}$, then there is a natural norm and inner product on $\mathcal{H}^{(n)}$ that makes it into a Hilbert space. Namely,

$$\left\| \begin{pmatrix} h_1 \\ \vdots \\ h_n \end{pmatrix} \right\|^2 = \|h_1\|^2 + \cdots + \|h_n\|^2$$

and

$$\left\langle \begin{pmatrix} h_1 \\ \vdots \\ h_n \end{pmatrix}, \begin{pmatrix} k_1, \\ \vdots \\ k_n \end{pmatrix} \right\rangle_{\mathcal{H}^{(n)}} = \langle h_1, k_1 \rangle_{\mathcal{H}} + \cdots + \langle h_n, k_n \rangle_{\mathcal{H}},$$

where

$$\begin{pmatrix} h_1 \\ \vdots \\ h_n \end{pmatrix} \text{ and } \begin{pmatrix} k_1 \\ \vdots \\ k_n \end{pmatrix}$$

are in $\mathcal{H}^{(n)}$. This Hilbert space is also often denoted $\ell_n^2(\mathcal{H})$. We prefer to regard elements of $\mathcal{H}^{(n)}$ as column vectors, for reasons that will become apparent shortly.

There is a natural way to regard an element of $M_n(B(\mathcal{H}))$ as a linear map on $\mathcal{H}^{(n)}$, by using the ordinary rules for matrix products. That is, we set

$$(T_{ij}) \begin{pmatrix} h_1 \\ \vdots \\ h_n \end{pmatrix} = \begin{pmatrix} \sum_{j=1}^{n} T_{1j} h_j \\ \vdots \\ \sum_{j=1}^{n} T_{nj} h_j \end{pmatrix},$$

for (T_{ij}) in $M_n(B(\mathcal{H}))$ and $\begin{pmatrix} h_1 \\ \vdots \\ h_n \end{pmatrix}$ in $\mathcal{H}^{(n)}$. It is easily checked (Exercise 1.1) that every element of $M_n(B(\mathcal{H}))$ defines a bounded linear operator on $\mathcal{H}^{(n)}$ and that this correspondence yields a one-to-one $*$-isomorphism of $M_n(B(\mathcal{H}))$ onto $B(\mathcal{H}^{(n)})$ (Exercise 1.2). Thus, the identification $M_n(B(\mathcal{H})) = B(\mathcal{H}^{(n)})$ gives us a norm that makes $M_n(B(\mathcal{H}))$ a C^*-algebra.

Now, given any C^*-algebra $\mathcal{A}$, one way that $M_n(\mathcal{A})$ can be viewed as a C^*-algebra is to first choose a one-to-one $*$-representation of $\mathcal{A}$ on some Hilbert space $\mathcal{H}$ so that $\mathcal{A}$ can be identified as a C^*-subalgebra of $B(\mathcal{H})$. This allows us to identify $M_n(\mathcal{A})$ as a $*$-subalgebra of $M_n(B(\mathcal{H}))$. It is straightforward to verify that the image of $M_n(\mathcal{A})$ under this representation is closed and hence a C^*-algebra.

Thus, by using a one-to-one *-representation of $\mathcal{A}$, we have *a* way to turn $M_n(\mathcal{A})$ into a C^*-algebra. But since the norm is unique on a C^*-algebra, we see that the norm on $M_n(\mathcal{A})$ defined in this fashion is independent of the particular representation of $\mathcal{A}$ that we chose. Since positive elements remain positive under $*$-isomorphisms, we see that the positive elements of $M_n(\mathcal{A})$ are also uniquely determined.

So we see that in addition to having a norm and an order, every C^*-algebra $\mathcal{A}$ carries along this extra "baggage" of canonically defined norms and orders on each $M_n(\mathcal{A})$. Remarkably, keeping track of how this extra structure behaves yields far more information than one might expect. The study of these *matrix norms* and *matrix orders* will be a central topic of this book.

For some examples of this structure, we first consider M_k. We can regard this as a C^*-algebra by identifying M_k with the linear transformations on k-dimensional (complex) Hilbert space, $\mathbb{C}^k$. There is a natural way to identify $M_n(M_k)$ with M_{nk}, namely, forget the additional parentheses. It is easy to see that, with this identification, the multiplication and $*$-operation on $M_n(M_k)$ become the usual multiplication and $*$-operation on M_{nk}, that is, the identification defines a $*$-isomorphism. Hence, the unique norm on $M_n(M_k)$ is just the norm obtained by this identification with M_{nk}. An element of $M_n(M_k)$ will be positive if and only if the corresponding matrix in M_{nk} is positive.

For a second example, let X be a compact Hausdorff space, and let $C(X)$ denote the continuous complex-valued functions on X. Setting $f^*(x) = \overline{f(x)}$, we have

$$\|f\| = \sup\{|f(x)|: x \in X\},$$

and defining the algebra operations pointwise makes $C(X)$ into a C^*-algebra. An element $F = (f_{i,j})$ of $M_n(C(X))$ can be thought of as a continuous M_n-valued function. Note that addition, multiplication, and the $*$-operation in $M_n(C(X))$ are just the pointwise addition, pointwise multiplication, and pointwise conjugate-transpose operations of these matrix-valued functions. If we set

$$\|F\| = \sup\{\|F(x)\|: x \in X\},$$

where by $\|F(x)\|$ we mean the norm in M_n, then it is easily seen that this defines a C^*-norm on $M_n(C(X))$, and thus is the unique norm in which $M_n(C(X))$ is a C^*-algebra. Note that the positive elements of $M_n(C(X))$ are those F for which $F(x)$ is a positive matrix for all x.

Now, given two C^*-algebras $\mathcal{A}$ and $\mathcal{B}$ and a map $\phi: \mathcal{A} \to \mathcal{B}$, we also obtain maps $\phi_n: M_n(\mathcal{A}) \to M_n(\mathcal{B})$ via the formula

$$\phi_n((a_{i,j})) = (\phi(a_{i,j})).$$

In general the adverb *completely* means that all of the maps $\{\phi_n\}$ enjoy some property.

For example, the map ϕ is called *positive* if it maps positive elements of $\mathcal{A}$ to positive elements of $\mathcal{B}$, and ϕ is called *completely positive* if every ϕ_n is a positive map.

In a similar fashion, if ϕ is a bounded map, then each ϕ_n will be bounded, and when $\|\phi\|_{cb} = \sup_n \|\phi_n\|$ is finite, we call ϕ a *completely bounded* map.

One's initial hope is perhaps that C^*-algebras are sufficiently nice that every positive map is completely positive and every bounded map is completely bounded. Indeed, one might expect that $\|\phi\| = \|\phi_n\|$ for all n. For these reasons, we begin with an example of a fairly nice map where those norms are different.

Let $\{E_{i,j}\}_{i,j=1}^2$ denote the system of matrix units for M_2 [that is, $E_{i,j}$ is 1 in the (i, j)th entry and 0 elsewhere], and let $\phi\colon M_2 \to M_2$ be the transpose map, so that $\phi(E_{i,j}) = E_{j,i}$. It is easy to verify (Exercise 1.9) that the transpose of a positive matrix is positive and that the norm of the transpose of a matrix is the same as the norm of the matrix, so ϕ is positive and $\|\phi\| = 1$. Now let's consider $\phi_2\colon M_2(M_2) \to M_2(M_2)$.

Note that the matrix of matrix units,

$$\begin{bmatrix} E_{11} & E_{12} \\ E_{21} & E_{22} \end{bmatrix} = \begin{bmatrix} 1 & 0 & 0 & 1 \\ 0 & 0 & 0 & 0 \\ 0 & 0 & 0 & 0 \\ 1 & 0 & 0 & 1 \end{bmatrix},$$

is positive, but that

$$\phi_2\left(\begin{bmatrix} E_{11} & E_{12} \\ E_{21} & E_{22} \end{bmatrix}\right) = \begin{bmatrix} \phi(E_{11}) & \phi(E_{12}) \\ \phi(E_{21}) & \phi(E_{22}) \end{bmatrix} = \begin{bmatrix} 1 & 0 & 0 & 0 \\ 0 & 0 & 1 & 0 \\ 0 & 1 & 0 & 0 \\ 0 & 0 & 0 & 1 \end{bmatrix}$$

is not positive. Thus, ϕ is a positive map but not completely positive. In a similar fashion, we have that

$$\left\| \begin{bmatrix} E_{11} & E_{21} \\ E_{12} & E_{22} \end{bmatrix} \right\| = 1,$$

while the norm of its image under ϕ_2 has norm 2. Thus, $\|\phi_2\| \geq 2$, so $\|\phi_2\| \neq \|\phi\|$. It turns out that ϕ is completely bounded, in fact, $\sup_n \|\phi_n\| = 2$, as we shall see later in this book.

To obtain an example of a map that's not completely bounded, we need to repeat the above example but on an infinite-dimensional space. So let $\mathcal{H}$ be a separable, infinite-dimensional Hilbert space with a countable, orthonormal basis, $\{e_n\}_{n=1}^\infty$. Every bounded, linear operator T on $\mathcal{H}$ can be thought of as an infinite matrix whose (i, j)th entry is the inner product $\langle Te_j, e_i\rangle$. One then defines a map ϕ from the C^*-algebra of bounded linear operators on $\mathcal{H}$, $B(\mathcal{H})$, to $B(\mathcal{H})$ by the transpose. Again ϕ will be positive and an isometry (Exercise 1.9), but $\|\phi_n\| \geq n$. To see this last claim, let $\{E_{i,j}\}_{i,j=1}^\infty$ be matrix units on $\mathcal{H}$, and for fixed n, let $A = (E_{j,i})$, that is, A is the element of $M_n(B(\mathcal{H}))$ whose (i, j)th entry is $E_{j,i}$. We leave it to the reader to verify that $\|A\| = 1$ (in fact, A is a partial isometry), but $\|\phi_n(A)\| = n$ (Exercise 1.8).

There is an alternative approach to the above constructions, via tensor products. A reader familiar with tensor products has perhaps realized that the algebra $M_n(\mathcal{A})$ that we've defined is readily identified with the tensor product algebra $M_n \otimes \mathcal{A}$. Recall that one makes the tensor product of two algebras into an algebra by defining $(a_1 \otimes b_1) \cdot (a_2 \otimes b_2) = (a_1 a_2) \otimes (b_1 b_2)$ and then extending linearly. If $\{E_{i,j}\}_{i,j=1}^n$ denotes the canonical basis for M_n, then an element $(a_{i,j})$ in $M_n(\mathcal{A})$ can be identified with $\sum_{i,j=1}^n a_{i,j} \otimes E_{i,j}$ in $M_n \otimes \mathcal{A}$. We leave it to the reader to verify (Exercise 1.10) that with this identification of $M_n(\mathcal{A})$ and $M_n \otimes \mathcal{A}$, the multiplication defined on $M_n(\mathcal{A})$ becomes the tensor product multiplication on $M_n \otimes \mathcal{A}$. Thus, this identification is an isomorphism of these algebras.

We shall on occasion return to this tensor product notation to simplify concepts.

Now that the reader has been introduced to the concepts of completely positive and completely bounded maps, we turn to the topic of dilations.

In general, the key idea behind a dilation is to realize an operator or a mapping into a space of operators as "part" of something simpler on a larger space.

The simplest case is the *unitary dilation of an isometry*. Let V be an isometry on $\mathcal{H}$, and let $P = I_{\mathcal{H}} - VV^*$ be the projection onto the orthocomplement of the range of V. If we define U on $\mathcal{H} \oplus \mathcal{H} = \mathcal{K}$ via

$$U = \begin{pmatrix} V & P \\ 0 & V^* \end{pmatrix},$$

then it is easily checked that $U^*U = UU^* = I_{\mathcal{K}}$, so that U is a unitary on $\mathcal{K}$. Moreover, if we identify $\mathcal{H}$ with $\mathcal{H} \oplus 0$, then

$$V^n = P_{\mathcal{H}} U^n|_{\mathcal{H}} \quad \text{for all} \quad n \geq 0.$$

Thus, any isometry V can be realized as the restriction of some unitary to one of its subspaces in a manner that also respects the powers of both operators.

In a similar fashion, one can construct an *isometric dilation of a contraction*. Let T be an operator on $\mathcal{H}$, $\|T\| \leq 1$, and let $D_T = (I - T^*T)^{1/2}$. Note that $\|Th\|^2 + \|D_T h\|^2 = \langle T^*Th, h\rangle + \langle D_T^2 h, h\rangle = \|h\|^2$.

We set

$$\ell^2(\mathcal{H}) = \left\{ (h_1, h_2, \ldots) : h_n \in \mathcal{H} \text{ for all } n, \ \sum_{n=1}^{\infty} \|h_n\|^2 < +\infty \right\}.$$

This is a Hilbert space with $\|(h_1, h_2, \dots)\|^2 = \sum_{n=1}^{\infty} \|h_n\|^2$, and inner product $\langle (h_1, h_2, \dots), (k_1, k_2, \dots) \rangle = \sum_{n=1}^{\infty} \langle h_n, k_n \rangle$.

We define $V \colon \ell^2(\mathcal{H}) \to \ell^2(\mathcal{H})$ via $V((h_1, h_2, \dots)) = (Th_1, D_T h_1, h_2, \dots)$. Since $\|V((h_1, h_2, \dots))\|^2 = \|Th_1\|^2 + \|D_T h_1\|^2 + \|h_2\|^2 + \cdots = \|(h_1, h_2, \dots)\|^2$, V is an isometry on $\ell^2(\mathcal{H})$. If we identify $\mathcal{H}$ with $\mathcal{H} \oplus 0 \oplus \cdots$, then it is clear that $T^n = P_{\mathcal{H}} V^n|_{\mathcal{H}}$ for all $n \geq 0$.

Combining these two constructions yields the unitary dilation of a contraction.

Theorem 1.1 (Sz.-Nagy's dilation theorem). *Let T be a contraction operator on a Hilbert space $\mathcal{H}$. Then there is a Hilbert space $\mathcal{K}$ containing $\mathcal{H}$ as a subspace and a unitary operator U on $\mathcal{K}$ such that*

$$T^n = P_{\mathcal{H}} U^n|_{\mathcal{H}}.$$

Proof. Let $\mathcal{K} = \ell^2(\mathcal{H}) \oplus \ell^2(\mathcal{H})$, and identify $\mathcal{H}$ with $(\mathcal{H} \oplus 0 \oplus \cdots) \oplus 0$. Let V be the isometric dilation of T on $\ell^2(\mathcal{H})$, and let U be the unitary dilation of V on $\ell^2(\mathcal{H}) \oplus \ell^2(\mathcal{H})$. Since $\mathcal{H} \subseteq \ell^2(\mathcal{H}) \oplus 0$, we have that $P_{\mathcal{H}} U^n|_{\mathcal{H}} = P_{\mathcal{H}} V^n|_{\mathcal{H}} = T^n$ for all $n \geq 0$. □

Whenever Y is an operator on a Hilbert space $\mathcal{K}$, $\mathcal{H}$ is a subspace of $\mathcal{K}$, and $X = P_{\mathcal{H}} Y|_{\mathcal{H}}$, then we call X a *compression* of Y.

There is a certain sense in which a "minimal" unitary dilation can be chosen, and this dilation is in some sense unique. We shall not need these facts now, but shall return to them in Chapter 4.

To see the power of this simple geometric construction, we now give Sz.-Nagy's proof of an inequality due to von Neumann.

Corollary 1.2 (von Neumann's inequality). *Let T be a contraction on a Hilbert space. Then for any polynomial p,*

$$\|p(T)\| \leq \sup\{|p(z)| \colon |z| \leq 1\}.$$

Proof. Let U be a unitary dilation of T. Since $T^n = P_{\mathcal{H}} U^n|_{\mathcal{H}}$ for all $n \geq 0$, it follows, by taking linear combinations, that $p(T) = P_{\mathcal{H}} p(U)|_{\mathcal{H}}$, and hence $\|p(T)\| \leq \|p(U)\|$. Since unitaries are normal operators, we have that $\|p(U)\| = \sup\{|p(\lambda)| \colon \lambda \in \sigma(U)\}$, where $\sigma(U)$ denotes the spectrum of U. Finally, since U is unitary, $\sigma(U)$ is contained in the unit circle and the result follows. □

In Chapter 2, we will give another proof of von Neumann's inequality, using some facts about positive maps, and then in Chapter 4 we will obtain Sz.-Nagy's dilation theorem as a consequence of von Neumann's inequality.

Exercises

1.1 Let (T_{ij}) be in $M_n(B(\mathcal{H}))$. Verify that the linear transformation it defines on $\mathcal{H}^{(n)}$ is bounded and that, in fact, $\|(T_{ij})\| \leq (\sum_{i,j=1}^{n} \|T_{ij}\|^2)^{1/2}$.

1.2 Let $\pi \colon M_n(B(\mathcal{H})) \to B(\mathcal{H}^{(n)})$ be the identification given in the text.
   (i) Verify that π is a one-to-one $*$-homomorphism.
   (ii) Let $E_j \colon \mathcal{H} \to \mathcal{H}^{(n)}$ be the map defined by setting $E_j(h)$ equal to the vector that has h for its jth component and is 0 elsewhere. Show that $E_j^* \colon \mathcal{H}^{(n)} \to \mathcal{H}$ is the map that sends a vector in $\mathcal{H}^{(n)}$ to its jth component.
   (iii) Given $T \in B(\mathcal{H}^{(n)})$, set $T_{ij} = E_i^* T E_j$. Show that $\pi((T_{ij})) = T$ and that consequently π is onto.

1.3 Let (T_{ij}) be in $M_n(B(\mathcal{H}))$. Prove that (T_{ij}) is a contraction if and only if for every choice of $2n$ unit vectors $x_1, \ldots, x_n, y_1, \ldots, y_n$ in $\mathcal{H}$, the scalar matrix $(\langle T_{ij} x_j, y_i \rangle)$ is a contraction.

1.4 Let (T_{ij}) be in $M_n(B(\mathcal{H}))$. Prove that (T_{ij}) is positive if and only if for every choice of n vectors $x_1, \ldots, x_n$ in $\mathcal{H}$ the scalar matrix $(\langle T_{ij} x_j, x_i \rangle)$ is positive.

1.5 Let $\mathcal{A}$ and $\mathcal{B}$ be unital C^*-algebras, and let $\pi \colon \mathcal{A} \to \mathcal{B}$ be a $*$-homomorphism with $\pi(1) = 1$. Show that π is completely positive and completely bounded and that $\|\pi\| = \|\pi_n\| = \|\pi\|_{\text{cb}} = 1$.

1.6 Let $\mathcal{A}$, $\mathcal{B}$, and $\mathcal{C}$ be C^*-algebras, and let $\phi \colon \mathcal{A} \to \mathcal{B}$ and $\psi \colon \mathcal{B} \to \mathcal{C}$ be (completely) positive maps. Show that $\psi \circ \phi$ is (completely) positive.

1.7 Let $\{E_{i,j}\}_{i,j=1}^{n}$ be matrix units for M_n, let $A = (E_{j,i})_{i,j=1}^{n}$, and let $B = (E_{i,j})_{i,j=1}^{n}$ be in $M_n(M_n)$. Show that A is unitary and that $\frac{1}{n} B$ is a rank one projection.

1.8 Let $\{E_{i,j}\}_{i,j=1}^{\infty}$ be a system of matrix units for $B(\mathcal{H})$, let $A = (E_{j,i})_{i,j=1}^{n}$, and let $B = (E_{i,j})_{i,j=1}^{n}$ be in $M_n(B(\mathcal{H}))$. Show that A is a partial isometry, and that $\frac{1}{n} B$ is a projection. Show that $\phi_n(A) = B$ and $\|\phi_n(A)\| = n$.

1.9 Let A be in M_n, and let A^t denote the transpose of A. Prove that A is positive if and only if A^t is positive, and that $\|A\| = \|A^t\|$. Prove that these same results hold for operators on a separable, infinite-dimensional Hilbert space, when we fix an orthonormal basis, regard operators as infinite matrices, and use this to define a transpose map.

1.10 Prove that the map $\pi \colon M_n(\mathcal{A}) \to M_n \otimes \mathcal{A}$ defined by $\pi((a_{i,j})) = \sum_{i,j=1}^{n} a_{i,j} \otimes E_{i,j}$ is an algebra isomorphism.

Chapter 2

Positive Maps

Before turning our attention to the completely positive or completely bounded maps, we begin with some results on positive maps that we shall need repeatedly. These results also serve to illustrate how many simplifications arise when one passes to the smaller class of completely positive maps.

If $\mathcal{S}$ is a subset of a C^*-algebra $\mathcal{A}$, then we set

$$S^* = \{a : a^* \in \mathcal{S}\},$$

and we call $\mathcal{S}$ *self-adjoint* when $\mathcal{S} = \mathcal{S}^*$. If $\mathcal{A}$ has a unit 1 and $\mathcal{S}$ is a self-adjoint subspace of $\mathcal{A}$ containing 1, then we call $\mathcal{S}$ an *operator system*. If $\mathcal{S}$ is an operator system and h is a self-adjoint element of S, then even though $f^+(h)$ and $f^-(h)$ need not belong to $\mathcal{S}$ (since these only belong to the norm-closed algebra generated by h), we can still write h as the difference of two positive elements in $\mathcal{S}$. Indeed,

$$h = \frac{1}{2}(\|h\| \cdot 1 + h) - \frac{1}{2}(\|h\| \cdot 1 - h).$$

If $\mathcal{S}$ is an operator system, $\mathcal{B}$ is a C^*-algebra, and $\phi \colon \mathcal{S} \to \mathcal{B}$ is a linear map, then ϕ is called a *positive map* provided that it maps positive elements of $\mathcal{S}$ to positive elements of $\mathcal{B}$. In this chapter, we develop some of the properties of positive maps. In particular, we shall be concerned with how the assumption of positivity is related to the norm of the map, and conversely, when assumptions about the norm of a map guarantee that it is positive. We give a fairly elementary proof of von Neumann's inequality (Corollary 2.7), which only uses these observations about positive maps and an elementary result from complex analysis due to Fejer and Riesz.

If ϕ is a positive, linear functional on an operator system $\mathcal{S}$, then it is easy to show that $\|\phi\| = \phi(1)$ (Exercise 2.3). When the range is a C^*-algebra the situation is quite different.

Proposition 2.1. *Let $\mathcal{S}$ be an operator system, and let $\mathcal{B}$ be a C^*-algebra. If $\phi\colon \mathcal{S} \to \mathcal{B}$ is a positive map, then ϕ is bounded and*

$$\|\phi\| \leq 2\|\phi(1)\|.$$

Proof. First note that if p is positive, then $0 \leq p \leq \|p\| \cdot 1$ and so $0 \leq \phi(p) \leq \|p\| \cdot \phi(1)$, from which it follows that $\|\phi(p)\| \leq \|p\| \cdot \|\phi(1)\|$ when $p \geq 0$.

Next note that if p_1 and p_2 are positive, then $\|p_1 - p_2\| \leq \max\{\|p_1\|, \|p_2\|\}$. If h is self-adjoint in $\mathcal{S}$, then using the above decomposition of h, we have

$$\phi(h) = \frac{1}{2}\phi(\|h\| \cdot 1 + h) - \frac{1}{2}\phi(\|h\| \cdot 1 - h),$$

which expresses $\phi(h)$ as a difference of two positive elements of $\mathcal{B}$. Thus,

$$\|\phi(h)\| \leq \frac{1}{2}\max\{\|\phi(\|h\| \cdot 1 + h)\|, \|\phi(\|h\| \cdot 1 - h)\|\} \leq \|h\| \cdot \|\phi(1)\|.$$

Finally, if a is an arbitrary element of $\mathcal{S}$, then $a = h + ik$ with $\|h\|, \|k\| \leq \|a\|$, $h = h^*$, $k = k^*$, and so

$$\|\phi(a)\| \leq \|\phi(h)\| + \|\phi(k)\| \leq 2\|a\| \cdot \|\phi(1)\|. \qquad \square$$

Let us reproduce an example of Arveson, which shows that 2 is the best constant in Proposition 2.1.

Example 2.2. Let $\mathbb{T}$ denote the unit circle in the complex plane, $C(\mathbb{T})$ the continuous functions on $\mathbb{T}$, z the coordinate function, and $\mathcal{S} \subseteq C(\mathbb{T})$ the subspace spanned by 1, z, and $\bar{z}$.

We define $\phi\colon \mathcal{S} \to M_2$ by

$$\phi(a + bz + c\bar{z}) = \begin{bmatrix} a & 2b \\ 2c & a \end{bmatrix}.$$

We leave it to the reader to verify that an element $a1 + bz + c\bar{z}$ of $\mathcal{S}$ is positive if and only if $c = \bar{b}$ and $a \geq 2|b|$. It is fairly standard that a self-adjoint element of M_2 is positive if and only if its diagonal entries and its determinant are nonnegative real numbers. Combining these two facts, it is clear that ϕ is a positive map. However,

$$2\|\phi(1)\| = 2 = \|\phi(z)\| \leq \|\phi\|,$$

so that $\|\phi\| = 2\|\phi(1)\|$.

The existence of unital, positive maps that are not contractive can be roughly attributed to two factors. One is the noncommutativity of the range, the other is the lack of sufficiently many positive elements in the domain. This first principle is illustrated in the exercises, and we concentrate here on properties of the domain that ensure that unital, positive maps are contractive.

Lemma 2.3. *Let $\mathcal{A}$ be a C^*-algebra with unit, and let $p_i, i = 1, \ldots, n$, be positive elements of $\mathcal{A}$ such that*

$$\sum_{i=1}^{n} p_i \leq 1.$$

If $\lambda_i, i = 1, \ldots, n$, are scalars with $|\lambda_i| \leq 1$, then

$$\left\| \sum_{i=1}^{n} \lambda_i p_i \right\| \leq 1.$$

Proof. Note that

$$\begin{bmatrix} \sum_{i=1}^{n} \lambda_i p_i & 0 & \cdots & 0 \\ 0 & 0 & \cdots & 0 \\ \vdots & \vdots & \ddots & \vdots \\ 0 & 0 & \cdots & 0 \end{bmatrix} = \begin{bmatrix} p_1^{1/2} & \ldots & p_n^{1/2} \\ 0 & \ldots & 0 \\ \vdots & & \vdots \\ 0 & \ldots & 0 \end{bmatrix}$$

$$\times \begin{bmatrix} \lambda_1 & 0 & \ldots & 0 \\ 0 & \ddots & \ddots & \vdots \\ \vdots & \ddots & \ddots & 0 \\ 0 & \ldots & 0 & \lambda_n \end{bmatrix} \cdot \begin{bmatrix} p_1^{1/2} & 0 & \ldots & 0 \\ \vdots & \vdots & & \vdots \\ p_n^{1/2} & 0 & \ldots & 0 \end{bmatrix}.$$

The norm of the matrix on the left is $\| \sum_{i=1}^{n} \lambda_i p_i \|$, while each of the three matrices on the right can be easily seen to have norm less than 1, by using the fact that $\|a^*a\| = \|aa^*\| = \|a\|^2$. □

Theorem 2.4. *Let $\mathcal{B}$ be a C^*-algebra with unit, let X be a compact Hausdorff space, with $C(X)$ the continuous functions on X, and let $\phi\colon C(X) \to \mathcal{B}$ be a positive map. Then $\|\phi\| = \|\phi(1)\|$.*

Proof. By scaling, we may assume that $\phi(1) \leq 1$. Let $f \in C(X)$, $\|f\| \leq 1$, and let $\varepsilon > 0$ be given. First, we note that, by a standard partition-of-unity argument, f may be approximated to within ε by a sum of the form given in Lemma 2.3. To see this, first choose a finite open covering $\{U_i\}_{i=1}^{n}$ of X such that $|f(x) - f(x_i)| < \varepsilon$ for x in U_i, and let $\{p_i\}$ be a partition of unity subordinate to the covering. That is, $\{p_i\}$ are nonnegative continuous functions satisfying $\sum_{i=1}^n p_i = 1$ and $p_i(x) = 0$ for $x \notin U_i$, $i = 1, \ldots, n$. Set $\lambda_i = f(x_i)$, and note that if $p_i(x) \neq 0$ for some i, then $x \in U_i$ and so $|f(x) - \lambda_i| < \varepsilon$. Hence, for any x,

$$\left|f(x) - \sum \lambda_i p_i(x)\right| = \left|\sum (f(x) - \lambda_i) p_i(x)\right|$$
$$\leq \sum |f(x) - \lambda_i| p_i(x) < \sum \varepsilon \cdot p_i(x) = \varepsilon.$$

Finally, by Lemma 2.3, $\left\|\sum \lambda_i \phi(p_i)\right\| \leq 1$, so that

$$\|\phi(f)\| \leq \left\|\phi\left(f - \sum \lambda_i p_i\right)\right\| + \left\|\sum \lambda_i \phi(p_i)\right\| < 1 + \varepsilon \cdot \|\phi\|,$$

and since ε was arbitrary, $\|\phi\| \leq 1$. □

As an application of Theorem 2.4, we now prove an inequality due to von Neumann, which will be central to many later results. First we need a preliminary lemma.

Lemma 2.5 (Fejer–Riesz). *Let $\tau(e^{i\theta}) = \sum_{n=-N}^{+N} a_n e^{in\theta}$ be a strictly positive function on the unit circle $\mathbb{T}$. Then there is a polynomial $p(z) = \sum_{n=0}^{N} p_n z^n$ such that*

$$\tau(e^{i\theta}) = |p(e^{i\theta})|^2.$$

Proof. First note that since τ is real-valued, $a_{-n} = \bar{a}_n$ and a_0 is real. We may assume $a_{-N} \neq 0$. Set $g(z) = \sum_{n=-N}^{+N} a_n z^{n+N}$, so that g is a polynomial of degree $2N$ with $g(0) \neq 0$. We have $g(e^{i\theta}) = \tau(e^{i\theta}) \cdot e^{iN\theta} \neq 0$. Notice that the antisymmetry of the coefficients of g implies

$$\overline{g(1/\bar{z})} = z^{-2N} g(z).$$

This means that the $2N$ zeros of g may be written as $z_1, \ldots, z_N, 1/\bar{z}_1, \ldots, 1/\bar{z}_N$.

We set $q(z) = (z - z_1) \cdots (z - z_N)$, $h(z) = (z - 1/\bar{z}_1) \cdots (z - 1/\bar{z}_N)$, and have that

$$g(z) = a_N q(z) h(z),$$

with

$$\overline{h(z)} = \frac{(-1)^N \bar{z}^N q(1/\bar{z})}{z_1 \cdots z_N}.$$

Thus,

$$\tau(e^{i\theta}) = e^{-iN\theta} g(e^{i\theta}) = |g(e^{i\theta})| = |a_N| \cdot |q(e^{i\theta})| \cdot |\overline{h(e^{i\theta})}|$$
$$= \left| \frac{a_N}{z_1 \cdots z_N} \right| \cdot |q(e^{i\theta})|^2,$$

so that $\tau(e^{i\theta}) = |p(e^{i\theta})|^2$, where $p(z) = |\frac{a_N}{z_1 \cdots z_N}|^{1/2} q(z)$. □

Writing $p(z) = \alpha_0 + \cdots + \alpha_N z^N$, we see that $\tau(e^{i\theta}) = \sum_{\ell,k=0}^{N} \alpha_\ell \bar{\alpha}_k e^{i(\ell-k)\theta}$, so that the coefficients of every strictly positive trigonometric polynomial have this special form.

The above results can be shown to hold for positive trigonometric polynomials as well as strictly positive trigonometric polynomials. To do this, one must carefully examine the roots of the polynomial g that lie on the circle.

Theorem 2.6. *Let T be an operator on a Hilbert space $\mathcal{H}$ with $\|T\| \leq 1$, and let $\mathcal{S} \subseteq C(\mathbb{T})$ be the operator system defined by*

$$\mathcal{S} = \{p(e^{i\theta}) + \overline{q(e^{i\theta})} \colon p, q \text{ are polynomials}\}.$$

Then the map $\phi \colon \mathcal{S} \to B(\mathcal{H})$ defined by $\phi(p + \bar{q}) = p(T) + q(T)^*$ is positive.

Proof. First, note that it is enough to prove that $\phi(\tau)$ is positive for every strictly positive τ. Indeed, if τ is only positive, then $\tau + \epsilon 1$ is strictly positive for every $\epsilon > 0$, and hence we have $\phi(\tau) + \epsilon I = \phi(\tau + \epsilon 1) \geq 0$ for every $\epsilon > 0$, and it follows that $\phi(\tau) \geq 0$. Let $\tau(e^{i\theta})$ be strictly positive in $\mathcal{S}$, so that $\tau(e^{i\theta}) = \sum_{\ell,k=0}^{+n} \alpha_\ell \bar{\alpha}_k e^{i(\ell-k)\theta}$. We must prove that

$$\phi(\tau) = \sum_{\ell,k=0}^{+n} \alpha_\ell \bar{\alpha}_k T(\ell - k)$$

is a positive operator, where we define

$$T(j) = \begin{cases} T^j, & j \geq 0, \\ T^{*-j}, & j < 0. \end{cases}$$

To this end, fix a vector x in our Hilbert space $\mathcal{H}$ and note that

$$\langle \phi(\tau)x, x \rangle = \left\langle \begin{bmatrix} I & T^* & \dots & T^{*n} \\ T & \ddots & \ddots & \vdots \\ \vdots & \ddots & \ddots & T^* \\ T^n & \dots & T & I \end{bmatrix} \begin{bmatrix} \bar{\alpha}_1 x \\ \vdots \\ \bar{\alpha}_n x \end{bmatrix}, \begin{bmatrix} \bar{\alpha}_1 x \\ \vdots \\ \bar{\alpha}_n x \end{bmatrix} \right\rangle, \quad (*)$$

where the matrix operator on the right is acting on the direct sum of n copies of the Hilbert space $\mathcal{H}^{(n)}$.

Thus, if we can show that the matrix operator is positive, we shall be done. To this end, set

$$R = \begin{bmatrix} 0 & \cdots & \cdots & \cdots & 0 \\ T & \ddots & & & \vdots \\ 0 & \ddots & \ddots & & \vdots \\ \vdots & \ddots & \ddots & \ddots & \vdots \\ 0 & \cdots & 0 & T & 0 \end{bmatrix},$$

and note that $R^{n+1} = 0$, $\|R\| \leq 1$.

Using I to also denote the identity operator on $\mathcal{H}^{(n)}$, we see that the matrix operator in $(*)$ can be written as

$$I + R + R^2 + \cdots + R^n + R^* + \cdots + R^{*n} = (I - R)^{-1} + (I - R^*)^{-1} - I.$$

To see that this latter operator is positive, fix h in $\mathcal{H}^{(n)}$, and let $h = (I - R)y$ for y in $\mathcal{H}^{(n)}$. One obtains

$$\begin{aligned} &\langle((I - R)^{-1} + (I - R^*)^{-1} - I)h, h\rangle \\ &\quad = \langle y, (I - R)y\rangle + \langle(I - R)y, y\rangle - \langle(I - R)y, (I - R)y\rangle \\ &\quad = \|y\|^2 - \|Ry\|^2 \geq 0, \end{aligned}$$

since R is a contraction. □

We can now give our second proof of von Neumann's inequality.

Corollary 2.7 (von Neumann's Inequality). *Let T be an operator on a Hilbert space with $\|T\| \leq 1$. Then for any polynomial p,*

$$\|p(T)\| \leq \|p\|,$$

where $\|p\| = \sup_{\theta} |p(e^{i\theta})|$.

Proof. Note that the operator system $\mathcal{S}$ defined in Theorem 2.6 is actually a $*$-algebra and separates points. Hence, by the Stone–Weierstrass theorem, $\mathcal{S}$ is dense in $C(\mathbb{T})$. By Proposition 2.1, the map ϕ is bounded and hence extends to $C(\mathbb{T})$. By Exercise 2.2, the extension will also be positive. Hence, ϕ is contractive by Theorem 2.4, from which the result follows. □

Excercises 2.15 and 2.16 present two other proofs of von Neumann's inequality.

We let $A(\mathbb{D})$ denote the functions that are analytic on $\mathbb{D}$ and continuous on $\mathbb{D}^-$. By the maximum modulus principle the supremum of such a function over $\mathbb{D}^-$ is attained on $\mathbb{T}$. Thus, we may regard $A(\mathbb{D})$ as a closed subalgebra of $C(\mathbb{T})$. Since the polynomials are dense in $A(\mathbb{D})$, the above inequality guarantees that the homomorphism $p \to p(T)$ extends to a homomorphism of $A(\mathbb{D})$, and we denote the image of an element f simply by $f(T)$, so that one has $\|f(T)\| \leq \|f\|$ for all f in $A(\mathbb{D})$.

Another consequence of Theorem 2.6 that we shall frequently use is that if a is an element of some unital C^*-algebra $\mathcal{A}$, $\|a\| \leq 1$, then there is a unital, positive map ϕ: $C(\mathbb{T}) \to \mathcal{A}$ with $\phi(p) = p(a)$. This observation is used in the following two results.

Corollary 2.8. *Let $\mathcal{B}, \mathcal{C}$ be C^*-algebras with unit, let $\mathcal{A}$ be a subalgebra of $\mathcal{B}$, $1 \in \mathcal{A}$, and let $\mathcal{S} = \mathcal{A} + \mathcal{A}^*$. If ϕ: $\mathcal{S} \to \mathcal{C}$ is positive, then $\|\phi(a)\| \leq \|\phi(1)\| \cdot \|a\|$ for all a in $\mathcal{A}$.*

Proof. Let a be in $\mathcal{A}$, $\|a\| \leq 1$. By Proposition 2.1 and Exercise 2.2, we may extend ϕ to a positive map on the closure $\mathcal{S}^-$ of $\mathcal{S}$. As remarked above, there is a positive map ψ: $C(\mathbb{T}) \to \mathcal{B}$ with $\psi(p) = p(a)$. Since $\mathcal{A}$ is an algebra, the range of ψ is actually contained in $\mathcal{S}^-$.

Clearly, the composition of positive maps is positive, so by Theorem 2.4,

$$\|\phi(a)\| = \|\phi \circ \psi(e^{i\theta})\| \leq \|\phi \circ \psi(1)\| \cdot \|e^{i\theta}\| = \|\phi(1)\|. \qquad \square$$

If $\phi(1) = 1$ in the above, then ϕ is a contraction on $\mathcal{A}$. It is somewhat surprising that ϕ need not be a contraction on all of $\mathcal{S}$. We shall see an example of this phenomenon in Chapter 5.

Corollary 2.9 (Russo–Dye). *Let $\mathcal{A}$ and $\mathcal{B}$ be C^*-algebras with unit, and let ϕ: $\mathcal{A} \to \mathcal{B}$ be a positive map. Then $\|\phi\| = \|\phi(1)\|$.*

Proof. Apply Corollary 2.8. $\square$

So far we have concentrated on positive maps without indicating how positive maps arise. We close this discussion with two such results.

Unlike our previous results, we shall see (Exercise 2.9) that the hypothesis that our map is unital is crucial to the next three results.

Lemma 2.10. *Let $\mathcal{A}$ be a C^*-algebra, $\mathcal{S} \subseteq \mathcal{A}$ an operator system, and $f: \mathcal{S} \to \mathbb{C}$ a linear functional with $f(1) = 1$, $\|f\| = 1$. If a is a normal element of $\mathcal{A}$ and $a \in \mathcal{S}$, then $f(a)$ will lie in the closed convex hull of the spectrum of a.*

Proof. Suppose not. Note that the convex hull of a compact set is the intersection of all closed disks containing the set. Thus, there will exist a λ and $r > 0$ such that $|f(a) - \lambda| > r$, while the spectrum $\sigma(a)$ of a satisfies

$$\sigma(a) \subseteq \{z: |z - \lambda| \leq r\}.$$

But then $\sigma(a - \lambda \cdot 1) \subseteq \{z: |z| \leq r\}$, and since norm and spectral radius agree for normal elements, $\|a - \lambda 1\| \leq r$, while $|f(a - \lambda \cdot 1)| > r$. This contradiction completes the proof. □

Since the convex hull of the spectrum of a positive operator is contained in the nonnegative reals, we see that Lemma 2.10 implies that such an f must be positive.

Proposition 2.11. *Let $\mathcal{S}$ be an operator system, $\mathcal{B}$ a unital C^*-algebra, and $\phi: \mathcal{S} \to \mathcal{B}$ a unital contraction. Then ϕ is positive.*

Proof. Since $\mathcal{B}$ can be represented on a Hilbert space, we may, without loss of generality, assume that $\mathcal{B} = \mathcal{B}(\mathcal{H})$ for some Hilbert space $\mathcal{H}$. Fix x in $\mathcal{H}$, $\|x\| = 1$. Setting $f(a) = \langle \phi(a)x, x \rangle$, we have that $f(a) = 1$, $\|f\| \leq \|\phi\|$. By Lemma 2.10, if a is positive, then $f(a)$ is positive, and consequently, since x was arbitrary, $\phi(a)$ is positive. □

Proposition 2.12. *Let $\mathcal{A}$ be a unital C^*-algebra, and let $\mathcal{M}$ be a subspace of $\mathcal{A}$ containing 1. If $\mathcal{B}$ is a unital C^*-algebra and $\phi: \mathcal{M} \to \mathcal{B}$ is a unital contraction, then the map $\tilde{\phi}: \mathcal{M} + \mathcal{M}^* \to \mathcal{B}$ given by*

$$\tilde{\phi}(a + b^*) = \phi(a) + \phi(b)^*,$$

is well defined and is the unique positive extension of ϕ to $\mathcal{M} + \mathcal{M}^*$.

Proof. If ϕ does extend to a positive map $\tilde{\phi}$, then by the self-adjointness of positive maps (Exercise 2.1), $\tilde{\phi}$ necessarily satisfies the above equation. So we must prove that this formula yields a well-defined, positive map.

Note that to prove that $\tilde{\phi}$ is well defined, it is enough to prove that if a and a^* belong to $\mathcal{M}$, then $\phi(a^*) = \phi(a)^*$. For this, set

$$\mathcal{S}_1 = \{a: a \in \mathcal{M} \text{ and } a^* \in \mathcal{M}\};$$

then $\mathcal{S}_1$ is an operator system, and ϕ is a unital, contractive map on $\mathcal{S}_1$ and hence positive by Proposition 2.11. Consequently by Exercise 2.1, ϕ is self-adjoint on $\mathcal{S}_1$ and so $\phi(a^*) = \phi(a)^*$ for a in $\mathcal{S}_1$. Thus, $\tilde{\phi}$ is well defined.

To see that $\tilde{\phi}$ is positive, it is sufficient to assume that $\mathcal{B} = \mathcal{B}(\mathcal{H})$, fix x in $\mathcal{H}$ with $\|x\| = 1$, set $\tilde{\rho}(a) = \langle \tilde{\phi}(a)x, x\rangle$, and prove that $\tilde{\rho}$ is positive. Let $\rho\colon \mathcal{M} \to \mathbb{C}$ be defined by $\rho(a) = \langle \phi(a)x, x\rangle$; then $\|\rho\| = 1$ and so, by the Hahn–Banach theorem, ρ extends to

$$\rho_1\colon \mathcal{M} + \mathcal{M}^* \to \mathbb{C} \quad \text{with} \quad \|\rho_1\| = 1.$$

But by Proposition 2.11, ρ_1 is positive, and so $\rho_1(a + b^*) = \rho(a) + \overline{\rho(b)} = \tilde{\rho}(a + b^*)$. Hence, $\tilde{\rho}$ is positive. □

Note that the above result also shows that there is a unique, norm-preserving Hahn–Banach extension of ρ to $\mathcal{M} + \mathcal{M}^*$.

Example 2.13. A positive map need not have a positive extension unless the range is $\mathbb{C}$ (Exercise 2.10). Indeed, if the positive map of Example 2.2 had a positive extension to $C(\mathbb{T})$, then by Corollary 2.9, this extension would be a contraction.

If $\mathcal{S}$ is an operator system, contained in a C^*-algebra $\mathcal{A}$, and ϕ is a linear functional on $\mathcal{S}$ with $\phi(1) = 1$, then by Exercise 2.3 and Proposition 2.11, we see that ϕ is contractive if and only if ϕ is positive. These maps are called *states* on $\mathcal{S}$.

Spectral Sets

There is a natural way that positive maps on operator systems arise in operator theory, and that is in the study of *spectral sets* for operators.

Let X be a compact set in the complex plane, and let $\mathcal{R}(X)$ be the subalgebra of $C(X)$ consisting of quotients p/q of polynomials, where the zeros of q lie off X. Note that two quotients of polynomials can be equal as functions on X even though they may be distinct as elements of the formal algebra of quotients. Indeed, in the extreme case where X is a singleton, $X = \{\lambda\}$, then $z - \lambda = 0$ as a function on X.

If T is in $\mathcal{B}(\mathcal{H})$, with the spectrum $\sigma(T)$ of T, contained in X, then for any quotient p/q as above we have an operator $p(T)q(T)^{-1}$. Thus, we can attempt to define a homomorphism $\rho\colon \mathcal{R}(X) \to \mathcal{B}(\mathcal{H})$ by setting $\rho(p/q) = p(T)q(T)^{-1}$. In the extreme case that $X = \{\lambda\}$, this homomorphism will be well defined if and only if $T = \lambda \cdot I$.

If ρ is well defined and $\|\rho\| \leq 1$, then X is called a *spectral set for T*. When ρ is well defined and only bounded with $\|\rho\| \leq K$, then X is called a *K-spectral set for T*.

We may also regard $\mathcal{R}(X)$ as a subalgebra of $C(\partial X)$, the continuous functions on the boundary of X. By the maximum modulus theorem, this endows $\mathcal{R}(X)$ with the same norm as if we regarded it as a subalgebra of $C(X)$.

We let $\mathcal{S} = \mathcal{R}(X) + \overline{\mathcal{R}(X)}$ regarded as a subset of $C(\partial X)$. Thus, $f_1 + \overline{g_1} = f_2 + \overline{g_2}$ if and only if they are equal as functions on ∂X, and are positive if and only if they are positive on ∂X.

The concept of spectral sets is partially motivated by von Neumann's inequality, which can be interpreted as saying that an operator T is a contraction if and only if the closed unit disk is a spectral set for T.

By Proposition 2.12, if X is a spectral set for T, then there is a well-defined, positive map $\tilde{\rho}\colon \mathcal{S} \to \mathcal{B}(\mathcal{H})$ given by $\tilde{\rho}(f + \bar{g}) = f(T) + g(T)^*$. Conversely, if the above map $\tilde{\rho}$ is well defined and positive, then by Corollary 2.8, X is a spectral set for T.

Nonunital C^*-Algebras

Let $\mathcal{A}$ be a nonunital C^*-algebra. We wish to make some comments on positive maps in this case. Let $\mathcal{B}$ be a C^*-algebra, and let $\phi\colon \mathcal{A} \to \mathcal{B}$ be a positive map. We claim that ϕ is automatically bounded. To see this, note that it is enough to prove that ϕ is bounded on $\mathcal{A}^+$. To this end, suppose that ϕ is not bounded; then there exists a sequence p_n in $\mathcal{A}^+$, $\|p_n\| \leq 1$, with $\|\phi(p_n)\| \geq n^3$. Let $p = \sum_n n^{-2} p_n$; then we have that $n^{-2} p_n \leq p$ and so

$$n \leq \|\phi(n^{-2} p_n)\| \leq \|\phi(p)\|$$

for all n, an obvious contradiction. Thus, ϕ is bounded.

The second observation is that every nonunital C^*-algebra $\mathcal{A}$ embeds into a unital C^*-algebra, $\mathcal{A}_1$ [73]. Furthermore, positive maps from $\mathcal{A}$ to $\mathcal{B}$ extend to positive maps from $\mathcal{A}_1$ to $\mathcal{B}_1$.

To see this second statement, first note that $\mathcal{A}$ is a closed, two-sided ideal in $\mathcal{A}_1$, and so the map $a + \lambda 1 \to \lambda$, for a in $\mathcal{A}$, is a $*$-homomorphism. Thus, if $a + \lambda \cdot 1$ is positive, then $\lambda \geq 0$. Now, let $\phi\colon \mathcal{A} \to \mathcal{B}$ be positive; then since ϕ is bounded, $\phi(p) \leq \|\phi\| \cdot 1$ for all positive p with $\|p\| \leq 1$. Define $\phi_1\colon \mathcal{A}_1 \to \mathcal{B}_1$ by $\phi_1(a + \lambda 1) = \phi(a) + \lambda q$ where $q = \|\phi\| \cdot 1$. This map is positive, since if $a + \lambda 1 \geq 0$, then $-\lambda^{-1} a \leq 1$, so that $\phi(-\lambda^{-1} a) \leq q$ or $0 \leq \phi(a) + \lambda q$.

Note that if we let $\mathcal{A} = \mathcal{B}$ denote the compact operators on an infinite-dimensional Hilbert space, then there is no extension of the identity map to a positive map from $\mathcal{A}_1$ to $\mathcal{B}$.

Toeplitz Matrices and Trigonometric Moments

We close this chapter with another application of the Fejer–Riesz lemma. A sequence of numbers $\{a_k\}_{k=-\infty}^{+\infty}$ is called a *trigonometric moment sequence* provided there exists a positive Borel measure μ on the circle (equivalently, on $[0, 2\pi]$) such that

$$a_k = \int_0^{2\pi} e^{ik\theta}\, d\mu(\theta)$$

for all k in $\mathbb{Z}$. Note that in this case $a_{-k} = \bar{a}_k$ and $a_0 = \mu([0, 2\pi]) \geq 0$. Hence, μ is automatically bounded. We call μ a *representing measure* for $\{a_k\}_{k=-\infty}^{+\infty}$. By the Riesz representation theorem, there is a one-to-one correspondence between positive, bounded Borel measures μ on the circle and positive linear maps $\phi\colon C(\mathbb{T}) \to \mathbb{C}$ given by $\phi(f) = \int f\, d\mu$. Thus, $\{a_k\}_{k=-\infty}^{+\infty}$ is a trigonometric moment sequence precisely when setting $\phi(z^k) = a_k$, $k \in \mathbb{Z}$, extends to give a positive linear functional on $C(\mathbb{T})$.

We shall call an infinite matrix $(b_{i,j})_{i,j=0}^{\infty}$ *formally positive* if each of the finite matrices $B_N = (b_{i,j})_{i,j=0}^{N}$ is positive. (Recall that in this book positive means positive semidefinite.) Note that we are making no requirements about boundedness of B.

Theorem 2.14. *The sequence $\{a_k\}_{k=-\infty}^{+\infty}$ is a trigonometric moment sequence if and only if the Toeplitz matrix $(a_{i-j})_{i,j=0}^{\infty}$ is formally positive.*

Proof. If $\{a_k\}_{k=-\infty}^{+\infty}$ is a moment sequence with measure μ, and $p_0, \ldots, p_N$ are scalars, then

$$\sum_{i,j=0}^{N} a_{i-j}\bar{p}_j p_i = \int |p(z)|^2\, d\mu \geq 0,$$

where $p(z) = p_0 + \cdots + p_N z^N$. Hence $(a_{i-j})_{i,j=0}^{\infty}$ is formally positive.

Conversely, assume that $(a_{i-j})_{i,j=0}^{\infty}$ is formally positive. Let $\mathcal{S} \subseteq C(\mathbb{T})$ be the span of z^k, $k \in \mathbb{Z}$, which is a dense operator system in $C(\mathbb{T})$. Define $\phi\colon \mathcal{S} \to \mathbb{C}$ by $\phi(z^k) = a_k$. We wish to prove that ϕ is a positive linear functional on $\mathcal{S}$. As in the proof of Theorem 2.6, it will be enough to assume that $\tau \in \mathcal{S}$ is strictly positive and prove that $\phi(\tau) \geq 0$.

By the Fejer–Riesz lemma, $\tau(z) = \sum_{i,j=0}^{N} p_i \bar{p}_j z^{(i-j)}$ for some scalars, $p_0, \ldots, p_N$. Hence $\phi(\tau) = \sum_{i,j=0}^{N} p_i \bar{p}_j a_{i-j} \geq 0$ by the positivity of the matrix $(a_{i-j})_{i,j=0}^{N}$.

Now by Exercise 2.2, the positive map ϕ extends by continuity to a positive map on $\mathcal{S}^{-} = C(\mathbb{T})$, and hence by the Riesz representation theorem we have a bounded, positive Borel measure with $a_k = \phi(z^k) = \int z^k\, d\mu$ for all k in $\mathbb{Z}$.

□

There is another characterization of trigonometric moment sequences that is often useful. If $\{a_k\}_{k=-\infty}^{+\infty}$ is a trigonometric moment sequence with representing measure μ, then it is easily seen that $|a_k| \leq \mu([0, 2\pi]) = a_0$ and hence the power series $f(z) = a_0/2 + \sum_{k=1}^{\infty} a_k z^k$ converges to give an analytic function on the unit disk $\mathbb{D}$. For $|z| < 1$ the Poisson kernel

$$P(z, \theta) = \frac{1}{1 - e^{i\theta} z} + \frac{1}{1 - e^{-i\theta} z} - 1$$

defines a positive function on the circle, and hence

$$f(z) + \overline{f(z)} = \int P(z, \theta)\, d\mu(\theta) \geq 0.$$

Thus, we have shown that if $\{a_k\}_{k=-\infty}^{+\infty}$ is a trigonometric moment sequence then $f(z)$ is analytic on the disk and maps the disk into the right half plane.

The converse of this result holds as well, and we sketch the proof. If f, as above, is analytic on $\mathbb{D}$ and has nonnegative real part, then $g_r(e^{i\theta}) = f(re^{i\theta}) + \overline{f(re^{i\theta})}$, $r < 1$, is continuous on $\mathbb{T}$ and nonnegative. Hence the operator of multiplication by g_r on $L^2(\mathbb{T})$ is a positive operator. But the matrix of this operator with respect to the standard orthonormal basis $\{e^{ik\theta}\}_{k=-\infty}^{+\infty}$ is $(a_{i-j} r^{|i|+|j|})_{i,j=-\infty}^{+\infty}$. Letting r tend to 1 yields the formal positivity of the Toeplitz matrix $(a_{i-j})_{i,j=0}^{+\infty}$.

Notes

For a survey of the theory of positive maps, see [222].

The idea of using von Neumann's inequality to prove the Russo–Dye result seems to have originated with Choi. The usual proof involves another important result of Russo and Dye, namely, that the extreme points of the unit ball of a unital C^*-algebra are the unitary elements [206] (see [173] for an elegant proof of this result).

Lemma 2.10 is a minor adaptation of [8].

Von Neumann's original proof of his inequality [152] first showed that the inequality was met for the Möbius transformations of the disk, and then reduced verifying the inequality for a general analytic function to this special case. A later proof by Heinz [114] is based on the classical Cauchy–Poisson formula. Most other presentations rely on Sz.-Nagy's dilation theorem (Theorem 1.1). The Fejer–Riesz lemma can be found in [89].

Foias [99] has shown how particular von Neumann's inequality is to the theory of Hilbert spaces. He proves that if a Banach space has the property that every contraction operator on the Banach space satisfies von Neumann's inequality, then that space is necessarily a Hilbert space.

A result of Ando [5] implies that von Neumann's inequality holds for pairs of commuting contractions, that is,

$$\|p(T_1, T_2)\| \leq \sup\{|p(z_1, z_2)|: |z_i| \leq 1, \quad i = 1, 2\},$$

where T_1 and T_2 are commuting contractions and p is an arbitrary polynomial in two variables. Thus, by Proposition 2.12, there is a positive map

$$\phi(p + \bar{q}) = p(T_1, T_2) + q(T_1, T_2)^*.$$

It would be interesting to know if a proof of this latter fact could be given along the lines of Theorem 2.6. Such a proof could perhaps shed some additional light on the rather paradoxical results of Crabbe and Davies [69] and of Varopoulos [236], that for three or more commuting contractions, the analogue of von Neumann's inequality fails. We shall examine these ideas in detail in Chapter 5.

Exercises

2.1 Let $\mathcal{S}$ be an operator system, $\mathcal{B}$ be a C^*-algebra, and $\phi: \mathcal{S} \to \mathcal{B}$ a positive map. Prove that ϕ is *self-adjoint*, i.e., that $\phi(x^*) = \phi(x)^*$.

2.2 Let $\mathcal{S}$ be an operator system, $\mathcal{B}$ be a C^*-algebra, and $\phi: \mathcal{S} \to \mathcal{B}$ be a positive map. Prove that ϕ extends to a positive map on the norm closure of $\mathcal{S}$.

2.3 Let $\mathcal{S}$ be an operator system, and let $\phi: \mathcal{S} \to \mathbb{C}$ be positive. Prove that $\|\phi\| \leq \phi(1)$. [Hint: Given a, chosen λ, $|\lambda| = 1$, such that $|\phi(a)| = \phi(\lambda a)$.]

2.4 Let $\mathcal{S}$ be an operator system, and let $\phi: \mathcal{S} \to C(X)$, where $C(X)$ denotes the continuous functions on a compact Hausdorff space X. Prove that if ϕ is positive, then $\|\phi\| \leq \|\phi(1)\|$.

2.5 (Schwarz inequality) Let $\mathcal{A}$ be a C^*-algebra, and let $\phi: \mathcal{A} \to \mathbb{C}$ be a positive linear functional. Prove that $|\phi(x^*y)|^2 \leq \phi(x^*x)\phi(y^*y)$.

2.6 Let T be an operator on a Hilbert space $\mathcal{H}$. The *numerical radius* of T is defined by $w(T) = \sup\{|\langle Tx, x\rangle|: x \in \mathcal{H}, \|x\| = 1\}$. Prove that if $\phi: \mathcal{S} \to \mathcal{B}(\mathcal{H})$ is positive and $\phi(1) = 1$, then $w(\phi(a)) \leq \|a\|$.

2.7 Let T be an operator on a Hilbert space. Prove that $w(T) \leq 1$ if and only if $2 + (\lambda T) + (\lambda T)^* \geq 0$ for all complex numbers λ with $|\lambda| = 1$.

2.8 Prove that $w(T)$ defines a norm on $\mathcal{B}(\mathcal{H})$, with $w(T) \leq \|T\| \leq 2w(T)$. Show that both inequalities are sharp.

2.9 Let $\mathcal{S}$ be an operator system, $\mathcal{B}$ a C^*-algebra, and $\phi: \mathcal{S} \to \mathcal{B}$ a linear map such that $\phi(1)$ is positive, $\|\phi(1)\| = \|\phi\|$. Give an example to show that ϕ need not be positive. In a similar vein, show that if $\mathcal{M}$ is as in Proposition 2.12, and $\phi(1)$ is positive with $\|\phi(1)\| = \|\phi\|$, then $\tilde{\phi}$ need not be well defined.

2.10 (Krein) Let $\mathcal{S}$ be an operator system contained in the C^*-algebra $\mathcal{A}$, and let $\phi: \mathcal{S} \to \mathbb{C}$ be positive. Prove that ϕ can be extended to a positive map on $\mathcal{A}$.

2.11 Prove that for the following element of M_{n+1},

$$\left\| \begin{bmatrix} a_0 & 0 & \dots & 0 \\ a_1 & a_0 & \ddots & \vdots \\ \vdots & \ddots & \ddots & 0 \\ a_n & \dots & a_1 & a_0 \end{bmatrix} \right\| \leq \inf_{r(z)} \{\|a_0 + a_1 z + \cdots + a_n z^n + r(z)\|\},$$

where $r(z)$ is a polynomial whose lowest-degree term is strictly greater than n, and the latter norm is the supremum norm over the unit disk.

2.12 (Jorgensen) Show that $C = \{p^2(t)\colon p$ is a polynomial with real coefficients$\}$ is dense in $C([0, 1])^+$. Let $\mathcal{S} = \{p + \bar{q}\colon p, q$ are polynomials$\}$, and set $\phi(p + \bar{q}) = p(2) + \overline{q(2)}$. Show that ϕ is positive on C and well defined on the dense subset $\mathcal{S}$ of $C([0, 1])$, but still does not extend to a positive map on $C([0, 1])$. Compare with Proposition 2.1 and Exercise 2.2.

2.13 Let X be a compact subset of $\mathbb{C}$. Prove that if X is a finite set, then X is a spectral set for T if and only if T is a normal operator with $\sigma(T)$ contained in X. Prove that if X is a subset of $\mathbb{R}$, then X is a spectral set for T if and only if $T = T^*$ and $\sigma(T)$ is contained in X. Prove that if X is contained in the unit circle, then X is a spectral set for T if and only if T is a unitary with $\sigma(T)$ contained in X.

2.14 (von Neumann) Let X be a compact subset of $\mathbb{C}$, with $R(X)$ dense in $C(\partial X)$. Prove that X is a spectral set for T if and only if T is normal and $\sigma(T)$ is contained in ∂X.

2.15 In this exercise we give an alternative proof of von Neumann's inequality. We assume that the reader has some familiarity with integration of operator-valued functions. Let $T \in \mathcal{B}(\mathcal{H})$ with $\|T\| < 1$, and let p and q denote arbitrary polynomials.
 (i) Let $P(t; T) = (1 - e^{-it}T)^{-1} + (1 - e^{it}T^*)^{-1} - 1$, and show that $P(t; T) \geq 0$ for all t,
 (ii) Show that $p(T) + q(T)^* = \frac{1}{2\pi} \int_0^{2\pi} (p(e^{it}) + \overline{q(e^{it})}) P(t; T)\, dt$,
 (iii) Deduce von Neumann's inequality.

2.16 (Wermer) In this exercise we give an alternative proof of von Neumann's inequality that is only valid for matrices. We assume that the reader is familiar with the singular-value decomposition of a matrix. Let

$T \in M_n$ with $\|T\| \leq 1$, and write $T = USV$ with U, V unitary and $S = \text{diag}(s_1, \ldots, s_n)$ a positive diagonal matrix, $0 \leq s_i \leq 1, i = 1, \ldots, n$. Define an analytic matrix-valued function $T(z_1, \ldots, z_n) = UZV$ where $Z = \text{diag}(z_1, \ldots, z_n), |z_i| \leq 1, i = 1, \ldots, n$. Fix a polynomial p.

(i) Let $x, y \in \mathbb{C}^n$, and let $f(z_1, \ldots, z_n) = \langle p(T(z_1, \ldots, z_n))x, y\rangle$. Deduce that f achieves its maximum modulus at a point where $|z_1| = \cdots = |z_n| = 1$. Note that at such a point $T(z_1, \ldots, z_n)$ is unitary.

(ii) Deduce that $\sup_{|z_i| \leq 1} \|p(T(z_1, \ldots, z_n))\|$ occurs at a point where $T(z_1, \ldots, z_n)$ is unitary.

(iii) Deduce that $\|p(T)\| \leq \sup \|p(W)\|$ over $W \in M_n$, unitary.

(iv) Show that for W unitary, $\|p(W)\| \leq \|p\|_\infty$.

(v) Deduce von Neumann's inequality for $T \in M_n$.

2.17 Let $T = \left(\begin{smallmatrix} a & c \\ 0 & b \end{smallmatrix}\right) \in M_2, a \neq b$.

(i) Show that for any polynomial p,

$$p(T) = \begin{pmatrix} p(a) & c(p(a) - p(b))/(a - b) \\ 0 & p(b) \end{pmatrix}.$$

(ii) Show that if $|a| \leq 1, |b| \leq 1, f \in A(\mathbb{D})$, and p_n is a sequence of polynomials with $\|f - p_n\|_\infty \to 0$, then

$$p_n(T) \to \begin{pmatrix} f(a) & c(f(a) - f(b))/(a - b) \\ 0 & f(b) \end{pmatrix}$$

entrywise and hence in norm.

(iii) Assume $|a|, |b| < 1$. Let $\varphi(z) = (z - b)/(1 - \bar{b}z)$. Show that $\|T\| \leq 1$ if and only if

$$|\varphi(a)|^2(|a - b|^2 + |c|^2) \leq |a - b|^2.$$

(iv) What can you say when $|a|, |b| \leq 1$?

2.18 Let $T = \left(\begin{smallmatrix} a & c \\ 0 & a \end{smallmatrix}\right) \in M_2$.

(i) Show that for any polynomial p,

$$p(T) = \begin{pmatrix} p(a) & cp'(a) \\ 0 & p(a) \end{pmatrix}.$$

(ii) Show that if $|a| < 1, f \in A(\mathbb{D})$, and p_n is a sequence of polynomials with $\|f - p_n\|_\infty \to 0$, then

$$p_n(T) \to \begin{pmatrix} f(a) & cf'(a) \\ 0 & f(a) \end{pmatrix}$$

entrywise and hence in norm.

(iii) Let $|a| < 1$ and $\varphi(z) = (z - a)/(1 - \bar{a}z)$. Show that $\|T\| \le 1$ if and only if $|c| \le 1 - |a|^2$.

(iv) What can you say when $|a| = 1$?

(v) Let $|a| < 1$, set $c = 1 - |a|^2$, and use von Neumann's inequality to deduce that if $\|f\| \le 1$, then $|f'(a)| \le (1 - |f(a)|^2)/(1 - |a|^2)$.

2.19 Let

$$T = \begin{pmatrix} a & b & c \\ 0 & a & b \\ 0 & 0 & a \end{pmatrix} \in M_3.$$

Obtain necessary and sufficient conditions for $\|T\| \le 1$.

2.20 (Korovkin) Let $f \in C([0, 1])$ and let $g_x(t) = (t - x)^2$

(i) Given $\epsilon > 0$, show that there exists a constant $c > 0$ depending only on ϵ and f such that

$$|f(t) - f(x)| \le \epsilon + cg_x(t) \quad \text{for all} \quad 0 \le t, x \le 1.$$

[Hint: There is a $\delta > 0$ such that $|f(t) - f(x)| \le \epsilon$ for all $|t - x| < \delta$.]

(ii) Let $\phi\colon C([0, 1]) \to C([0, 1])$ be a positive map with $\phi(1) = 1$. Show that

$$-\epsilon - c\phi(g_x)(x) \le \phi(f)(x) - f(x) \le \epsilon + c\phi(g_x)(x),$$

and deduce that $\|\phi(f) - f\| \le \epsilon + c \sup_x |\phi(g_x)(x)|$.

(iii) Let $\phi_n\colon C([0, 1]) \to C([0, 1])$ be a sequence of positive maps. Prove that if $\|\phi_n(f_i) - f_i\| \to 0$ as $n \to \infty$ for $f_i(t) = t^i$, $i = 0, 1$, and 2, then $\|\phi_n(f) - f\| \to 0$ as $n \to \infty$ for all $f \in C([0, 1])$.

2.21 The *Bernstein maps* $\phi_n\colon C([0, 1]) \to C([0, 1])$ are defined by

$$\phi_n(f)(t) = \sum_{k=0}^{n} \binom{n}{k} f\left(\frac{k}{n}\right) t^k (1 - t)^{n-k}.$$

(i) Verify that the Bernstein maps are positive maps with $\phi_n(1) = 1$, $\phi_n(t) = t$, $\phi_n(t^2) = t^2 + \frac{t-t^2}{n}$. [Hint: Use $\binom{n}{k}\frac{k}{n} = \binom{n-1}{k-1}$.]

(ii) Deduce that $\|\phi_n(f) - f\|_\infty \to 0$ for all $f \in C([0, 1])$

(iii) Deduce the Weierstrass theorem, i.e., prove that the polynomials are dense in $C([0, 1])$.

2.22 A sequence $\{a_n\}_{n=0}^{\infty}$ of complex numbers is called a *Hausdorff moment sequence* if there exists a positive (finite) Borel measure μ on [0,1] such that $a_n = \int_0^1 t^n \, d\mu(t)$ for all n. Set $b_{n,m} = \sum_{k=0}^{n} \binom{n}{k} (-1)^k a_{k+m}$, for all $n, m \ge 0$.

(i) Assuming the existence of such a measure μ, show that $b_{n,m} = \int_0^1 t^m(t-1)^n \, d\mu(t)$, and deduce that necessarily $b_{n,m} \geq 0$ for all $n, m \geq 0$.

(ii) Let $\mathcal{P} \subseteq C([0, 1])$ denote the span of the polynomials, and define $\phi: \mathcal{P} \to \mathbb{C}$ by setting $\phi(t^n) = a_n$. Show that if $b_{n,m} \geq 0$ for all $n, m \geq 0$ then ϕ is a positive map. (Caution: You do not know ϕ is continuous! You must use Exercises 2.20 *and* 2.21.)

(iii) Prove that $\{a_n\}_{n=0}^{\infty}$ is a Hausdorff moment sequence if and only if $b_{n,m} \geq 0$ for all $n, m \geq 0$.

Chapter 3
Completely Positive Maps

Let $\mathcal{A}$ be a C^*-algebra, and let $\mathcal{M}$ be a subspace. Then we shall call $\mathcal{M}$ an *operator space*. Clearly, $M_n(\mathcal{M})$ can be regarded as a subspace of $M_n(\mathcal{A})$, and we let $M_n(\mathcal{M})$ have the norm structure that it inherits from the (unique) norm structure on the C^*-algebra $M_n(\mathcal{A})$. We make no attempt at this time to define a norm structure on $M_n(\mathcal{M})$ without reference to $\mathcal{A}$. Thus, one thing that distinguishes $\mathcal{M}$ from an ordinary normed space is that it comes naturally equipped with norms on $M_n(\mathcal{M})$ for all $n \geq 1$. Later in this book we shall give a more axiomatic definition of operator spaces, at which time we shall begin to refer to subspaces of C^*-algebras as *concrete operator spaces*. For now we simply stress that by an operator space $\mathcal{M}$ we mean a concrete subspace of a C^*-algebra, together with this extra "baggage" of a well-defined sequence of norms on $M_n(\mathcal{M})$. Similarly, if $\mathcal{S} \subseteq \mathcal{A}$ is an operator system, then we endow $M_n(\mathcal{S})$ with the norm and order structure that it inherits as a subspace of $M_n(\mathcal{A})$.

As before, if $\mathcal{B}$ is a C^*-algebra and $\phi \colon \mathcal{S} \to \mathcal{B}$ is a linear map, then we define $\phi_n \colon M_n(\mathcal{S}) \to M_n(\mathcal{B})$ by $\phi_n((a_{i,j})) = (\phi(a_{i,j}))$. We call ϕ *n-positive* if ϕ_n is positive, and we call ϕ *completely positive* if ϕ is n-positive for all n. We call ϕ *completely bounded* if $\sup_n \|\phi_n\|$ is finite, and we set

$$\|\phi\|_{\mathrm{cb}} = \sup_n \|\phi_n\|.$$

Note that $\|\cdot\|_{\mathrm{cb}}$ is a norm on the space of completely bounded maps. We use the terms *completely isometric* and *completely contractive* to indicate that each ϕ_n is isometric and that $\|\phi\|_{\mathrm{cb}} \leq 1$, respectively. We note that if ϕ is n-positive, then ϕ is k-positive for $k \leq n$. Also, $\|\phi_k\| \leq \|\phi_n\|$ for $k \leq n$ (Exercise 3.1).

In this chapter we investigate some of the elementary properties of these classes of maps and prove some theorems about when positive maps are automatically completely positive. We begin by relating some of the results of the previous chapter to these concepts.

Lemma 3.1. *Let $\mathcal{A}$ be a C^*-algebra with unit, and let a and b belong to $\mathcal{A}$. Then:*

(i) $\|a\| \leq 1$ *if and only if*

$$\begin{bmatrix} 1 & a \\ a^* & 1 \end{bmatrix}$$

is positive in $M_2(\mathcal{A})$.

(ii) $\left[\begin{smallmatrix} 1 & a \\ a^* & b \end{smallmatrix}\right]$ *is positive in $M_2(\mathcal{A})$ if and only if $a^*a \leq b$.*

Proof. We represent $\mathcal{A}$ on a Hilbert space $\mathcal{H}$ via $\pi\colon \mathcal{A} \to \mathcal{B}(\mathcal{H})$ and set $A = \pi(a)$.

If $\|A\| \leq 1$, then for any vectors x, y in $\mathcal{H}$,

$$\left\langle \begin{bmatrix} I & A \\ A^* & I \end{bmatrix} \begin{bmatrix} x \\ y \end{bmatrix}, \begin{bmatrix} x \\ y \end{bmatrix} \right\rangle = \langle x, x\rangle + \langle Ay, x\rangle + \langle x, Ay\rangle + \langle y, y\rangle$$

$$\geq \|x\|^2 - 2\|A\|\|y\|\|x\| + \|y\|^2 \geq 0.$$

Conversely, if $\|A\| > 1$, then there exist unit vectors x and y such that $\langle Ay, x\rangle < -1$ and the above inner product will be negative.

The proof of (ii) is similar and is left as an exercise [Exercise 3.2(ii)]. □

Proposition 3.2. *Let $\mathcal{S}$ be an operator system, $\mathcal{B}$ a C^*-algebra with unit, and $\phi\colon \mathcal{S} \to \mathcal{B}$ a unital, 2-positive map. Then ϕ is contractive.*

Proof. Let $a \in \mathcal{S}$, $\|a\| \leq 1$. Then

$$\phi_2 \begin{bmatrix} 1 & a \\ a^* & 1 \end{bmatrix} = \begin{bmatrix} 1 & \phi(a) \\ \phi(a)^* & 1 \end{bmatrix}$$

is positive and hence $\|\phi(a)\| \leq 1$. □

Proposition 3.3 (Schwarz inequality for 2-positive maps). *Let $\mathcal{A}, \mathcal{B}$ be unital C^*-algebras, and let $\phi\colon \mathcal{A} \to \mathcal{B}$ be a unital 2-positive map. Then $\phi(a)^*\phi(a) \leq \phi(a^*a)$ for all a in $\mathcal{A}$.*

Proof. We have that $\left[\begin{smallmatrix} 1 & a \\ 0 & 0 \end{smallmatrix}\right]^* \left[\begin{smallmatrix} 1 & a \\ 0 & 0 \end{smallmatrix}\right] = \left[\begin{smallmatrix} 1 & a \\ a^* & a^*a \end{smallmatrix}\right] \geq 0$ and hence

$$\begin{bmatrix} 1 & \phi(a) \\ \phi(a)^* & \phi(a^*a) \end{bmatrix} \geq 0.$$

By Lemma 3.1(ii), we have the result. □

Proposition 3.4. *Let $\mathcal{A}$ and $\mathcal{B}$ be C^*-algebras with unit, let $\mathcal{M}$ be a subspace of $\mathcal{A}$, $1 \in \mathcal{M}$, and let $\mathcal{S} = \mathcal{M} + \mathcal{M}^*$. If $\phi\colon \mathcal{M} \to \mathcal{B}$ is unital and 2-contractive (i.e., $\|\phi_2\| \le 1$), then the map $\tilde{\phi}\colon \mathcal{S} \to \mathcal{B}$ given by $\tilde{\phi}(a + b^*) = \phi(a) + \phi(b)^*$ is 2-positive and contractive.*

Proof. Since ϕ is contractive, $\tilde{\phi}$ is well defined by Proposition 2.12. Note that $M_2(\mathcal{S}) = M_2(\mathcal{M}) + M_2(\mathcal{M})^*$ and that $(\tilde{\phi})_2 = (\tilde{\phi_2})$. Since ϕ_2 is contractive, we have again by Proposition 2.12, that $\tilde{\phi}_2$ is positive and so $\tilde{\phi}$ is contractive by Proposition 3.2. □

Proposition 3.5. *Let $\mathcal{A}$ and $\mathcal{B}$ be C^*-algebras with unit, let $\mathcal{M}$ be a subspace of $\mathcal{A}$, $1 \in \mathcal{M}$, and let $\mathcal{S} = \mathcal{M} + \mathcal{M}^*$. If $\phi\colon \mathcal{M} \to \mathcal{B}$ is unital and completely contractive, then $\tilde{\phi}\colon \mathcal{S} \to \mathcal{B}$ is completely positive and completely contractive.*

Proof. We have that $\tilde{\phi}_n$ is positive, since ϕ_n is unital and contractive, and $\tilde{\phi}_n$ is contractive, since $\tilde{\phi}_{2n} = (\tilde{\phi}_n)_2$ is positive. □

We've glossed over one point in the above proof. Namely, we've identified $M_{2n}(\mathcal{S})$ with $M_2(M_n(\mathcal{S}))$. It's obvious how to do this: one simply "erases" the additional brackets in an element of $M_2(M_n(\mathcal{S}))$. One must check, however, that the norms one defines are the same in each instance. One has that $M_2(M_n(\mathcal{S}))$ inherits its norm from $M_2(M_n(\mathcal{A}))$, while $M_{2n}(\mathcal{S})$ inherits its norm from $M_{2n}(\mathcal{A})$. However, this "erasure" operation defines a $*$-isomorphism between $M_2(M_n(\mathcal{A}))$ and $M_{2n}(\mathcal{A})$; thus the norms are indeed the same.

Now that we've seen some of the advantages of considering maps in these two classes, let's begin by describing some maps that belong to them. Let $\mathcal{A}$ and $\mathcal{B}$ be C^*-algebras. First, note that if $\pi\colon \mathcal{A} \to \mathcal{B}$ is a $*$-homomorphism, then π is completely positive and completely contractive, since each map, $\pi_n\colon M_n(\mathcal{A}) \to M_n(\mathcal{B})$, is a $*$-homomorphism, and $*$-homomorphisms are both positive and contractive. For a second class of maps, fix x and y in $\mathcal{A}$ and define $\phi\colon \mathcal{A} \to \mathcal{A}$ by $\phi(a) = xay$. Note that if $(a_{i,j})$ is in $M_n(\mathcal{A})$, then

$$
\begin{aligned}
\|\phi_n((a_{i,j}))\| &= \|(xa_{i,j}y)\| \\
&= \left\| \begin{bmatrix} x & 0 & \dots & 0 \\ 0 & \ddots & \ddots & \vdots \\ \vdots & \ddots & \ddots & 0 \\ 0 & \dots & 0 & x \end{bmatrix} \begin{bmatrix} a_{11} & \dots & a_{1n} \\ \vdots & & \vdots \\ a_{n1} & \dots & a_{nn} \end{bmatrix} \begin{bmatrix} y & 0 & \dots & 0 \\ 0 & \ddots & \ddots & \vdots \\ \vdots & \ddots & \ddots & 0 \\ 0 & \dots & 0 & y \end{bmatrix} \right\| \\
&\le \|x\| \cdot \|(a_{i,j})\| \cdot \|y\|.
\end{aligned}
$$

Thus, ϕ is completely bounded, and $\|\phi\|_{cb} \le \|x\| \cdot \|y\|$. A similar calculation shows that if $x = y^*$, then ϕ is completely positive.

Combining these two examples, we obtain the prototypical example of a completely bounded map. Namely, let $\mathcal{H}_1$ and $\mathcal{H}_2$ be Hilbert spaces, let $v_i\colon \mathcal{H}_1 \to \mathcal{H}_2$, $i = 1, 2$, be bounded operators, and let $\pi\colon \mathcal{A} \to \mathcal{B}(\mathcal{H}_2)$ be a $*$-homomorphism. Define a map $\phi\colon \mathcal{A} \to \mathcal{B}(\mathcal{H}_1)$ via $\phi(a) = v_2^*\pi(a)v_1$. Then ϕ is completely bounded with $\|\phi\|_{cb} \le \|v_1\| \cdot \|v_2\|$, and if $v_1 = v_2$, then ϕ is completely positive. We shall prove in Chapter 8 that all completely bounded maps have this form.

In each of the above examples, we see that the completely positive maps are all completely bounded. This is always the case.

Proposition 3.6. *Let $\mathcal{S} \subseteq \mathcal{A}$ be an operator system, let $\mathcal{B}$ be a C^*-algebra, and let $\phi\colon \mathcal{S} \to \mathcal{B}$ be completely positive. Then ϕ is completely bounded and $\|\phi(1)\| = \|\phi\| = \|\phi\|_{cb}$.*

Proof. Clearly, we have that $\|\phi(1)\| \le \|\phi\| \le \|\phi\|_{cb}$, so it is sufficient to show $\|\phi\|_{cb} \le \|\phi(1)\|$. To this end, let $A = (a_{i,j})$ be in $M_n(\mathcal{S})$ with $\|A\| \le 1$, and let I_n be the unit of $M_n(\mathcal{A})$, i.e., the diagonal matrix with 1's on the diagonal. Since

$$\begin{bmatrix} I_n & A \\ A^* & I_n \end{bmatrix}$$

is positive, we have that

$$\phi_{2n}\left(\begin{bmatrix} I_n & A \\ A^* & I_n \end{bmatrix}\right) = \begin{bmatrix} \phi_n(I_n) & \phi_n(A) \\ \phi_n(A)^* & \phi_n(I_n) \end{bmatrix}$$

is positive. Thus, by Exercise 3.2(iii), $\|\phi_n(A)\| \le \|\phi_n(I_n)\| = \|\phi(1)\|$, which completes the proof. □

Schur Products and Tensor Products

As an application of the above result, we study the *Schur product* of matrices. If $A = (a_{i,j})$, $B = (b_{i,j})$ are elements of M_n, then we define the Schur product by

$$A * B = (a_{i,j} \cdot b_{i,j}).$$

For fixed A, this gives rise to a linear map,

$$S_A\colon M_n \to M_n, \quad \text{via} \quad S_A(B) = A * B.$$

In order to study this map, we need to recall a few facts about tensor products of matrices. Let A be in M_n and B be in M_m, so that A and B can be thought of as linear transformations on $\mathbb{C}^n$ and $\mathbb{C}^m$, respectively. Then $A \otimes B$ is the linear transformation on $\mathbb{C}^n \otimes \mathbb{C}^m \simeq \mathbb{C}^{nm}$, which is defined by setting $A \otimes B(x \otimes y) = Ax \otimes By$ and extending linearly. Writing $A \otimes B = (A \otimes I)(I \otimes B)$, it is easy to see that $\|A \otimes B\| = \|A\| \cdot \|B\|$.

Let $\{e_1, \ldots, e_n\}$ and $\{f_1, \ldots, f_m\}$ be the canonical orthonormal bases for $\mathbb{C}^n$ and $\mathbb{C}^m$, respectively. If we order our basis for $\mathbb{C}^{nm}$ by $e_1 \otimes f_1, e_1 \otimes f_2, \ldots, e_1 \otimes f_m, e_2 \otimes f_1, \ldots, e_n \otimes f_m$, then the matrix for $A \otimes B$ with respect to this ordered basis is given in block form by

$$\begin{bmatrix} a_{11}B & \ldots & a_{1n}B \\ \vdots & & \vdots \\ a_{n1}B & \ldots & a_{nn}B \end{bmatrix}.$$

This last matrix is often referred to as the *Kronecker product* of A and B.

On the other hand, if we order our basis for $\mathbb{C}^{nm}$ by $e_1 \otimes f_1, e_2 \otimes f_1, \ldots, e_n \otimes f_1, e_1 \otimes f_2, e_2 \otimes f_2, \ldots, e_n \otimes f_m$, then the matrix for $A \otimes B$ with respect to this ordered basis is given in block form by

$$\begin{bmatrix} b_{11}A & \ldots & b_{1m}A \\ \vdots & & \vdots \\ b_{m1}A & \ldots & b_{mmA} \end{bmatrix},$$

which is the Kronecker product of B and A.

Since the two block matrices given above represent the same linear transformation, they are unitarily equivalent. The unitary matrix that implements this unitary equivalence is simply the permutation matrix corresponding to the given reordering of the basis vectors.

Note that one obtained the (i, j)th block of the second matrix, $b_{ij}A$, by taking the (i, j) entry of the (k, ℓ)th block, $a_{k,\ell}B$, of the first matrix. We shall encounter this rearrangement of matrix entries repeatedly in this book. We'll refer to it as the *canonical shuffle*.

Now let A and B be in M_n, and define an isometry $V\colon \mathbb{C}^n \to \mathbb{C}^n \otimes \mathbb{C}^n$ by $V(e_i) = e_i \otimes e_i$. A simple calculation shows that

$$V^*(A \otimes B)V = A * B.$$

To see this, note that

$$\begin{aligned} \langle V^*(A \otimes B)Ve_j, e_i\rangle &= \langle A \otimes B(e_j \otimes e_j), (e_i \otimes e_i)\rangle \\ &= \langle Ae_j, e_i\rangle \cdot \langle Be_j, e_i\rangle = a_{i,j} \cdot b_{i,j} = \langle A * Be_j, e_i\rangle. \end{aligned}$$

Thus, $\|S_A(B)\| = \|V^*(A \otimes B)V\| \le \|A\| \cdot \|B\|$ and so $\|S_A\| \le \|A\|$.

Similarly, if $(B_{i,j})$ is in $M_k(M_n)$, then

$$(S_A)_k((B_{i,j})) = (V^*(A \otimes B_{i,j})V)$$

$$= \begin{bmatrix} V^* & 0 & \dots & 0 \\ 0 & \ddots & \ddots & \vdots \\ \vdots & \ddots & \ddots & 0 \\ 0 & \dots & 0 & V^* \end{bmatrix} A \otimes \begin{bmatrix} B_{11} & \dots & B_{1n} \\ \vdots & & \vdots \\ B_{n1} & \dots & B_{nn} \end{bmatrix} \begin{bmatrix} V & 0 & \dots & 0 \\ 0 & \ddots & \ddots & \vdots \\ \vdots & \ddots & \ddots & 0 \\ 0 & \dots & 0 & V \end{bmatrix}$$

and so, $\|(S_A)_k\| \le \|A\|$ also. Thus, $\|S_A\|_{\text{cb}} \le \|A\|$.

This estimate is not very good in general. For example, the identity map is the Schur product against the matrix of all 1's, and this matrix has norm n.

If A is positive, then we shall prove below that S_A is completely positive. Hence, for positive matrices we can obtain $\|S_A\|_{\text{cb}}$ explicitly, by invoking Proposition 3.5. Namely,

$$\|S_A\| = \|S_A(I)\| = \|S_A\|_{\text{cb}} = \max\{a_{i,i} \colon i = 1, \dots, n\}.$$

It is more difficult to calculate $\|S_A\|_{\text{cb}}$ when A is not positive. We return to this topic in Chapter 8. Clearly, if one decomposes $A = (P_1 - P_2) + i(P_3 - P_4)$ with P_i positive, then $S_A = (S_{P_1} - S_{P_2}) + i(S_{P_3} - S_{P_4})$. Thus, $\|S_A\|_{\text{cb}} \le \|S_{P_1}\|_{\text{cb}} + \|S_{P_2}\|_{\text{cb}} + \|S_{P_3}\|_{\text{cb}} + \|S_{P_4}\|_{\text{cb}}$, and each of the right hand terms is given by the maximum diagonal entry of the corresponding matrix. However, we shall see in Chapter 8 that this estimate can be far from $\|S_A\|_{\text{cb}}$.

The following characterizes when a Schur product map is completely positive.

Theorem 3.7. *Let $A = (a_{ij})$ be in M_n. Then the following are equivalent:*

(i) A is positive,
(ii) $S_A \colon M_n \to M_n$ is positive,
(iii) $S_A \colon M_n \to M_n$ is completely positive.

Proof. Clearly (iii) implies (ii). Note that the matrix J, all of whose entries are 1, is positive, and that $S_A(J) = A$. Hence, (ii) implies (i). It remains to prove that (i) implies (iii).

First note that if A and B are positive then $A \otimes B$ is positive. To see this, note that $A \otimes B = (A^{1/2} \otimes B^{1/2})^2$. Now if $B \in M_n$ is positive, then

$$S_A(B) = V^*(A \otimes B)V = \left[\left(A^{1/2} \otimes B^{1/2}\right)V\right]^* \left[\left(A^{1/2} \otimes B^{1/2}\right)V\right]$$

is positive. Hence, (i) implies (ii). To see that (i) implies (iii), let $B = (B_{ij}) \in$

$M_k(M_n)$ be positive, write $B = (X_{ij})^*(X_{ij})$, and observe that

$$(S_A)_k(B) = (V^*(A \otimes B_{ij})V) = Y^*Y,$$

where $Y = ((A^{1/2} \otimes X_{ij})V)$. □

There is an analogous theory of Schur product maps on $B(\ell^2)$ where we regard bounded operators as infinite matrices. If we demand that A be a bounded positive operator, then the arguments above can be used to show that S_A is completely positive. However, the requirement that A be bounded is very restrictive. For example, the identity map on $B(\ell^2)$ is the Schur product against the infinite matrix of all 1's, and this is not the matrix of a bounded operator. We shall leave these more delicate questions until we return to Schur product maps in Chapter 8.

We now return to results about general completely positive maps. The next result shows that for linear functionals, the adverb "completely" introduces nothing new.

Proposition 3.8. *Let $\mathcal{S}$ be an operator space, and let $f\colon \mathcal{S} \to \mathbb{C}$ be a bounded linear functional. Then $\|f\|_{\text{cb}} = \|f\|$. Furthermore, if $\mathcal{S}$ is an operator system and f is positive, then f is completely positive.*

Proof. Let $(a_{i,j})$ be in $M_n(\mathcal{S})$, and let $x = (x_1, \ldots, x_n)$, $y = (y_1, \ldots, y_n)$ be unit vectors in $\mathbb{C}^n$. We have that

$$|\langle f_n((a_{i,j}))x, y\rangle| = \left|\sum_{i,j} f(a_{i,j})x_j\bar{y}_i\right| = \left|f\left(\sum_{i,j} a_{i,j}x_j\bar{y}_i\right)\right|$$
$$\leq \|f\| \cdot \left\|\sum_{i,j} a_{i,j}x_j\bar{y}_i\right\|.$$

Thus, we must show that this latter element has norm less than $\|(a_{i,j})\|$. To see this, note that the above sum is the (1,1) entry of the product

$$\begin{bmatrix} \bar{y}_1 \cdot 1 & \ldots & \bar{y}_n \cdot 1 \\ 0 & \ldots & 0 \\ \vdots & & \vdots \\ 0 & \ldots & 0 \end{bmatrix} \begin{bmatrix} a_{11} & \ldots & a_{1n} \\ \vdots & & \vdots \\ a_{n1} & \ldots & a_{nn} \end{bmatrix} \begin{bmatrix} x_1 \cdot 1 & 0 & \ldots & 0 \\ \vdots & \vdots & & \vdots \\ x_n \cdot 1 & 0 & \ldots & 0 \end{bmatrix},$$

and that the outer two factors each have norm equal to one, since x and y were chosen to be unit vectors.

To prove that f is completely positive whenever f is positive reduces to showing that $\langle f_n((a_{i,j}))x, x\rangle = f(\sum_{i,j} a_{i,j}x_j\bar{x}_i)$ is positive whenever $(a_{i,j})$ is positive. But using the above product with $x = y$, we see that the summation that f is being evaluated at is the (1,1) entry of a positive matrix and hence is positive. □

Let X be a compact Hausdorff space, and let $C(X)$ be the C^*-algebra of continuous functions on X. Note that every element $F = (f_{i,j})$ of $M_n(C(X))$ can be thought of as a continuous matrix-valued function and that multiplication and the $*$-operation in $M_n(C(X))$ are just the pointwise multiplication and $*$-operation of the matrix-valued functions. Thus, one way to make $M_n(C(X))$ into a C^*-algebra is to set $\|F\| = \sup\{\|F(x)\|: x \in X\}$, and by the uniqueness of C^*-norms, this is the only way. With these observations, the following is a direct consequence of Proposition 3.8.

Theorem 3.9. *Let $\mathcal{S}$ be an operator space, and let $\phi: \mathcal{S} \to C(X)$ be a bounded linear map. Then $\|\phi\|_{\text{cb}} = \|\phi\|$. Furthermore, if $\mathcal{S}$ is an operator system and ϕ is positive, then ϕ is completely positive.*

Proof. Let $x \in X$, and define $\phi^x: \mathcal{S} \to \mathbb{C}$ by $\phi^x(a) = \phi(a)(x)$. By the above observations,

$$\|\phi_n\| = \sup\left\{\left\|\phi_n^x\right\|: x \in X\right\} = \sup\{\|\phi^x\|: x \in X\} = \|\phi\|.$$

Similarly, $\phi_n((a_{i,j}))$ is positive if and only if $\phi_n^x((a_{i,j}))$ is positive for all $x \in X$, from which the second statement follows. □

Thus, when the range C^*-algebra is commutative, the concepts of bounded and completely bounded coincide, as do positive and completely positive. A commutative domain is also enough to ensure that positive maps are completely positive, as we shall prove shortly. However, we show in Chapter 14 that it is not enough to guarantee that bounded maps are completely bounded.

Lemma 3.10. *Let $(p_{i,j})$ be a positive scalar matrix, and let q be a positive element of some C^*-algebra $\mathcal{B}$. Then $(q \cdot p_{i,j})$ is positive in $M_n(\mathcal{B})$.*

Proof. Straightforward. □

Theorem 3.11 (Stinespring). *Let $\mathcal{B}$ be a C^*-algebra, and let $\phi: C(X) \to \mathcal{B}$ be positive. Then ϕ is completely positive.*

Proof. Let $P(x)$ be positive in $M_n(C(X))$. We must prove that $\phi_n(P)$ is positive. Given $\varepsilon > 0$ and arguing as in Theorem 2.4, we obtain a partition of unity $\{u_\ell(x)\}$

and positive matrices $P_\ell = (p^\ell_{i,j})$ such that

$$\left\| P(x) - \sum_\ell u_\ell(x) P_\ell \right\| < \varepsilon.$$

But $\phi_n(u_\ell \cdot P_\ell) = \phi_n((u_\ell \cdot p^\ell_{i,j})) = (\phi(u_\ell) \cdot p^\ell_{i,j})$, which is positive by Lemma 3.10. Thus, $\phi_n(P)$, to within $\varepsilon\|\phi_n\|\|P\|$, is a sum of positive elements. Since $M_n(\mathcal{B})^+$ is a closed set, we have that $\phi_n(P)$ is positive. □

The original proof of Theorem 3.11 [221] is quite different.

As an immediate application of Theorem 3.11 we have the following matrix-valued version of von Neumann's inequality.

Corollary 3.12. *Let T be an operator on a Hilbert space $\mathcal{H}$ with $\|T\| \leq 1$, and let $(p_{i,j})$ be an $n \times n$ matrix of polynomials. Then*

$$\|(p_{i,j}(T))\|_{B(\mathcal{H}^{(n)})} \leq \sup\{\|(p_{ij}(z))\|_{M_n} \colon |z| = 1\}.$$

Proof. The map given by $\phi(p + \bar{q}) = p(T) + q(T)^*$ extends to a positive map $\phi \colon C(\mathbb{T}) \to B(\mathcal{H})$. By Theorem 3.11 this map is completely positive, and so by Proposition 3.6, $\|\phi\|_{\text{cb}} = \|\phi(1)\| = 1$. Hence,

$$\|(p_{ij}(T))\|_{B(\mathcal{H}^{(n)})} = \|\phi_n((p_{ij}))\| \leq \|(p_{ij})\|_{M_n(C(\mathbb{T}))},$$

and the result follows. □

We've seen that for a commutative domain or range, positivity implies complete positivity. We shall see that if the domain or range isn't too badly noncommutative, then slightly stronger hypotheses than positivity will be enough to imply complete positivity. For now we restrict the domain.

Lemma 3.13. *Let $\mathcal{A}$ be a C^*-algebra. Then every positive element of $M_n(\mathcal{A})$ is a sum of n positive elements of the form $(a_i^* a_j)$ for some $\{a_1, \ldots, a_n\} \subseteq A$.*

Proof. We remark that if we let R be the element of $M_n(\mathcal{A})$ whose kth row is $a_1, \ldots, a_n$ and whose other entries are 0, then $R^*R = (a_i^* a_j)$, so such an element is positive. Now let P be positive, so $P = B^*B$, and write $B = R_1 + \cdots + R_n$, where R_k is the kth row of B and 0 elsewhere.

We have that $P = B^*B = R_1^*R_1 + \cdots + R_n^*R_n$, since $R_i^*R_j = 0$ when $i \neq j$. □

We note that by the above lemma, to verify that $\phi\colon \mathcal{A} \to \mathcal{B}$ is n-positive it is sufficient to check that $(\phi(a_i^* a_j))$ is positive for all $\{a_1, \ldots, a_n\}$ in $\mathcal{A}$.

Theorem 3.14 (Choi). *Let $\mathcal{B}$ be a C^*-algebra, let $\phi\colon M_n \to \mathcal{B}$, and let $\{E_{i,j}\}_{i,j=1}^n$ denote the standard matrix units for M_n. The following are equivalent:*

(i) ϕ is completely positive.
(ii) ϕ is n-positive.
(iii) $(\phi(E_{i,j}))_{i,j=1}^n$ is positive in $M_n(\mathcal{B})$.

Proof. Clearly, (i) implies (ii), and since $(E_{i,j})_{i,j=1}^n$ is positive, (ii) implies (iii). Thus, we shall prove that (iii) implies (i).

For this it is sufficient to assume that $\mathcal{B} = B(\mathcal{H})$. Fix k, and let $x_1, \ldots, x_k$ belong to $\mathcal{H}$ and $B_1, \ldots, B_k$ belong to M_n. By the above lemma, it is sufficient to prove that $\sum_{i,j} \langle \phi(B_i^* B_j) x_j, x_i \rangle$ is positive.

Write $B_\ell = \sum_{r,s=1}^n b_{r,s,\ell} E_{r,s}$ so that

$$B_i^* B_j = \sum_{r,s,t=1}^n \bar{b}_{r,s,i} b_{r,t,j} E_{s,t}.$$

Set $y_{t,r} = \sum_{j=1}^k b_{r,t,j} x_j$; then

$$\begin{aligned}\sum_{i,j} \langle \phi(B_i^* B_j) x_j, x_i \rangle &= \sum_{r=1}^n \sum_{s,t=1}^n \left\langle \phi(E_{s,t}) \left(\sum_{i,j} \bar{b}_{r,s,i} b_{r,t,j} x_j \right), x_i \right\rangle \\ &= \sum_{r=1}^n \sum_{s,t} \langle \phi(E_{s,t}) y_{t,r}, y_{s,r} \rangle.\end{aligned}$$

But for each r, this last sum is positive, since $(\phi(E_{s,t}))_{s,t=1}^n$ is positive. Thus, we've expressed our original sum as the sum of n positive quantities. This completes the proof. □

By combining the results of this chapter with the technique used in the proof of Theorem 2.6, some fairly deep operator theory results can be readily obtained. Recall that if T is in $B(\mathcal{H})$, then we define the *numerical radius* of T by

$$w(T) = \sup\{|\langle Tx, x\rangle| \colon \|x\| \leq 1, x \in \mathcal{H}\}.$$

The following result is essentially due to Berger [15].

Theorem 3.15. *Let T be in $B(\mathcal{H})$, and let $\mathcal{S} \subseteq C(\mathbb{T})$ be the operator system defined by $\mathcal{S} = \{p + \bar{q} \colon p, q \text{ polynomials}\}$. The following are equivalent:*

(i) $w(T) \leq 1$.

(ii) The map $\phi\colon \mathcal{S} \to B(\mathcal{H})$ *defined by*

$$\phi(p + \bar{q}) = p(T) + q(T)^* + (p(0) + \overline{q(0)})I$$

is positive.

Proof. We first show that (i) implies (ii). Let R_n be the $n \times n$ operator matrix whose subdiagonal entries are T and whose remaining entries are 0. It is not difficult to show that $w(R_n) \leq w(T)$ (Exercise 3.13).

Mimicking the first part of the proof of Theorem 2.6, we see that the above map ϕ is positive, provided that the operator matrices

$$\begin{bmatrix} 2 & T^* & \dots & T^{*n} \\ T & \ddots & \ddots & \vdots \\ \vdots & \ddots & \ddots & T^* \\ T^n & \dots & T & 2 \end{bmatrix} \tag{$*$}$$

are positive for all n.

Note that as in Theorem 2.6, $R_n^{n+1} = 0$ and so the matrix $(*)$ can be written as $(I - R_n)^{-1} + (I - R_n^*)^{-1}$. Fix a vector $x = (I - R_n)y$, and compute

$$\langle((I - R_n)^{-1} + (I - R_n^*)^{-1})x, x\rangle = 2\|y\|^2 - 2\operatorname{Re}\langle R_n y, y\rangle.$$

Thus, $(*)$ is positive if and only if $w(R_n) \leq 1$ (Exercise 3.13 (iv)).

If $w(T) \leq 1$, then $w(R_n) \leq 1$ and so $(*)$ is positive, which implies that ϕ is positive.

Conversely, if ϕ is positive, then, since $\mathcal{S}$ is dense in $C(\mathbb{T})$, ϕ will be completely positive by Theorem 3.11. Note that the matrix

$$\begin{bmatrix} 1 & \bar{z} & \dots & \bar{z}^n \\ z & 1 & \ddots & \vdots \\ \vdots & \ddots & \ddots & \bar{z} \\ z^n & \dots & z & 1 \end{bmatrix} = \begin{bmatrix} 1 & 0 & \dots & 0 \\ 0 & z & \ddots & \vdots \\ \vdots & \ddots & \ddots & 0 \\ 0 & \dots & 0 & z^n \end{bmatrix} \tag{$**$}$$

$$\times \begin{bmatrix} 1 & \dots & 1 \\ \vdots & & \vdots \\ 1 & \dots & 1 \end{bmatrix} \begin{bmatrix} 1 & 0 & \dots & 0 \\ 0 & \bar{z} & \ddots & \vdots \\ \vdots & \ddots & \ddots & 0 \\ 0 & \dots & 0 & \bar{z}^n \end{bmatrix}$$

is positive in $M_n(C(\mathbb{T}))$, and so its image under ϕ_n will be positive. But the image of $(**)$ under ϕ_n is $(*)$. Thus, $(*)$ is positive for all n, and hence $w(R_n) \leq 1$.

Let $x \in \mathcal{H}$, $\|x\| = 1$, and let $y = (x \oplus \cdots \oplus x)/\sqrt{n}$ be a unit vector in the direct sum of n copies of $\mathcal{H}$. We have that

$$1 \geq |\langle R_n y, y\rangle| = \frac{n-1}{n} |\langle Tx, x\rangle|,$$

from which it follows that $w(T) \leq \frac{n}{n-1}$ for all n.

Thus, $w(T) \leq 1$, which completes the proof of the theorem. □

Note that if $w(T) \leq 1$ and ϕ is as above, then since ϕ is positive, $\|\phi\| = \|\phi(1)\| = 2$. Thus, if p is a polynomial, then $\|p(T)\| = \|\phi(p) - p(0)I\| \leq 3\|p\|$. In particular, if $w(T) \leq 1$, then the functional calculus can be extended from polynomials to the disk algebra $A(\mathbb{D})$.

Corollary 3.16 (Berger–Kato–Stampfli). *Let T be in $\mathcal{B}(\mathcal{H})$ with $w(T) \leq 1$, and let f be in $A(\mathbb{D})$ with $f(0) = 0$. Then $w(f(T)) \leq \|f\|$.*

Proof. It is sufficient to assume that f is a polynomial and that $\|f\| \leq 1$. Let ϕ be the map of Theorem 3.14 for T. We must show that the map $\psi(p + \bar{q}) = p(f(T)) + q(f(T))^* + (p(0) + \overline{q(0)})I$ is positive. But if $p + \bar{q}$ is positive, then $p \circ f + \overline{q \circ f}$ is positive, and thus,

$$\psi(p + \bar{q}) = p(f(T)) + q(f(T))^* + (p(0) + \overline{q(0)})I = \phi(p \circ f + \overline{q \circ f})$$

is positive. □

Corollary 3.17 (Berger). *Let T be in $B(\mathcal{H})$. Then $w(T^n) \leq w(T)^n$.*

Proof. We may assume $w(T) = 1$, but then applying Corollary 3.16, with $f(z) = z^n$, yields the result. □

Actually, the proof of Theorem 3.15 yields a matrix-valued generalization of the Berger–Kato–Stampfli theorem. If $w(T) \leq 1$ and $F = (f_{ij})$ is in $M_n(A(\mathbb{D}))$ with $F(0) = 0$, then $w(F(T)) \leq \|F\|$.

Module Mappings and Multiplicative Domains

In later chapters an increasingly important role will be played by module actions and module mappings. Several key facts about module mappings are consequences of the Schwarz inequality, and so we present them here.

Let $\mathcal{A}$ be a unital C^*-algebra, and assume that $\mathcal{C}$ is a subalgebra with $1_{\mathcal{C}} = 1_{\mathcal{A}}$. Then we can regard $\mathcal{A}$ as a left $\mathcal{C}$-module with module action $c \circ a = ca$. Similarly, we can regard $\mathcal{A}$ as a right $\mathcal{C}$-module or a $\mathcal{C}$-bimodule. If $\mathcal{B}$ is another C^*-algebra containing $\mathcal{C}$ as such a subalgebra and $\phi\colon \mathcal{A} \to \mathcal{B}$ is linear, then we call ϕ a *left $\mathcal{C}$-module map* provided $\phi(ca) = c\phi(a)$ for every $a \in \mathcal{A}$ and $c \in \mathcal{C}$. We define *right $\mathcal{C}$-module maps* and *$\mathcal{C}$-bimodule maps* analogously.

One of our main interests is in finding conditions that guarantee a map ϕ is a module map.

Theorem 3.18. *Let $\mathcal{A}$ and $\mathcal{B}$ be unital C^*-algebras, and let $\phi\colon \mathcal{A} \to \mathcal{B}$ be completely positive with $\phi(1) = 1$. We have the following:*

(i) $\{a \in \mathcal{A}\colon \phi(a)^*\phi(a) = \phi(a^*a)\} = \{a \in \mathcal{A}\colon \phi(ba) = \phi(b)\phi(a)$ *for all* $b \in \mathcal{A}\}$ *is a subalgebra of $\mathcal{A}$, and ϕ is a homomorphism when restricted to this set.*

(ii) $\{a \in \mathcal{A}\colon \phi(a)\phi(a)^* = \phi(aa^*)\} = \{a \in \mathcal{A}\colon \phi(ab) = \phi(a)\phi(b)$ *for all* $b \in \mathcal{A}\}$ *is a subalgebra of $\mathcal{A}$, and ϕ is a homomorphism when restricted to this set.*

(iii) $\{a \in \mathcal{A}\colon \phi(a)^*\phi(a) = \phi(a^*a)$ *and* $\phi(a)\phi(a)^* = \phi(aa^*)\} = \{a \in \mathcal{A}\colon \phi(ab) = \phi(a)\phi(b)$ *and* $\phi(ba) = \phi(b)\phi(a)$ *for all* $b \in \mathcal{A}\}$ *is a C^*-subalgebra of $\mathcal{A}$, and ϕ is a $*$-homomorphism when restricted to this set.*

Proof. We prove (i). The proofs of (ii) and (iii) are similar.

First, if a belongs to the set on the right, then choosing $b = a^*$ yields $\phi(a^*a) = \phi(a^*)\phi(a) = \phi(a)^*\phi(a)$, since ϕ is self-adjoint. Thus, a belongs to the set on the left.

So assume $\phi(a)^*\phi(a) = \phi(a^*a)$, and apply the Schwarz inequality to the map $\phi^{(2)}$ and the matrix $\left[\begin{smallmatrix} a & b^* \\ 0 & 0 \end{smallmatrix}\right]$. We obtain

$$\begin{bmatrix} \phi(a) & \phi(b^*) \\ 0 & 0 \end{bmatrix}^* \begin{bmatrix} \phi(a) & \phi(b^*) \\ 0 & 0 \end{bmatrix} \leq \phi^{(2)}\left(\begin{pmatrix} a^*a & a^*b^* \\ ba & bb^* \end{pmatrix}\right)$$

and so

$$\begin{bmatrix} \phi(a^*a) - \phi(a)^*\phi(a) & \phi(a^*b^*) - \phi(a)^*\phi(b^*) \\ \phi(b)\phi(a) - \phi(ba) & \phi(b)\phi(b^*) - \phi(bb^*) \end{bmatrix} \geq 0.$$

Since the (1,1) entry of this matrix is 0, it follows that the (2,1) entry must be 0 as well. Thus, $\phi(b)\phi(a) = \phi(ba)$, and we have that the two sets are equal.

The remaining claims are trivial. □

Corollary 3.19. *Let $\mathcal{A}$, $\mathcal{B}$, and $\mathcal{C}$ be unital C^*-algebras, and assume that $\mathcal{C}$ is a C^*-subalgebra of both $\mathcal{A}$ and $\mathcal{B}$ with $1_{\mathcal{C}} = 1_{\mathcal{A}} = 1_{\mathcal{B}}$. If $\phi\colon \mathcal{A} \to \mathcal{B}$ is completely positive and $\phi(c) = c$ for every $c \in \mathcal{C}$, then ϕ is a $\mathcal{C}$-bimodule map.*

We call the set given in Theorem 3.18(i) the *right multiplicative domain of* ϕ and denote it by $\mathcal{R}$. Note that if we define a right $\mathcal{R}$-module action on $\mathcal{B}$ via

$b \circ a = b\phi(a)$ for $b \in \mathcal{B}$, $a \in \mathcal{R}$, then ϕ is a right $\mathcal{R}$-module map. Similarly, we call the set in Theorem 3.18(ii) the *left multiplicative domain of* ϕ, denoting it by $\mathcal{L}$, and the set in Theorem 3.18(iii) the *multiplicative domain of* ϕ, denoting it by $\mathcal{C}$. We have that $\mathcal{L} = \mathcal{R}^*$ and $\mathcal{C} = \mathcal{R} \cap \mathcal{R}^*$ and that ϕ is a $\mathcal{C}$-bimodule map.

It is perhaps worth remarking that the above proof only used 4-positivity of the map ϕ.

Notes

Lemma 3.1 is an observation used in the work of Choi and Effros [48].

Proposition 3.5 appears in Arveson [6], which is the source for many of the applications of complete positivity to operator theory.

Theorem 3.11 is due to Stinespring [221], where the term "completely positive" is introduced and used. Stinespring's proof was measure-theoretic.

Theorem 3.14 can be found in Arveson [9] and Choi [41, 43].

Theorem 3.15, Corollary 3.16, and Corollary 3.17 can be found in Berger [15], Kato [130], and Berger and Stampfli [17]. For some related work see [16], [118], and [239]. These ideas were further generalized by the theory of $\mathcal{C}_p$ operators in Sz.-Nagy and Foias [231] (see Exercises 4.16 and 8.10). For an elementary proof of Corollary 3.16, see [172]. Further results in the direction of Exercise 3.10 can be found in Tomiyama's survey article [235].

Exercise 3.6(iii) appears in Choi [43] with a different proof. Exercise 3.9(v) is Walter's 3×3 matrix trick [240], which yields a simple proof of 3.9(vi). Exercise 3.11 is an unpublished result of Smith [215].

Theorem 3.18 and Corollary 3.19 are due to Choi [42].

Exercises

3.1 Prove that $\|\phi_n\| \le \|\phi_k\|$ for $n \le k$ and that if ϕ_k is positive, then ϕ_n is positive.

3.2 Let P, Q, A be operators on some Hilbert space $\mathcal{H}$ with P and Q positive.
  (i) Show that $\left[\begin{smallmatrix} P & A \\ A^* & Q \end{smallmatrix}\right] \ge 0$ if and only if $|\langle Ax, y\rangle|^2 \le \langle Py, y\rangle \cdot \langle Qx, x\rangle$ for all x, y in $\mathcal{H}$.
  (ii) Prove Lemma 3.1(ii).
  (iii) Show that if $\left[\begin{smallmatrix} P & A \\ A^* & Q \end{smallmatrix}\right] \ge 0$, then for any x in $\mathcal{H}$ we have that

$$0 \le \langle (P + A + A^* + Q)x, x\rangle \le (\sqrt{\langle Px, x\rangle} + \sqrt{\langle Qx, x\rangle})^2$$

  and hence $\|P + AA^* + Q\| \le (\|P\|^{1/2} + \|Q\|^{1/2})^2$.
  (iv) Show that if $\left[\begin{smallmatrix} P & A \\ A^* & P \end{smallmatrix}\right] \ge 0$, then $A^*A \le \|P\| \cdot P$ and in particular $\|A\| \le \|P\|$.

3.3 Prove a nonunital version of Proposition 3.2.

3.4 (Modified Schwarz inequality for 2-positive maps) Let $\mathcal{A}$ and $\mathcal{B}$ be C^*-algebras, and $\phi\colon \mathcal{A} \to \mathcal{B}$ 2-positive. Prove that $\phi(a)^*\phi(a) \leq \|\phi(1)\|\phi(a^*a)$ and that $\|\phi(a^*b)\|^2 \leq \|\phi(a^*a)\| \cdot \|\phi(b^*b)\|$. (Hint: Consider $\left[\begin{smallmatrix} 1 & a \\ 0 & 0 \end{smallmatrix}\right]$ and $\left[\begin{smallmatrix} a & b \\ 0 & 0 \end{smallmatrix}\right]$.)

3.5 Let $\mathcal{A}$ be a C^*-algebra with unit. Show that the maps Tr, $\sigma\colon M_n(\mathcal{A}) \to \mathcal{A}$ defined by $\mathrm{Tr}((a_{i,j})) = \sum_i a_{i,i}$ and $\sigma((a_{i,j})) = \sum_{i,j} a_{i,j}$ are completely positive maps. Deduce that if $\|(a_{i,j})\| \leq 1$, then $\|\sum_{i,j} a_{i,j}\| \leq n$.

3.6 (Choi) Let $\mathcal{A}$ be a C^*-algebra, let λ be a complex number with $|\lambda| = 1$, let U_λ be the unitary element of $M_n(\mathcal{A})$ that is diagonal with $u_{i,i} = \lambda^i$, and let Diag: $M_n(\mathcal{A}) \to M_n(\mathcal{A})$ be defined by $\mathrm{Diag}((a_{i,j})) = (b_{i,j})$, where $b_{i,j} = 0$, for $i \neq j$ and $b_{i,i} = a_{i,i}$.
   (i) Show that $U_\lambda^*(a_{i,j})U_\lambda = (\lambda^{j-i}a_{i,j})$.
   (ii) By considering the nontrivial nth roots of unity, show the map Φ: $M_n(\mathcal{A}) \to M_n(\mathcal{A})$ defined by $\Phi(A) = n\,\mathrm{Diag}(A) - A$ is completely positive.
   (iii) Show that the map Φ: $M_n \to M_n$ defined by $\Phi(A) = (n-1)\,\mathrm{Diag}(A) - A$ is not completely positive.

3.7 Let $\mathcal{A}$ and $\mathcal{B}$ be C^*-algebras with unit, and let $\phi_1, \phi_2\colon \mathcal{A} \to \mathcal{B}$ be bounded linear maps with $\phi_1 \pm \phi_2$ completely positive. Prove that $\|\phi_2\|_{\mathrm{cb}} \leq \|\phi_1(1)\|$. [Hint: Use Lemma 3.1 applied to a and $-a$ and Exercise 3.2(iv).]

3.8 Let $\mathcal{A}$ be a C^*-algebra with unit. Define T_1, T_2: $M_n(\mathcal{A}) \to M_n(\mathcal{A})$ by $T_1((a_{i,j})) = (b_{i,j})$ where $b_{i,i} = \sum_\ell a_{\ell,\ell}$, $b_{i,j} = 0$, $i \neq j$, and $T_2((a_{i,j})) = (c_{i,j})$ where $c_{i,j} = a_{j,i}$. Fix k and ℓ, $k \neq \ell$, and define $U_{k,\ell}^{\pm}$ to be 1 in the (k, ℓ) entry and ± 1 in the (ℓ, k) entry and 0 elsewhere.
   (i) Show that

$$T_1(A) - T_2(A) = \frac{1}{2} \sum_{k \neq \ell} U_{k,\ell}^{-*} A U_{k,\ell}^{-}.$$

   (ii) Show that

$$T_1(A) + T_2(A) = \frac{1}{2} \sum_{k \neq \ell} U_{k,\ell}^{+*} A U_{k,\ell}^{+} + \mathrm{Diag}(A).$$

   (iii) Deduce that $T_1 \pm T_2$ are completely positive and that $\|T_2\|_{\mathrm{cb}} \leq n$.
   (iv) By considering $\mathcal{A} = \mathbb{C}$, show that $\|T_2\|_{\mathrm{cb}} = n$.

3.9 Let $\mathcal{A}$ be a C^*-algebra, and let $\mathcal{A}^{\mathrm{op}}$ denote the set $\mathcal{A}$ with the same norm and $*$-operation, but with a multiplication defined by $a \circ b = ba$.
   (i) Prove that $\mathcal{A}^{\mathrm{op}}$ is a C^*-algebra.
   (ii) Prove that M_2 and M_2^{op} are $*$-isomorphic via the transpose map.
   (iii) Show that the identity map from $\mathcal{A}$ to $\mathcal{A}^{\mathrm{op}}$ is always positive.

(iv) Prove that the identity map from M_2 to M_2^{op} is not 2-positive.

(v) (Walter) Let U, V, and X be elements of $\mathcal{A}$ with U, V unitary. Prove that

$$\begin{bmatrix} I & U & X \\ U^* & I & V \\ X^* & V^* & I \end{bmatrix} \geq 0 \quad \text{if and only if} \quad X = UV.$$

(vi) Prove that the identity map from $\mathcal{A}$ to $\mathcal{A}^{\text{op}}$ is completely positive if and only if $\mathcal{A}$ is commutative.

3.10 (Tomiyama) Let $\mathcal{A}$ and $\mathcal{B}$ be unital C^*-algebras, and let $(a_{i,j})$ be in $M_n(\mathcal{A})$.

(i) Prove that $\|(a_{i,j})\| \leq \|(\|a_{i,j}\|)\| \leq \left(\sum_{i,j} \|a_{i,j}\|^2\right)^{1/2} \leq n\|(a_{i,j})\|$, and give examples to show that all of these inequalities are sharp.

(ii) Let $\mathcal{M}$ be an operator space in $\mathcal{A}$, and let $\phi\colon \mathcal{M} \to \mathcal{B}$ be bounded. Prove that $\|\phi_n\| \leq n\|\phi\|$.

3.11 (Smith) Let $\mathcal{M}$ be an operator space, and let $\phi\colon \mathcal{M} \to M_n$ be bounded. Prove that $\|\phi\|_{\text{cb}} \leq n\|\phi\|$. [Hint: Write

$$\phi(a) = \sum_{i,j=1}^{n} \phi_{i,j}(a) \otimes E_{i,j},$$

where $\phi_{i,j}\colon \mathcal{M} \to \mathbb{C}$.]

3.12 Let X be a compact subset of $\mathbb{C}$, and let $R(X)$ be the quotients of polynomials with poles off X. We may regard $R(X)$ as a subalgebra of $C(\partial X)$ or of $C(X)$. Prove that with respect to these two embeddings, the identity map from $R(X)$ to $R(X)$ is completely isometric. Use this result to deduce that the real part of a function in $R(X)$ is positive on X if and only if it is positive on ∂X.

3.13 Let S_n denote the cyclic forward shift on $\mathbb{C}^n$. That is, $S_n e_j = e_{j+1}$ (mod n), where $e_0, \ldots, e_{n-1}$ is the canonical basis for $\mathbb{C}^n$.

(i) Show that S_n is unitarily equivalent to a diagonal matrix whose entries are the nth roots of unity.

(ii) Let T be in $\mathcal{B}(\mathcal{H})$. Show that $w(T) = w(T \otimes S_n)$.

(iii) Let R_n be the $n \times n$ operator matrix whose subdiagonal entries are T and that is 0 elsewhere. Show that $w(R_n) \leq w(T \otimes S_n)$. (Hint: Consider $x_\lambda = \lambda x_1 \oplus \cdots \oplus \lambda^n x_n$ with $|\lambda| = 1$.)

(iv) Show that $\text{Re}\langle R_n y, y\rangle \leq 1$ for all $\|y\| = 1$ if and only if $w(R_n) \leq 1$.

3.14 In this exercise we outline an alternative proof of Theorem 3.14. Let $T \in B(\mathcal{H})$ with $w(T) < 1$, and let p and q be arbitrary polynomials.

(i) Show directly that $\sigma(T)$ is contained in the open unit disk. (Hint: Recall Exercise 2.7.)

(ii) Let

$$Q(t;T) = (1 - e^{-it}T)^{-1} + (1 - e^{it}T^*)^{-1},$$

and show that $Q(t;T) \geq 0$ for all t.

(iii) Show that

$$p(T) + q(T)^* + (p(0) + \overline{q(0)})I = \frac{1}{2\pi}\int_0^{2\pi}(p(e^{it}) + \overline{q(e^{it})})Q(t;T)\,dt.$$

(iv) Deduce Theorem 3.14.

3.15 Let $\{A_n\}_{n=-\infty}^{+\infty}$ be a sequence in $B(\mathcal{H})$. Prove that $\phi(z^n) = A_n$ extends linearly to a completely positive map $\phi\colon C(\mathbb{T}) \to B(\mathcal{H})$ if and only if the $n \times n$ Toeplitz matrix (A_{i-j}) is positive for all n. A sequence $\{A_n\}_{n=-\infty}^{+\infty}$ satisfying these conditions is called an *operator-valued trigonometric moment sequence*.

3.16 Prove that the equivalent conditions of Exercise 3.15 are also equivalent to the operator-valued power series $F(z) = A_0/2 + \sum_{k=1}^{\infty} A_k z^k$ converging on $\mathbb{D}$ and satisfying $F(z) + F(z)^* \geq 0$ for $|z| < 1$.

3.17 Let $\{A_n\}_{n=0}^{\infty}$ be a sequence in $B(\mathcal{H})$. Prove that $\phi(t^n) = A_n$ extends linearly to a completely positive map $\phi\colon C([0,1]) \to B(\mathcal{H})$ if and only if $B_{n,m} = \sum_{k=0}^{n}\binom{n}{k}(-1)^k A_{m+k} \geq 0$ for all $n, m \geq 0$. We call such a sequence an *operator-valued Hausdorff moment sequence*.

3.18 Let $\mathcal{A}$ be a unital C^*-algebra, and let (b_{ij}) be in $M_n(\mathcal{A})$. Prove that (b_{ij}) is positive if and only if for every n-tuple $(a_1, \ldots, a_n)$ of elements of $\mathcal{A}$, we have that $\sum_{i,j=1}^{n} a_i b_{ij} a_j^*$ is a positive element of $\mathcal{A}$.

Chapter 4

Dilation Theorems

We saw our first example of a dilation theorem in Chapter 1. Sz.-Nagy's dilation theorem (Theorem 1.1) showed that every contraction operator on a Hilbert space $\mathcal{H}$ was the compression of a unitary operator on a Hilbert space $\mathcal{K}$ that contains $\mathcal{H}$. In this chapter we focus on dilation theorems that characterize various classes of maps into $B(\mathcal{H})$ as compressions to $\mathcal{H}$ of "nicer" maps into $B(\mathcal{K})$, where $\mathcal{K}$ is a Hilbert space containing $\mathcal{H}$. One of the most general dilation theorems of this type is Stinespring's theorem, which characterizes completely positive maps from C^*-algebras into $B(\mathcal{H})$ in terms of $*$-homomorphisms into $B(\mathcal{K})$ for some other Hilbert space $\mathcal{K}$.

Theorem 4.1 (Stinespring's dilation theorem). *Let $\mathcal{A}$ be a unital C^*-algebra, and let $\phi\colon \mathcal{A} \to \mathcal{B}(\mathcal{H})$ be a completely positive map. Then there exists a Hilbert space $\mathcal{K}$, a unital $*$-homomorphism $\pi\colon \mathcal{A} \to B(\mathcal{K})$, and a bounded operator $V\colon \mathcal{H} \to \mathcal{K}$ with $\|\phi(1)\| = \|V\|^2$ such that*

$$\phi(a) = V^*\pi(a)V.$$

Proof. Consider the algebraic tensor product $\mathcal{A} \otimes \mathcal{H}$, and define a symmetric bilinear function $\langle , \rangle$ on this space by setting

$$\langle a \otimes x, b \otimes y \rangle = \langle \phi(b^*a)x, y \rangle_{\mathcal{H}}$$

and extending linearly, where $\langle , \rangle_{\mathcal{H}}$ is the inner product on $\mathcal{H}$.

The fact that ϕ is completely positive ensures that $\langle , \rangle$ is positive semidefinite, since

$$\left\langle \sum_{j=1}^{n} a_j \otimes x_j, \sum_{i=1}^{n} a_i \otimes x_i \right\rangle = \left\langle \phi_n((a_i^* a_j)) \begin{pmatrix} x_1 \\ \vdots \\ x_n \end{pmatrix}, \begin{pmatrix} x_1 \\ \vdots \\ x_n \end{pmatrix} \right\rangle_{\mathcal{H}^{(n)}} \geq 0,$$

where $\langle , \rangle_{\mathcal{H}^{(n)}}$ denotes the inner product on the direct sum $\mathcal{H}^{(n)}$ of n copies of $\mathcal{H}$, given by

$$\left\langle \begin{pmatrix} x_1 \\ \vdots \\ x_n \end{pmatrix}, \begin{pmatrix} y_1 \\ \vdots \\ y_n \end{pmatrix} \right\rangle_{\mathcal{H}^{(n)}} = \langle x_1, y_1 \rangle_{\mathcal{H}} + \cdots + \langle x_n, y_n \rangle_{\mathcal{H}}.$$

Positive semidefinite bilinear forms satisfy the Cauchy–Schwarz inequality,

$$|\langle u, v \rangle|^2 \le \langle u, u \rangle \cdot \langle v, v \rangle.$$

Thus, we have that

$$\{u \in \mathcal{A} \otimes \mathcal{H} \mid \langle u, u \rangle = 0\} = \{u \in \mathcal{A} \otimes \mathcal{H} \mid \langle u, v \rangle = 0 \text{ for all } v \in \mathcal{A} \otimes \mathcal{H}\}$$

is a subspace, $\mathcal{N}$, of $\mathcal{A} \otimes \mathcal{H}$. The induced bilinear form on the quotient space $\mathcal{A} \otimes \mathcal{H}/\mathcal{N}$ defined by

$$\langle u + \mathcal{N}, v + \mathcal{N} \rangle = \langle u, v \rangle$$

will be an inner product. We let $\mathcal{K}$ denote the Hilbert space that is the completion of the inner product space $\mathcal{A} \otimes \mathcal{H}/\mathcal{N}$.

If $a \in \mathcal{A}$, define a linear map $\pi(a)\colon \mathcal{A} \otimes \mathcal{H} \to \mathcal{A} \otimes \mathcal{H}$ by

$$\pi(a)\left(\sum a_i \otimes x_i\right) = \sum (aa_i) \otimes x_i.$$

A matrix factorization shows that the following inequality in $M_n(\mathcal{A})^+$ is satisfied:

$$(a_i^* a^* a a_j) \le \|a^* a\| \cdot (a_i^* a_j),$$

and consequently,

$$\begin{aligned} &\left\langle \pi(a)\left(\sum a_j \otimes x_j\right), \pi(a)\left(\sum a_i \otimes x_i\right)\right\rangle \\ &\quad = \sum_{i,j} \langle \phi(a_i^* a^* a a_j) x_j, x_i \rangle_{\mathcal{H}} \le \|a^* a\| \cdot \sum_{i,j} \langle \phi(a_i^* a_j) x_j, x_i \rangle_{\mathcal{H}} \\ &\quad = \|a\|^2 \cdot \left\langle \sum a_j \otimes x_j, \sum a_i \otimes x_i \right\rangle. \end{aligned}$$

Thus, $\pi(a)$ leaves $\mathcal{N}$ invariant and consequently induces a quotient linear transformation on $\mathcal{A} \otimes \mathcal{H}/\mathcal{N}$, which we still denote by $\pi(a)$. The above inequality also shows that $\pi(a)$ is bounded with $\|\pi(a)\| \le \|a\|$. Thus, $\pi(a)$ extends to a bounded linear operator on $\mathcal{K}$, which we still denote by $\pi(a)$.

It is straightforward to verify that the map $\pi\colon \mathcal{A} \to B(\mathcal{K})$ is a unital $*$-homomorphism.

Now define $V\colon \mathcal{H} \to \mathcal{K}$ via

$$V(x) = 1 \otimes x + \mathcal{N}.$$

Then V is bounded, since

$$\|Vx\|^2 = \langle 1 \otimes x, 1 \otimes x \rangle = \langle \phi(1)x, x \rangle_{\mathcal{H}} \leq \|\phi(1)\| \cdot \|x\|^2.$$

Indeed, it is clear that $\|V\|^2 = \sup\{\langle \phi(1)x, x \rangle_{\mathcal{H}}\colon \|x\| \leq 1\} = \|\phi(1)\|$.

To complete the proof, we only need observe that

$$\langle V^*\pi(a)Vx, y \rangle_{\mathcal{H}} = \langle \pi(a)1 \otimes x, 1 \otimes y \rangle_{\mathcal{K}} = \langle \phi(a)x, y \rangle_{\mathcal{H}}$$

for all x and y, and so $V^*\pi(a)V = \phi(a)$. □

There are several remarks to be made. First, any map of the form $\phi(a) = V^*\pi(a)V$ is easily seen to be completely positive. Thus, Stinespring's dilation theorem characterizes the completely positive maps from any C^*-algebra into the algebra of bounded operators on any Hilbert space.

Second, note that if ϕ is unital, then V is an isometry. In this case we may identify $\mathcal{H}$ with the subspace $V\mathcal{H}$ of $\mathcal{K}$. With this identification, V^* becomes the projection of $\mathcal{K}$ onto $\mathcal{H}$, $P_{\mathcal{H}}$. Thus, we see that

$$\phi(a) = P_{\mathcal{H}}\pi(a)|_{\mathcal{H}}.$$

If T is in $B(\mathcal{K})$, then the operator $P_{\mathcal{H}}T|_{\mathcal{H}}$ in $B(\mathcal{H})$ is called the *compression of T to $\mathcal{H}$*. If we decompose $\mathcal{K} = \mathcal{H} \oplus \mathcal{H}^\perp$ and, using this decomposition, regard T as a 2×2 operator matrix, then the compression of T to $\mathcal{H}$ is just the (1,1) entry of this operator matrix for T. Thus, when $\phi(1) = 1$, Stinespring's theorem shows that every completely positive map into $B(\mathcal{H})$ is the compression to $\mathcal{H}$ of a *-homomorphism into a Hilbert space that contains $\mathcal{H}$.

The third point to be made is that Stinespring's theorem is really the natural generalization of the Gelfand–Naimark–Segal (GNS) representation of states. Indeed, if $\mathcal{H} = \mathbb{C}$ is one-dimensional so that $B(\mathbb{C}) = \mathbb{C}$, then an isometry $V\colon \mathbb{C} \to \mathcal{K}$ is determined by $V1 = x$ and we have

$$\phi(a) = \phi(a)1 \cdot 1 = V^*\pi(a)V1 \cdot 1 = \langle \pi(a)V1, V1 \rangle_{\mathcal{K}} = \langle \pi(a)x, x \rangle.$$

In fact, rereading the above proof with $\mathcal{H} = \mathbb{C}$ and $\mathcal{A} \otimes \mathbb{C} = \mathcal{A}$, the reader will find a proof of the GNS representation of states.

Finally, we note that if $\mathcal{H}$ and $\mathcal{A}$ are separable, then the space $\mathcal{K}$ constructed above will be separable as well. Similarly, if $\mathcal{H}$ and $\mathcal{A}$ are finite-dimensional, then $\mathcal{K}$ is finite-dimensional.

We now turn our attention to considering the uniqueness of the Stinespring representation. We shall call a triple $(\pi, V, \mathcal{K})$ as obtained in Stinespring's theorem a *Stinespring representation for* ϕ. Given a Stinespring representation $(\pi, V, \mathcal{K})$, let $\mathcal{K}_1$ be the closed linear span of $\pi(\mathcal{A})V\mathcal{H}$. It is easily verified that $\mathcal{K}_1$ reduces $\pi(\mathcal{A})$ so that the restriction of π to $\mathcal{K}_1$ defines a $*$-homomorphism, $\pi_1\colon \mathcal{A} \to B(\mathcal{K}_1)$.

Clearly, $V\mathcal{H} \subseteq \mathcal{K}_1$, so we have that $\phi(a) = V^*\pi_1(a)V$, i.e., that $(\pi_1, V, \mathcal{K}_1)$ is also a Stinespring representation. It enjoys one additional property, namely, that $\mathcal{K}_1$ is the closed linear span of $\pi_1(\mathcal{A})V\mathcal{H}$. Whenever the space of the representation enjoys this additional property, we call the triple a *minimal Stinespring representation*. The following result summarizes the importance of this minimality condition.

Proposition 4.2. *Let $\mathcal{A}$ be a C^*-algebra, let $\phi\colon \mathcal{A} \to B(\mathcal{H})$ be completely positive, and let*

$$(\pi_i, V_i, \mathcal{K}_i), \qquad i = 1, 2,$$

be two minimal Stinespring representations for ϕ. Then there exists a unitary $U\colon \mathcal{K}_1 \to \mathcal{K}_2$ satisfying $UV_1 = V_2$ and $U\pi_1U^* = \pi_2$.

Proof. If U exists, then necessarily,

$$U\left(\sum_i \pi_1(a_i)V_1h_i\right) = \sum_i \pi_2(a_i)V_2h_i.$$

Thus, it will be sufficient to verify that the above formula yields a well-defined isometry from $\mathcal{K}_1$ to $\mathcal{K}_2$, since by the minimality condition, U will have dense range and hence be onto.

To this end, note that

$$\left\|\sum_i \pi_1(a_i)V_1h_i\right\|^2 = \sum_{i,j}\langle V_1^*\pi_1(a_i^*a_j)V_1h_j, h_i\rangle$$

$$= \sum_{i,j}\langle \phi(a_i^*a_j)h_j, h_i\rangle = \left\|\sum_i \pi_2(a_i)V_2h_i\right\|^2,$$

so U is isometric and consequently well defined, which is all that we needed to show. □

We now show how some other dilation theorems can be deduced from Stinespring's result.

The following is a slightly refined version of Theorem 1.1. It is interesting to compare the proofs.

Theorem 4.3 (Sz.-Nagy's dilation theorem). *Let $T \in B(\mathcal{H})$ with $\|T\| \le 1$. Then there exists a Hilbert space $\mathcal{K}$ containing $\mathcal{H}$ as a subspace and a unitary U on $\mathcal{K}$ with the property that $\mathcal{K}$ is the smallest closed reducing subspace for U containing $\mathcal{H}$ such that*

$$T^n = P_{\mathcal{H}} U^n|_{\mathcal{H}} \quad \text{for all nonnegative integers} \quad n.$$

Moreover, if $(U', \mathcal{K}')$ is another pair with the above properties, then there is a unitary $V: \mathcal{K} \to \mathcal{K}'$ such that $Vh = h$ for $h \in \mathcal{H}$ and $VUV^* = U'$.

Proof. By Theorem 2.6 and Exercise 2.2, the map $\phi(p + \bar{q}) = p(T) + q(T)^*$, where p and q are polynomials, extends to a positive map of $C(\mathbb{T})$ into $B(\mathcal{H})$. This map is completely positive by Theorem 3.11.

Let $(\pi, V, \mathcal{K})$ be a minimal Stinespring representation of ϕ, and recall that since $\phi(1) = 1$, we may identify $V\mathcal{H}$ with $\mathcal{H}$. Setting $\pi(z) = U$, where z is the coordinate function, we have that U is unitary and that

$$T^n = \phi(z^n) = P_{\mathcal{H}} \pi(z^n)|_{\mathcal{H}} = P_{\mathcal{H}} U^n|_{\mathcal{H}}.$$

The minimality condition on $(\pi, V, \mathcal{K})$ is equivalent to requiring that the span of

$$\{U^n \mathcal{H}: n = 0, \pm 1, \pm 2, \dots\}$$

be dense in $\mathcal{K}$, which is equivalent to the requirement that there be no closed reducing subspace for U containing $\mathcal{H}$ other than $\mathcal{K}$ itself.

The final statement of the theorem is a consequence of the uniqueness of a minimal Stinespring representation up to unitary equivalence, Proposition 4.2. □

The techniques used to prove Theorem 4.3 can be used to prove a far more general result. Let $X \subseteq \mathbb{C}$ be a compact set, and let $\mathcal{R}(X)$ be the algebra of rational functions on X. An operator T is said to have a *normal ∂X-dilation* if there is a Hilbert space $\mathcal{K}$ containing $\mathcal{H}$ as a subspace and a normal operator N on $\mathcal{K}$ with $\sigma(N) \subseteq \partial X$ such that

$$r(T) = P_{\mathcal{H}} r(N)|_{\mathcal{H}}$$

for all r in $\mathcal{R}(X)$. We shall call N a *minimal normal ∂X-dilation of T*, provided that $\mathcal{K}$ is the smallest reducing subspace for N that contains $\mathcal{H}$.

Clearly, when T has a normal ∂X dilation N,

$$\|r(T)\| \le \|r(N)\| \le \sup\{|r(z)|: z \in \partial X\},$$

and so a necessary condition for T to have a normal ∂X-dilation is that X is a spectral set for T. It is a long-standing problem to determine if this condition is also sufficient [231]. That is, if X is a spectral set for T, then does it necessarily follow that T has a normal ∂X-dilation?

No compact subsets of $\mathbb{C}$ are known where this fails to be true. Yet the collection of sets X for which this statement is known to always hold is somewhat limited. It has been verified that this condition is sufficient for X an annulus [1], but the answer to this question is still unknown even when X is a nice region with two holes and T is a finite matrix. On the other hand, if T is restricted to be a 2×2 matrix, then this statement is known to be true [63] for every set X.

If $\mathcal{S} = \mathcal{R}(X) + \overline{\mathcal{R}(X)}$ is dense in $C(\partial X)$, then $\mathcal{R}(X)$ is called a *Dirichlet algebra on* ∂X. For example, if $\mathbb{C}/X$ has only finitely many components and the interior of X is simply connected, then $\mathcal{R}(X)$ is a Dirichlet algebra on ∂X. See [68] for this and further topological conditions on X that ensure that $\mathcal{R}(X)$ is a Dirichlet algebra on ∂X.

Theorem 4.4 (Berger–Foias–Lebow). *Let $\mathcal{R}(X)$ be a Dirichlet algebra on ∂X. If X is a spectral set for T, then T has a minimal normal ∂X-dilation. Moreover, any two minimal normal ∂X-dilations for T are unitarily equivalent, via a unitary which leaves $\mathcal{H}$ invariant.*

Proof. Let $\rho\colon \mathcal{R}(X) \to B(\mathcal{H})$ be the unital contraction defined by $\rho(r) = r(T)$, so that $\tilde{\rho}\colon \mathcal{S} \to B(\mathcal{H})$ is positive, where $\mathcal{S} = \mathcal{R}(X) + \overline{\mathcal{R}(X)}$. Since $\mathcal{S}$ is dense in $C(\partial X)$ and positive maps are bounded, $\tilde{\rho}$ extends to a positive map ϕ on $C(\partial X)$. But by Theorem 3.11, ϕ is completely positive. The remainder of the proof proceeds as in Theorem 4.3. □

When $\mathcal{R}(X)$ is not a Dirichlet algebra, minimal normal ∂X-dilations of an operator need not be unitarily equivalent [165].

To state the next dilation theorem, we need to introduce some notation. Let X be a compact Hausdorff space, and let $\mathcal{B}$ be the σ-algebra of Borel sets on X. A *$B(\mathcal{H})$-valued measure on X* is a map $E\colon \mathcal{B} \to B(\mathcal{H})$ that is weakly countably additive, that is, if $\{B_i\}$ is a countable collection of disjoint Borel sets with union B, then

$$\langle E(B)x, y\rangle = \sum_i \langle E(B_i)x, y\rangle$$

for all x, y in $\mathcal{H}$. The measure is *bounded* provided that

$$\sup\{\|E(B)\|\colon B \in \mathcal{B}\} < \infty,$$

and we let $\|E\|$ denote this supremum. The measure is *regular* provided that for all x, y in $\mathcal{H}$, the complex measure given by

$$\mu_{x,y}(B) = \langle E(B)x, y\rangle \qquad (*)$$

is regular.

Given a regular bounded $B(\mathcal{H})$-valued measure E, one obtains a bounded, linear map

$$\phi_E \colon C(X) \to B(\mathcal{H})$$

via

$$\langle \phi_E(f)x, y\rangle = \int f \, d\mu_{x,y}. \qquad (**)$$

Conversely, given a bounded, linear map $\phi\colon C(X) \to B(\mathcal{H})$, if one defines regular Borel measures $\{\mu_{x,y}\}$ for each x and y in $\mathcal{H}$ by the above formula $(**)$, then for each Borel set B, there exists a unique, bounded operator $E(B)$, defined by the formula $(*)$, and the map $B \to E(B)$ defines a bounded, regular $B(\mathcal{H})$-valued measure. Thus, we see that there is a one-to-one correspondence between the bounded, linear maps of $C(X)$ into $B(\mathcal{H})$ and the regular bounded $B(\mathcal{H})$-valued measures. Such measures are called

(i) *spectral* if $E(M \cap N) = E(M) \cdot E(N)$,
(ii) *positive* if $E(M) \geq 0$,
(iii) *self-adjoint* if $E(M)^* = E(M)$,

for all Borel sets M and N.

Note that if E is spectral and self-adjoint, then $E(M)$ must be an orthogonal projection and hence E is also positive.

The following proposition, whose proof we leave to the reader (Exercise 4.10), summarizes the relationships between the above properties of measures and the properties of their corresponding linear maps.

Proposition 4.5. *Let E be a regular bounded $B(\mathcal{H})$-valued measure, and let $\phi\colon C(X) \to \mathcal{B}(\mathcal{H})$ be the corresponding linear map. Then:*

(i) ϕ is a homomorphism if and only if E is spectral;
(ii) ϕ is positive if and only if E is positive;
(iii) ϕ is self-adjoint if and only if E is self-adjoint;
(iv) ϕ is a $*$-homomorphism if and only if E is self-adjoint and spectral.

The correspondence between these measures and linear maps leads to a dilation result for these measures.

Theorem 4.6 (Naimark). *Let E be a regular, positive, $B(\mathcal{H})$-valued measure on X. Then there exists a Hilbert space $\mathcal{K}$, a bounded linear operator $V: \mathcal{H} \to \mathcal{K}$, and a regular, self-adjoint, spectral, $B(\mathcal{K})$-valued measure F on X, such that*

$$E(B) = V^* F(B) V.$$

Proof. Let ϕ: $C(X) \to B(\mathcal{H})$ be the positive, linear map corresponding to E. Then ϕ is completely positive by Theorem 3.11, and so we may apply Stinespring's theorem to obtain a $*$-homomorphism $\pi: C(X) \to B(\mathcal{K})$ and a bounded, linear operator $V: \mathcal{H} \to \mathcal{K}$ such that $\phi(f) = V^*\pi(f)V$ for all f in $C(X)$. If we let F be the $B(\mathcal{K})$-valued measure corresponding to π, then it is easy to verify that F has the desired properties. □

As another application of Stinespring's theorem, we shall give a characterization of the completely positive maps between two matrix algebras. In contrast, there is an entire panoply of classes of positive maps between matrix algebras and much that is not known about the relationships between these various classes (see, for example [46] and [244]).

We begin by remarking that if π: $M_n \to B(\mathcal{K})$ is a $*$-homomorphism, then up to unitary equivalence, $\mathcal{K}$ decomposes as an orthogonal, direct sum of n-dimensional subspaces such that π is the identity representation on each of the subspaces (Exercise 4.11).

Now let ϕ: $M_n \to M_k = B(\mathbb{C}^k)$, and let $(\pi, V, \mathcal{K})$ be a minimal Stinespring representation for ϕ. By the construction of the space $\mathcal{K}$ given in Theorem 4.1, we see that $\dim(\mathcal{K}) \le \dim(M_n \otimes \mathbb{C}^k) = n^2 k$.

Thus, up to unitary equivalence, we may decompose $\mathcal{K}$ into the direct sum of fewer than nk subspaces, each of dimension n, such that π: $M_n \to B(\mathcal{K})$ is the identity representation on each one. So let us write $\mathcal{K} = \sum_{i=1}^{\ell} \oplus C_i^n$, $\ell \le nk$, and let P_i denote the projection of $\mathcal{K}$ onto $\mathbb{C}_i^n$. We have that for any A in M_n,

$$P_i \pi(A)|_{\mathbb{C}_i^n} = A.$$

Now, if we let V_i: $\mathbb{C}^k \to \mathbb{C}_i^n$ be defined by $V_i = P_i V$, then

$$\phi(A) = V^*\pi(A)V = \sum_{i,j=1}^{\ell} V_i^*\pi(A)V_j = \sum_{i=1}^{\ell} V_i^* A V_i,$$

after identifying each $\mathbb{C}_i^n$ with $\mathbb{C}^n$.

We summarize these calculations in the following:

Proposition 4.7 (Choi). *Let ϕ: $M_n \to M_k$ be completely positive. Then there exist fewer than nk linear maps, V_i: $\mathbb{C}^k \to \mathbb{C}^n$, such that $\phi(A) = \sum_i V_i^* A V_i$ for all A in M_n.*

There is another dilation theorem due to Naimark whose proof is closely related to the proof of Stinespring's theorem. Let G be a group and let $\phi\colon G \to B(\mathcal{H})$. We call ϕ *completely positive definite* if for every finite set of elements $g_1, \ldots, g_n$ of G, the operator matrix $(\phi(g_i^{-1} g_j))$ is positive.

If G is a topological group, then we call a map $\phi\colon G \to B(\mathcal{H})$ *weakly continuous* provided $\langle \phi(g_\lambda)x, y\rangle \to \langle \phi(g)x, y\rangle$ for every pair of vectors x, y in $\mathcal{H}$, and every net $\{g_\lambda\}$ in G converging to g. Similarly, ϕ is called *strongly continuous* provided $\|\phi(g_\lambda)x - \phi(g)x\| \to 0$, and $*$*-strongly continuous* provided $\|\phi(g_\lambda)x - \phi(g)x\| \to 0$ and $\|\phi(g_\lambda)^* x - \phi(g)x\| \to 0$.

Theorem 4.8 (Naimark). *Let G be a topological group, and let $\phi\colon G \to B(\mathcal{H})$ be weakly continuous and completely positive definite. Then there exists a Hilbert space $\mathcal{K}$, a bounded operator $V\colon \mathcal{H} \to \mathcal{K}$, and a $*$-strongly continuous unitary representation $\rho\colon G \to B(\mathcal{H})$ such that*

$$\phi(g) = V^* \rho(g) V.$$

Consequently, ϕ is automatically $*$-strongly continuous.

Proof. Consider the vector space $C_0(G, \mathcal{H})$ of finitely supported functions from G to $\mathcal{H}$ and define a bilinear function on this space by

$$\langle f_1, f_2\rangle = \sum_{g,g'} \langle \phi(g^{-1}g') f_1(g'), f_2(g)\rangle_{\mathcal{H}}.$$

As in the proof of Stinespring's theorem, we have that $\langle f, f\rangle \geq 0$ and that the set $\mathcal{N} = \{f \mid \langle f, f\rangle = 0\}$ is a subspace of $C_0(G, \mathcal{H})$. We let $\mathcal{K}$ be the completion of $C_0(G, \mathcal{H})/\mathcal{N}$ with respect to the induced inner product.

For h in $\mathcal{H}$, define Vh by

$$(Vh)(g) = \begin{cases} h & \text{if } g = e, \\ 0 & \text{if } g \neq e, \end{cases}$$

where e denotes the identity of G, and let $\rho\colon G \to B(\mathcal{K})$ be left translation, that is,

$$(\rho(g)f)(g') = f(g^{-1}g').$$

It is straightforward to check that V is bounded and linear, that ρ is a unitary representation, and that $\phi(g) = V^* \rho(g) V$.

Now we show that ρ is weakly continuous. Let $\{g_\lambda\}$ be a net in G that converges to g_0. Since ρ is a unitary representation, it will suffice to show that $\rho(g_\lambda)$ converges weakly to $\rho(g_0)$ on a dense subspace.

Let f_1, f_2 be in $C_0(G, \mathcal{H})$. Then

$$\begin{aligned}\langle \rho(g_\lambda) f_1, f_2 \rangle &= \sum_{g,g'} \langle \phi(g^{-1}g') f_1(g_\lambda^{-1}g'), f_2(g) \rangle_{\mathcal{H}} \\ &= \sum_{g,g''} \langle \phi(g^{-1}g_\lambda g'') f_1(g''), f_2(g) \rangle_{\mathcal{H}},\end{aligned}$$

which, since both sums involved are finite, converges to

$$\sum_{g,g''} \langle \phi(g^{-1}g_0 g'') f_1(g''), f_2(g) \rangle_{\mathcal{H}} = \langle \rho(g_0) f_1, f_2 \rangle.$$

Thus, we see that $\rho(g_\lambda)$ converges weakly to $\rho(g_0)$.

Now it is easily checked that if a net of unitaries converges in the weak operator topology to another unitary, then it converges $*$-strongly. Thus, ρ is $*$-strongly continuous. The $*$-strong continuity of ϕ now follows from the representation and the $*$-strong continuity of ρ. □

It is useful to consider the special case of the above theorem when $G = \mathbb{Z}$, the group of integers. Setting $\phi(n) = A_n$, we see that ϕ is completely positive definite if and only if $(A_{n_i - n_j})_{i,j=1}^k$ is positive for every choice of finitely many integers $n_1, \ldots, n_k$ with k arbitrary. Taking $n_i = i$, we see that this implies the formal positivity of $(A_{i-j})_{i,j=0}^\infty = T$. A little careful reflection shows that an arbitrary choice of $n_1, \ldots, n_k$ is simply restricting T to the entries represented by the set $\{n_1, \ldots, n_k\}$ and then permuting these entries.

Thus, a completely positive definite function on $\mathbb{Z}$ is just an operator-valued trigonometric moment sequence (Exercise 3.15). Naimark's dilation theorem now tells us that there exists a unitary $U = \rho(1)$ and V such $A_n = V^* U^n V$ for all n.

As with Stinespring's representation, there is a minimality condition that guarantees the uniqueness of this representation up to unitary equivalence (Exercise 4.12).

A map $\phi\colon G \to B(\mathcal{H})$ will be called *positive definite* if for every choice of n elements $g_1, \ldots, g_n$ of G, and scalars, $\alpha_1, \ldots, \alpha_n$, the operator

$$\sum_{i,j} \bar{\alpha}_i \alpha_j \phi(g_i^{-1} g_j)$$

is positive. We remark that this is equivalent to requiring that for every x in $\mathcal{H}$, the map $\phi_x\colon G \to \mathbb{C}$, given by $\phi_x(g) = \langle \phi(g)x, x \rangle$, be completely positive definite.

We caution the reader that our terminology is not standard. What we have chosen to call completely positive definite is usually called simply positive definite, and the concept that we have introduced above and called positive definite is usually not introduced at all. Our rationale for this slight deviation in notation will be clear in the section "Group C^*-Algebras" at the end of this chapter, where we will show a correspondence between the two classes of maps on G that are defined above and maps on a certain C^*-algebra associated with G, $C^*(G)$. Not surprisingly, this correspondence will carry (completely) positive definite maps on G to (completely) positive maps on $C^*(G)$.

We begin by describing this correspondence in one case of particular interest. Let $\mathbb{Z}^n$ be the Cartesian product of n copies of the integers, and let $\mathbb{T}^n$ be the Cartesian product of n copies of the circle. Let $J = (j_1, \dots, j_n)$ be in $\mathbb{Z}^n$, and let z_j be the jth coordinate function on $\mathbb{T}^n$. We set $z^J = z_1^{j_1} \cdots z_n^{j_n}$. Note that there is a one-to-one correspondence between unitary representations $\rho\colon \mathbb{Z}^n \to B(\mathcal{H})$ and $*$-homomorphisms $\pi\colon C(\mathbb{T}^n) \to B(\mathcal{H})$, given by $\pi(z_j) = \rho(e_j)$, where e_j is the n-tuple that is 1 in the jth entry and 0 in the remaining entries.

Proposition 4.9. *Let $\phi\colon \mathbb{Z}^n \to B(\mathcal{H})$ be (completely) positive definite. Then there is a uniquely defined, (completely) positive map $\psi\colon C(\mathbb{T}^n) \to B(\mathcal{H})$, given by $\psi(z^J) = \phi(J)$. Conversely, if the (completely) positive map ψ is given, then the above formula defines a (completely) positive definite function ϕ on $\mathbb{Z}^n$.*

Proof. First, we consider the case where ϕ is completely positive definite. Let $(\rho, V, \mathcal{K})$ be the Naimark dilation of ϕ, so that $\phi(J) = V^*\rho(J)V$, and let $\pi\colon C(\mathbb{T}^n) \to B(\mathcal{H})$ be the $*$-homomorphism associated with ρ. If we set $\psi(f) = V^*\pi(f)V$, then we obtain a map $\psi\colon C(\mathbb{T}^n) \to B(\mathcal{H})$, which is completely positive. Moreover, ψ satisfies $\psi(z^J) = V^*\pi(z^J)V = V^*\rho(J)V = \phi(J)$.

The proof of the converse in the completely positive case is identical.

Now, suppose that ϕ is only positive definite. If we fix x in $\mathcal{H}$ and set $\phi_x(J) = \langle \phi(J)x, x\rangle$, then ϕ_x is a completely positive definite function on $\mathbb{Z}^n$. Thus, by the above there is a positive map $\psi_x\colon C(\mathbb{T}^n) \to \mathbb{C}$ with $\psi_x(z^J) = \phi_x(J)$. For fixed f in $C(\mathbb{T}^n)$, the function $x \to \psi_x(f)$ as x varies over $\mathcal{H}$ is a bounded, quadratic function (see Exercise 4.18), and hence there exists a bounded operator $\psi(f)$ such that $\langle \psi(f)x, x\rangle = \psi_x(f)$. This defines a linear map $\psi\colon C(\mathbb{T}^n) \to B(\mathcal{H})$, which is easily seen to be positive.

The converse in the positive case is similar. □

Corollary 4.10. *For $\mathbb{Z}^n$, the sets of positive definite and completely positive definite operator-valued functions coincide.*

Proof. Clearly, every completely positive definite map is positive definite. Now let $\phi\colon \mathbb{Z}^n \to B(\mathcal{H})$ be positive definite. Then $\psi\colon C(\mathbb{T}^n) \to B(\mathcal{H})$, given $\psi(z^J) = \phi(J)$ is a positive linear map. By Theorem 3.11, ψ is completely positive and hence ϕ is completely positive definite. □

Again it is useful to understand what this corollary says in the case of Z. Setting $\phi(n) = A_n$, we have seen that ϕ is completely positive definite if and only if $(A_{i-j})_{i,j=0}^{\infty}$ is formally positive. But ϕ is positive definite if and only if ϕ_x is positive definite for all x, which is if and only if $(\langle A_{i-j}x, x\rangle)_{i,j=0}^{\infty}$ is formally positive for all x. Thus, for Toeplitz operator matrices, formal positivity of (A_{i-j}) is equivalent to formal positivity of $(\langle A_{i-j}x, x\rangle)$ for every vector x.

Corollary 4.10 is a generalization of this observation to multivariable Toeplitz operators. We now wish to discuss the analogue for more general groups.

Group C^*-Algebras

The above results are part of a more general duality. Let G be a locally compact group, and let dg be a (left) Haar measure on G. The space $L^1(G)$ of integrable functions, on G can be made into a $*$-algebra by defining

$$f_1 \times f_2(g') = \int f_1(g) f_2(g^{-1}g')\, dg,$$

and a $*$-operation by

$$f^*(g) = \Delta(g)^{-1}\overline{f(g^{-1})},$$

where $\Delta(\cdot)$ is the modular function. It is then possible to endow $L^1(G)$ with a norm such that the completion of $L^1(G)$ is a C^*-algebra, denoted by $C^*(G)$ (see [73] or [173]).

There is a one-to-one correspondence between weakly continuous, unitary representations $\rho\colon G \to B(\mathcal{H})$, and $*$-homomorphisms $\pi\colon C^*(G) \to B(\mathcal{H})$, given by

$$\pi(f) = \int f(g)\rho(g)\, dg$$

when f is in $L^1(G)$. See [173] for a development of this theory.

In a similar fashion, there are one-to-one correspondences between the weakly continuous, (completely) positive definite, operator-valued functions on G and the (completely) positive maps on $C^*(G)$, given by

$$\psi(f) = \int f(g)\phi(g)\, dg$$

for f in $L^1(G)$. The proof that the above formula defines a one-to-one correspondence between these two classes of maps is similar to the proof of Proposition 4.9 and is left as an exercise (Exercise 4.16).

Proposition 4.9 follows from the above correspondence and the fact [173] that $C^*(Z^n) = C(\mathbb{T}^n)$.

Paralleling Corollary 4.10, we see that when G is commutative the positive definite and completely positive definite functions correspond.

Notes

Naimark's two dilation theorems ([150] and [151]) preceded Stinespring's dilation theorem [221]. Stinespring [221] defined completely positive maps on C^*-algebras, proved that positive maps on $C(X)$ were completely positive (Theorem 3.11), and then proved Theorem 4.1 as a generalization of Naimark's dilation theorem for positive, operator-valued measures.

Arveson [6] realized the important role that the theory of completely positive maps can play in the study of normal ∂X-dilations and gave the proofs of Theorems 4.3 and 4.4 that are presented here.

Other proofs of Sz.-Nagy's dilation theorem have used the theory of positive definite functions on Z or the "geometric" technique that we presented in Chapter 1, where the unitary and the space that it acts on are explicitly constructed ([228] and [210]). Two beautiful results of the geometric dilation techniques are Ando's dilation theorem for commuting pairs of contractions [5] and the Sz.-Nagy–Foias commutant lifting theorem [231]. We will present these topics in Chapter 5.

The correspondence between bounded, regular, operator-valued measures on a compact Hausdorff space X and bounded, linear maps on $C(X)$ is discussed in Hadwin [109].

Proposition 4.7 was proved by Choi [43], who also developed the theory of multiplicative domains [42] (Exercise 4.2).

See Bunce [37] for more on Korovkin-type theorems (Exercise 4.9).

Exercise 4.16 is from Berger [15].

The material distinguishing positive definite and completely positive definite functions on groups seems to be new.

Exercises

4.1 Use Stinespring's representation theorem to prove that $\|V\|^2 = \|\phi\|_{\text{cb}}$ when ϕ is completely positive. Also, use the representation theorem to prove that $\phi(a)^*\phi(a) \leq \|\phi(1)\|^2\phi(a^*a)$.

4.2 (Multiplicative domains) In this exercise, we present an alternative proof of Theorem 3.18. Let $\mathcal{A}$ be a C^*-algebra with unit, and let $\phi\colon \mathcal{A} \to B(\mathcal{H})$ be completely positive, $\phi(1) = 1$, with minimal Stinespring representation $(\pi, V, \mathcal{K})$.

   (i) Prove that $\phi(a)^*\phi(a) = \phi(a^*a)$ if and only if $V\mathcal{H}$ is an invariant subspace for $\pi(a)$.

(ii) Use this to give another proof that $\{a \in \mathcal{A}\colon \phi(a)^*\phi(a) = \phi(a^*a)\} = \{a \in \mathcal{A}\colon \phi(ba) = \phi(b)\phi(a) \text{ for all } b \in \mathcal{A}\}$. Recall that this set is the *right multiplicative domain* of ϕ.

(iii) Similarly, show that $\phi(a)^*\phi(a) = \phi(a^*a)$ and $\phi(a)\phi(a)^* = \phi(aa^*)$ if and only if $V\mathcal{H}$ is a reducing subspace for $\pi(a)$. Deduce that the set of all such elements is a C^*-subalgebra of $\mathcal{A}$. Recall that this subalgebra is the *multiplicative domain of* ϕ.

4.3 (Bimodule Maps) Let $\mathcal{A}$, $\mathcal{B}$, and $\mathcal{C}$ be C^*-algebras with unit, and suppose that $\mathcal{C}$ is contained in both $\mathcal{A}$ and $\mathcal{B}$, with $1_{\mathcal{C}} = 1_{\mathcal{A}}$ and $1_{\mathcal{C}} = 1_{\mathcal{B}}$. A linear map $\phi\colon \mathcal{A} \to \mathcal{B}$ is called a $\mathcal{C}$*-bimodule* map if $\phi(c_1ac_2) = c_1\phi(a)c_2$ for all c_1, c_2 in $\mathcal{C}$. Let $\phi\colon \mathcal{A} \to \mathcal{B}$ be completely positive.

(i) If $\phi(1) = 1$, then prove that ϕ is a $\mathcal{C}$-bimodule map if and only if $\phi(c) = c$ for all c in $\mathcal{C}$.

(ii) Prove, in general, that ϕ is a $\mathcal{C}$-bimodule map if and only if $\phi(c) = c \cdot \phi(1)$ for all c in $\mathcal{C}$. Moreover, in this case $\phi(1)$ commutes with $\mathcal{C}$.

4.4 Let $\mathcal{D}_n$ be the C^*-subalgebra of diagonal matrices in M_n. Prove that a linear map $\phi\colon M_n \to M_n$ is a $\mathcal{D}_n$-bimodule map if and only if ϕ is the Schur product map S_T for some matrix T.

4.5 Let $\phi\colon \mathcal{A} \to \mathcal{A}$ be a completely positive projection with $\phi(1) = 1$.

(i) Show that if $\phi(a) = a$, then $\phi(ax) = \phi(a\phi(x))$ for every x.

(ii) Show that $\mathcal{B} = \{a\colon \phi(a) = a\}$ is a C^*-algebra, in the product $a \circ b = \phi(ab)$, but that in general, $\mathcal{B}$ is distinct from the multiplicative domain of ϕ.

(iii) Show that ϕ is a $(\mathcal{B}, \circ)$-bimodule map.

4.6 Let $\phi\colon G \to B(\mathcal{H})$ be completely positive definite. Prove that ϕ is weakly continuous if and only if ϕ is strongly continuous.

4.7 Let $\phi\colon G \to M_n$ be continuous. Prove that ϕ is completely positive definite if and only if there exists a Hilbert space $\mathcal{H}$ and continuous functions $x_i\colon G \to \mathcal{H}$, $i = 1, \ldots, n$, such that $\phi(g^{-1}g') = (\langle x_j(g'), x_i(g)\rangle)$.

4.8 A semigroup G with an involution $g \to g^*$ satisfying $(g_1g_2)^* = g_2^*g_1^*$, $1^* = 1$ is called a $*$*-semigroup*. A function $\phi\colon G \to B(\mathcal{H})$ is called *completely positive definite* if $(\phi(g_i^*g_j))$ is positive for every set of finitely many elements $g_1, \ldots, g_n$ of G, and *bounded* if $(\phi(g_i^*g^*gg_j)) \le M_g(\phi(g_i^*g_j))$ where M_g is a constant depending only on g. Show that every group is a $*$-semigroup if we set $g^* = g^{-1}$. Prove a version of Naimark's dilation theorem for $*$-semigroups.

4.9 Let $\mathcal{A}$ be a C^*-algebra with unit, and let $\phi_n\colon \mathcal{A} \to B(\mathcal{H})$ be a sequence of completely positive maps such that $\phi_n(1) \to 1$ in the weak operator topology.

(i) Prove that $\{a\colon\ \phi_n(a)^*\phi_n(a) - \phi_n(a^*a) \to 0$ and $\phi_n(a)\phi_n(a)^* - \phi_n(aa^*) \to 0$ in the weak operator topology$\}$ is a C^*-subalgebra of $\mathcal{A}$.

(ii) (Korovkin) Prove that if $\phi_n\colon C([0, 1]) \to C([0, 1])$ is a sequence of positive maps such that $\phi_n(1)$, $\phi_n(t)$, and $\phi_n(t^2)$ converge in norm to $1, t$, and t^2, respectively, then $\phi_n(f)$ converges in norm to f, for all f.

4.10 Prove Proposition 4.5.

4.11 Let $\pi\colon M_n \to B(\mathcal{K})$ be a unital $*$-homomorphism. Prove that up to unitary equivalence, $\mathcal{K} = \mathcal{H}_1 \oplus \cdots \oplus \mathcal{H}_n$, with $\mathcal{H}_i = \mathcal{H}, i = 1, \ldots, n$, such that $\pi(E_{i,j})$ is the identity map from $\mathcal{H}_j$ to $\mathcal{H}_i$. Show that up to unitary equivalence, π is the direct sum of $\dim(\mathcal{H})$ copies of the identity map.

4.12 Give a minimality condition for the Naimark representation of completely positive definite functions on a group, and prove uniqueness of minimal representations up to unitary equivalence.

4.13 Let $t \to A(t)$, $t \geq 0$, be a weakly continuous semigroup of contraction operators, $A(0) = I$. For $t < 0$, set $A(t) = A(-t)^*$. Prove that this extended map is completely positive definite in $\mathbb{R}$. What does Naimark's dilation theorem imply? (Hint: Recall the proof of Theorem 2.6.)

4.14 (Trigonometric moments) Let $\{A_n\}_{n=-\infty}^{+\infty}$ be a sequence of bounded linear operators on a Hilbert space $\mathcal{H}$. Prove that $\{A_n\}_{n=-\infty}^{+\infty}$ is an operator-valued trigonometric moment sequence if and only if there exists a Hilbert space $\mathcal{K}$, a unitary operator U on $\mathcal{K}$, and a bounded linear operator $V\colon \mathcal{H} \to \mathcal{K}$ such that $A_n = V^*U^nV$ for all n.

4.15 (Hausdorff moments) Let $\{A_n\}_{n=0}^{+\infty}$ be a sequence of bounded linear operators on a Hilbert space $\mathcal{H}$. Prove that $\{A_n\}_{n=0}^{+\infty}$ is an operator-valued Hausdorff moment sequence if and only if there exists a Hilbert space $\mathcal{K}$, a positive contraction P on $\mathcal{K}$, and a bounded linear operator $V\colon \mathcal{H} \to \mathcal{K}$ such that $A_n = V^*P^nV$ for all $n \geq 0$.

4.16 Verify the claims of the subsection on group C^*-algebras.

4.17 (Berger) Let T be an operator on a Hilbert space $\mathcal{H}$. Prove that $w(T) \leq 1$ if and only if there exists a Hilbert space $\mathcal{K}$ containing $\mathcal{H}$ and a unitary operator U on $\mathcal{K}$ such that $T^n = 2P_{\mathcal{H}}U^n|_{\mathcal{H}}$ for all $n \geq 1$.

4.18 Let $\mathcal{H}$ be a Hilbert space, and let $\gamma\colon \mathcal{H} \to \mathbb{C}$ be a function. We call γ *quadratic* provided that $\gamma(\lambda x) = |\lambda|^2\gamma(x)$ and $\gamma(x + y) + \gamma(x - y) = 2(\gamma(x) + \gamma(y))$ for all λ in $\mathbb{C}$ and for every x and y in $\mathcal{H}$. If, in addition, there exists a constant M such that $|\gamma(x)| \leq M\|x\|^2$ for all x in $\mathcal{H}$, then we call γ *bounded.* Prove that γ is a bounded, quadratic function on $\mathcal{H}$ if and only if there exists T in $B(\mathcal{H})$ such that $\gamma(x) = \langle Tx, x\rangle$, and that T is unique.

Chapter 5

Commuting Contractions on Hilbert Space

In this chapter we study families of commuting contractions on Hilbert space. For pairs of commuting contractions there is an analogue of Sz.-Nagy's unitary dilation theorem due to Ando, and consequently a two-variable analogue of von Neumann's inequality. Unlike the case of Sz.-Nagy's and von Neumann's theorems, there is really only one proof known of Ando's dilation theorem, and this proof is geometric. It is similar in spirit to the proof of Sz.-Nagy's theorem given in Chapter 1. Given this fact, it is somewhat surprising that the analogues of von Neumann's inequality fail for three or more commuting contractions. This difference between the cases with two and with three (or more) variables is still not very well understood and is the source of a number of open questions.

We begin with a dilation theorem for sets of commuting isometries.

Theorem 5.1. *Let $\{V_1, \dots, V_n\}$ be a set of commuting isometries on a Hilbert space $\mathcal{H}$. Then there is a Hilbert space $\mathcal{K}$ containing $\mathcal{H}$ and a set of commuting unitaries $\{U_1, \dots, U_n\}$ on $\mathcal{K}$ such that*

$$V_1^{m_1} \cdots V_n^{m_n} = P_{\mathcal{H}} U_1^{m_1} \cdots U_n^{m_n}\big|_{\mathcal{H}}$$

for all sets $\{m_1, \dots, m_n\}$ of nonnegative integers.

Proof. Let U_1 on $\mathcal{K}_1$ be the minimal unitary dilation of V_1 given by Theorem 4.3. Recall that the span of $\{U_1^n \mathcal{H}\colon n \in \mathbb{Z}\}$ is dense in $\mathcal{K}_1$.

For $i \neq 1$ we claim that there is a well-defined isometry $W_i\colon \mathcal{K}_1 \to \mathcal{K}_1$ given by the formula

$$W_i \left(\sum_{n=-N}^{+N} U_1^n h_n \right) = \sum_{n=-N}^{+N} U_1^n V_i h_n.$$

To see this note that

$$\begin{aligned}\left\|\sum_{n=-N}^{+N} U_1^n V_i h_n\right\|^2 &= \sum_{n\geq m}\langle U_1^{n-m} V_i h_n, V_i h_m\rangle + \sum_{n<m}\langle V_i h_n, U^{m-n} V_i h_m\rangle \\ &= \sum_{n\geq m}\langle V_1^{n-m} V_i h_n, V_i h_m\rangle + \sum_{n<m}\langle V_i h_n, V_1^{m-n} V_i h_n\rangle \\ &= \sum_{n\geq m}\langle V_i V_1^{n-m} h_n, V_i h_m\rangle + \sum_{n<m}\langle V_i h_n, V_i V_1^{m-n} h_n\rangle \\ &= \sum_{n\geq m}\langle V_1^{n-m} h_n, h_m\rangle + \sum_{n<m}\langle h_n, V_1^{m-n} h_n\rangle \\ &= \sum_{n\geq m}\langle U_1^{n-m} h_n, h_m\rangle + \sum_{n<m}\langle h_n, U_1^{m-n} h_n\rangle \\ &= \left\|\sum_{n=-N}^{+N} U_1^n h_n\right\|^2 .\end{aligned}$$

This equality of the norms proves that W_i is well defined and an isometry. Note that if V_i is unitary, then W_i is onto a dense subspace of $\mathcal{K}_1$ and hence is also unitary.

It is easy to see that $\{U_1, W_2, \ldots, W_n\}$ commute and that

$$V_1^{m_1} \cdots V_n^{m_n} = P_{\mathcal{H}} U_1^{m_1} W_2^{m_2} \cdots W_n^{m_n}\big|_{\mathcal{H}}.$$

Now continue by next taking the unitary dilation of W_2 on $\mathcal{K}_2$ and extending $U_1, W_3, \ldots, W_n$ to be isometries on $\mathcal{K}_2$. Since U_1 is unitary, its extension will also be a unitary on $\mathcal{K}_2$. Thus, after n such dilations and extensions we shall obtain an n-tuple of unitaries on a space $\mathcal{K}_n$ with the desired properties. □

We arrive at the first of several results that rely on the above construction.

Corollary 5.2. *Let $\{V_1, \ldots, V_n\}$ be a set of commuting isometries on a Hilbert space $\mathcal{H}$, and let p_{ij}, $i, j = 1, \ldots, m$, be polynomials in n variables. Then*

$$\begin{aligned}&\|(p_{i,j}(V_1, \ldots, V_n))\|_{B(\mathcal{H}^{(m)})} \\ &\qquad \leq \sup\{\|(p_{i,j}(z_1, \ldots, z_n))\|_{M_m} : |z_k| \leq 1,\ 1 \leq k \leq n\}.\end{aligned}$$

Proof. For $U_1, \ldots, U_n$ as in Theorem 5.1 we have

$$p_{i,j}(V_1, \ldots, V_n) = P_{\mathcal{H}} p_{ij}(U_1, \ldots, U_n)|_{\mathcal{H}}$$

and hence

$$\|(p_{i,j}(V_1, \ldots, V_n))\| \leq \|(p_{i,j}(U_1, \ldots, U_n))\|,$$

and the proof is completed by using the $*$-isomorphism of $C^*(\{U_1, \ldots, U_n\})$ with $C(X)$ for some compact subset X of the n-torus $\mathbb{T}^n$. □

Corollary 5.3. *Let $\{V_1, \ldots, V_n\}$ be a set of commuting isometries on a Hilbert space $\mathcal{H}$. Then $(V_j^* V_i) \geq (V_i V_j^*) \geq 0$.*

Proof. Let $\{U_1, \ldots, U_n\}$ be as in Theorem 5.1, and decompose $\mathcal{K} = \mathcal{H} \oplus \mathcal{H}^\perp$. Since each V_i is an isometry, $U_i \mathcal{H} \subseteq \mathcal{H}$, and hence relative to this decomposition,

$$U_i = \begin{pmatrix} V_i & X_i \\ 0 & Y_i \end{pmatrix}.$$

Since $U_j^* U_i = U_i U_j^*$, computing the (1, 1) entry of their operator matrix yields $V_j^* V_i = V_i V_j^* + X_i X_j^*$. Hence,

$$\begin{aligned}(V_j^* V_i) &= (V_i V_j^*) + (X_i X_j^*) \\ &= (V_1^*, \ldots, V_n^*)^* \cdot (V_1^*, \ldots, V_n^*) + (X_1^*, \ldots, X_n^*)^* \cdot (X_1^*, \ldots, X_n^*),\end{aligned}$$

and both inequalities follow. □

These inequalities allow us to generalize Theorem 5.1 considerably.

Let G be an abelian group. We shall write the group operation as addition. We call $\mathcal{P} \subseteq G$ a *spanning cone* provided:

(i) $0 \in \mathcal{P}$,
(ii) if $g_1, g_2 \in \mathcal{P}$, then $g_1 + g_2 \in \mathcal{P}$,
(iii) if $g \in G$, then there exists $g_1, g_2 \in \mathcal{P}$ such that $g = g_1 - g_2$.

Some key examples of abelian groups with spanning cones are $G = \mathbb{Z}^n$ with $\mathcal{P} = (\mathbb{Z}^+)^n$ and any subgroup $G \subseteq (\mathbb{R}, +)$ with $\mathcal{P} = G \cap \mathbb{R}^+$.

Note that conditions (i) and (ii) make $\mathcal{P}$ a semigroup.

We call $\rho \colon \mathcal{P} \to B(\mathcal{H})$ a *semigroup homomorphism* if $\rho(0) = I$ and $\rho(g_1 + g_2) = \rho(g_1)\rho(g_2)$.

Theorem 5.4. *Let G be an abelian group with spanning cone $\mathcal{P}$, and let $\rho \colon \mathcal{P} \to B(\mathcal{H})$ be a semigroup homomorphism such that $\rho(g)$ is an isometry for every $g \in \mathcal{P}$. Then there exists a Hilbert space $\mathcal{K}$ containing $\mathcal{H}$ and a unitary representation $\pi \colon G \to B(\mathcal{K})$ such that $\rho(g) = P_{\mathcal{H}} \pi(g)|_{\mathcal{H}}$ for every $g \in \mathcal{P}$.*

Proof. Let $g_1 - g_2 = g_3 - g_4$ with $g_i \in \mathcal{P}, i = 1, 2, 3, 4$. Then

$$\begin{aligned}\rho(g_2)^*\rho(g_1) &= \rho(g_2)^*\rho(g_3)^*\rho(g_3)\rho(g_1) = \rho(g_2 + g_3)^*\rho(g_1 + g_3)\\ &= \rho(g_1 + g_4)^*\rho(g_1 + g_3) = \rho(g_4)^*\rho(g_1)^*\rho(g_1)\rho(g_3)\\ &= \rho(g_4)^*\rho(g_3).\end{aligned}$$

This calculation shows that there is a well-defined map $\phi\colon G \to B(\mathcal{H})$ given by setting $\phi(g) = \rho(g_2)^*\rho(g_1)$ where $g = g_1 - g_2$ with g_1, g_2 in $\mathcal{P}$.

We claim that ϕ is completely positive definite. To see this, choose $g_1, \ldots, g_n$ in G, and write $g_i = p_i - q_i$ where $p_1, \ldots, p_n, q_1, \ldots, q_n$ are in $\mathcal{P}$. We have that

$$\begin{aligned}(\phi(-g_i + g_j)) &= (\phi(-p_i + q_i + p_j - q_j)) = (\rho(p_i + g_j)^*\rho(q_i + p_j))\\ &= (\rho(p_i)^*\rho(q_j)^*\rho(q_i)\rho(p_j)) = D^*(\rho(q_j)^*\rho(q_i))D,\end{aligned}$$

where D is the diagonal matrix whose (j, j) entry is $\rho(p_j)$. Setting $\rho(q_i) = V_i$ and applying Corollary 5.3, it is now easily seen that $(\phi(-g_i + g_j)) \geq 0$. Hence ϕ is completely positive definite. By Naimark's dilation theorem there exists a Hilbert space $\mathcal{K}$, an operator $V\colon \mathcal{H} \to \mathcal{K}$, and a unitary representation $\pi\colon G \to B(\mathcal{K})$ such that $\phi(g) = V^*\pi(g)V$. Since $I = \phi(0) = V^*V$, we have that V is an isometry, and the proof is completed by identifying $\mathcal{H}$ with $V\mathcal{H}$. □

We now return to arbitrary commuting contractions.

Theorem 5.5 (Ando's dilation theorem). *Let T_1 and T_2 be commuting contractions on a Hilbert space $\mathcal{H}$. Then there exists a Hilbert space $\mathcal{K}$ that contains $\mathcal{H}$ as a subspace, and commuting unitaries U_1, U_2 on $\mathcal{K}$, such that*

$$T_1^n T_2^m = P_{\mathcal{H}} U_1^n U_2^m|_{\mathcal{H}}$$

for all nonnegative integers n, m.

Proof. By Theorem 5.1 it will be enough to find a pair of commuting isometries V_1, V_2 such that

$$T_1^n T_2^m = P_{\mathcal{H}} V_1^n V_2^m|_{\mathcal{H}} \qquad (*)$$

The concrete isometric dilations of Chapter 1, $V_i\colon \ell^2(\mathcal{H}) \to \ell^2(\mathcal{H})$ via

$$V_i((h_1, h_2, \ldots)) = (T_i h_1, D_i h_1, h_2, \ldots),\ D_i = (I - T_i^* T_i)^{1/2} \qquad i = 1, 2,$$

satisfy $(*)$, but these isometries need not commute. Indeed, we have

$$V_1V_2((h_1, h_2, \dots)) = (T_1T_2h_1, D_1T_2h_1, D_2h_1, h_2, \dots)$$

while

$$V_2V_1((h_1, h_2, \dots)) = (T_2T_1h_1, D_2T_1h_1, D_1h_1, h_2, \dots).$$

Assume for the moment that there exists a unitary $U: \mathcal{H} \oplus \mathcal{H} \to \mathcal{H} \oplus \mathcal{H}$ such that $U((D_1T_2h, D_2h)) = (D_2T_1h, D_1h)$. Let $W: \ell^2(\mathcal{H}) \to \ell^2(\mathcal{H})$ be the unitary defined by

$$W((h_1, h_2, \dots)) = (h_1, U((h_2, h_3)), U((h_4, h_5)), \dots).$$

Note that WV_1, V_2W^{-1} are still isometries satisfying $(*)$ and that

$$\begin{aligned} &WV_1V_2W^{-1}((h_1, h_2, \dots)) \\ &\quad = W(T_1T_2h_1, D_1T_2h_1, D_2h_1, U^{-1}((h_2, h_3)), U^{-1}((h_4, h_5)), \dots) \\ &\quad = (T_1T_2h_1, U((D_1T_2h_1, D_2h_1)), h_2, h_3, \dots) \\ &\quad = V_2V_1((h_1, h_2, \dots)) \\ &\quad = (V_2W^{-1})(WV_1)((h_1, h_2, \dots)), \end{aligned}$$

and so WV_1, V_2W^{-1} commute.

To obtain the desired unitary U, we note that

$$\begin{aligned} \|D_1T_2h\|^2 + \|D_2h\|^2 &= \langle [T_2^*(I - T_1^*T_1)T_2 + (I - T_2^*T_2)]h, h\rangle \\ &= \langle [T_1^*(I - T_2^*T_2)T_1 + (I - T_1^*T_1)]h, h\rangle \\ &= \|D_2T_1h\|^2 + \|D_1h\|^2. \end{aligned}$$

Thus, setting $U((D_1T_2h, D_2h)) = (D_2T_1h, D_1h)$ defines an isometric map between two subspaces of $\mathcal{H} \oplus \mathcal{H}$. When the codimensions of these two subspaces agree, then this isometry extends to the desired unitary on $\mathcal{H} \oplus \mathcal{H}$. This is the case for example when the dimension of $\mathcal{H}$ is finite.

When $\mathcal{H}$ is infinite-dimensional, however, these two codimensions can be different, and a more complicated argument is needed.

One begins by redefining the basic isometries, via

$$V_i((h_1, h_2, \dots)) = (T_ih_1, D_ih_1, 0, h_2, \dots).$$

Comparing V_1V_2 with V_2V_1, one now finds that a unitary $U: \mathcal{H}^{(4)} \to \mathcal{H}^{(4)}$ is needed satisfying

$$U((D_1T_2h, 0, D_2h, 0)) = (D_2T_1h, 0, D_1h, 0).$$

The extra 0's guarantee that when $\mathcal{H}$ is infinite-dimensional then both subspaces are of the same infinite codimension and so the desired U exists. Defining

$$W((h_1, h_2, \ldots)) = (h_1, U((h_2, h_3, h_4, h_5)), U((h_6, h_7, h_8, h_9)), \ldots),$$

one finds that WV_1 and V_2W^{-1} commute and satisfy $(*)$. □

Corollary 5.6. *Let T_1 and T_2 be commuting contractions on a Hilbert space $\mathcal{H}$, and let $p_{i,j}$, $i, j = 1, \ldots, m$, be polynomials in two variables. Then*

$$\|(p_{i,j}(T_1, T_2))\|_{B(\mathcal{H}^{(m)})} \leq \sup\{\|(p_{i,j}(z_1, z_2))\|_{M_m} \colon |z_1| \leq 1,\ |z_2| \leq 1\}.$$

Proof. The proof is similar to Corollary 5.2. □

We shall refer to the above result as the *two-variable von Neumann inequality*.

Surprisingly, Ando's construction cannot be generalized to more than two commuting contractions. For if it could, then one could prove an n-variable von Neumann inequality, and the analogue of von Neumann's inequality fails for three or more commuting contractions. There are several such counterexamples in the literature. The following appeared in [236] and is perhaps the simplest.

Example 5.7 (Kaijser–Varopoulos). Consider the following operators on $\mathbb{C}^5$:

$$A_1 = \begin{pmatrix} 0 & 0 & 0 & 0 & 0 \\ 1 & 0 & 0 & 0 & 0 \\ 0 & 0 & 0 & 0 & 0 \\ 0 & 0 & 0 & 0 & 0 \\ 0 & 1/\sqrt{3} & -1/\sqrt{3} & -1/\sqrt{3} & 0 \end{pmatrix},$$

$$A_2 = \begin{pmatrix} 0 & 0 & 0 & 0 & 0 \\ 0 & 0 & 0 & 0 & 0 \\ 1 & 0 & 0 & 0 & 0 \\ 0 & 0 & 0 & 0 & 0 \\ 0 & -1/\sqrt{3} & 1/\sqrt{3} & -1/\sqrt{3} & 0 \end{pmatrix},$$

and

$$A_3 = \begin{pmatrix} 0 & 0 & 0 & 0 & 0 \\ 0 & 0 & 0 & 0 & 0 \\ 0 & 0 & 0 & 0 & 0 \\ 1 & 0 & 0 & 0 & 0 \\ 0 & -1/\sqrt{3} & -1/\sqrt{3} & 1/\sqrt{3} & 0 \end{pmatrix}.$$

It is easily checked that $\|A_i\| \leq 1$ and that $A_i A_j = A_j A_i$, for $1 \leq i, j \leq 3$. If one considers the polynomial

$$p(z_1, z_2, z_3) = z_1^2 + z_2^2 + z_3^2 - 2z_1z_2 - 2z_1z_3 - 2z_2z_3,$$

then $\|p\|_\infty = \sup\{|p(z_1, z_2, z_3)|\colon |z_i| \leq 1\} = 5$, as a little calculus shows. However,

$$p(A_1, A_2, A_3) = \begin{pmatrix} 0 & 0 & 0 & 0 & 0 \\ 0 & 0 & 0 & 0 & 0 \\ 0 & 0 & 0 & 0 & 0 \\ 0 & 0 & 0 & 0 & 0 \\ 3\sqrt{3} & 0 & 0 & 0 & 0 \end{pmatrix},$$

and since $3\sqrt{3} > 5$, the analogue of von Neumann's inequality fails.

Of course, once it fails for $n = 3$, it also fails for all $n \geq 3$.

In spite of the many examples of commuting contractions for which von Neumann's inequality fails, it is still unknown whether or not it could hold up to some constant. That is, it is not known if for each n there exists a finite constant K_n such that for any commuting contractions $T_1, \ldots, T_n$ and any polynomial p in n variables one has

$$\|p(T_1, \ldots, T_n)\| \leq K_n \|p\|_\infty,$$

where $\|p\|_\infty$ denotes the supremum of p over $|z_i| \leq 1$, $1 \leq i \leq n$. It is generally believed that there does not exist such a constant.

Some lower bounds on the possible values of K_n are known. For example, it is known that $K_n \geq \frac{\sqrt{n}}{11}$ [75].

In a similar vein, one can consider the problem for matrices of polynomials, i.e., seek constants C_n such that for any commuting contractions $T_1, \ldots, T_n$, one has

$$\|(p_{i,j}(T_1, \ldots, T_n))\| \leq C_n \|(p_{i,j})\|_\infty,$$

where $\|(p_{i,j})\|_\infty$ denotes the supremum of the norm of the matrix over $|z_i| \leq 1$, $1 \leq i \leq n$, and the sizes of the matrices are arbitrarily large. Clearly, $C_n \geq K_n$, and so if there is no finite constant K_n, then it should be easier to prove that there is no finite constant C_n; but this is also still unknown.

These questions lead naturally to our first abstractly defined operator algebra. Let $\mathcal{P}_n$ denote the algebra of polynomials in n variables. Given a polynomial p, we set

$$\|p\|_u = \sup\|p(T_1, \ldots, T_n)\|,$$

where the supremum is taken over the family of all n-tuples of commuting contractions on all Hilbert spaces.

It is easy to see that $\|p\|_u$ is finite, since it is bounded by the sum of the absolute values of the coefficients of p, and that this quantity defines a norm on $\mathcal{P}_n$.

For a fixed polynomial p, there is always an n-tuple of contractions where this supremum is acheived. To see this, first choose a sequence of n-tuples $(T_{1,k}, \ldots, T_{n,k})$ such that $\|p\|_u = \sup_k\|p(T_{1,k}, \ldots, T_{n,k})\|$. If we then let $T_i = \sum_k \oplus T_{i,k}$, $i = 1, \ldots, n$, be the n-tuple of commuting contractions on the direct sum of the corresponding Hilbert spaces, then $\|p(T_1, \ldots, T_n)\| = \|p\|_u$.

Now, if for each polynomial p we choose a commuting n-tuple of contractions where $\|p\|_u$ is attained and form the direct sum of all such n-tuples, then we have a single n-tuple of commuting contractions $(T_1, \ldots, T_n)$ on a Hilbert space $\mathcal{H}$ such that $\|p\|_u = \|p(T_1, \ldots, T_n)\|$ for every polynomial. Consequently, the map $\pi : \mathcal{P}_n \to B(\mathcal{H})$ defined by $\pi(p) = p(T_1, \ldots, T_n)$ is an isometric homomorphism of $(\mathcal{P}_n, \|\cdot\|_u)$ into $B(\mathcal{H})$.

In a similar fashion, if for $(p_{i,j})$ in $M_k(\mathcal{P}_n)$ we set

$$\|(p_{i,j})\|_{u,k} = \sup\|(p_{i,j}(T_1, \ldots, T_n))\|,$$

then we obtain a sequence of norms on the matrices over $\mathcal{P}_n$. We call $(\mathcal{P}_n, \|\cdot\|_{u,k})$ the *universal operator algebra for n-tuples of commuting contractions.*

To justify this terminology, note that, arguing as above, we can choose a single Hilbert space $\mathcal{H}$ and a single n-tuple of commuting contractions $(T_1, \ldots, T_n)$ on that space such that $\|(p_{i,j}))\|_{u,k} = \|(p_{i,j}(T_1, \ldots, T_n))\|$ for every k and every $k \times k$ matrix of polynomials. The corresponding map π, will now be a completely isometric homomorphism of the matrix-normed algebra $\mathcal{P}_n$ into $B(\mathcal{H})$. Hence, $(\mathcal{P}_n, \|\cdot\|_{u,k})$ can be realized completely isometrically as a concrete algebra of operators.

We now turn our attention to further applications of Ando's theorem. We begin with some re-formulations due to Sz.-Nagy and Foias [231].

Theorem 5.8 (Commutant lifting theorem). *Let T be a contraction on a Hilbert space, and let $(U, \mathcal{K})$ be the minimal unitary dilation of T. If R commutes with T, then there exists an operator S commuting with U such that $\|R\| = \|S\|$ and $RT^n = P_{\mathcal{H}} SU^n|_{\mathcal{H}}$ for $n \geq 0$.*

Proof. By scaling we may assume that $\|R\| = 1$. Hence, by Ando's theorem there exists commuting unitaries U_1, U_2 on a Hilbert space $\mathcal{K}_1$ containing $\mathcal{H}$ such that $T^n R^m = P_{\mathcal{H}} U_1^n U_2^m|_{\mathcal{H}}$. Let $\mathcal{K}$ be the closed linear span of vectors of the form $U_1^n h, n \in \mathbb{Z}, h \in \mathcal{H}$. Then $\mathcal{K}$ is a reducing subspace for U_1, and $U = P_{\mathcal{K}} U_1|_{\mathcal{K}}$ is the minimal unitary dilation of T.

Relative to $\mathcal{K}_1 = \mathcal{K} \oplus \mathcal{K}^{\perp}$ we have operator matrices

$$U_1 = \begin{pmatrix} U & 0 \\ 0 & V \end{pmatrix}, \quad U_2 = \begin{pmatrix} S & B \\ C & D \end{pmatrix}.$$

Since U_1 and U_2 commute, it follows that U and S commute and that $P_{\mathcal{K}} U_2 U_1^n|_{\mathcal{K}} = SU^n$. Thus

$$P_{\mathcal{H}} SU^n|_{\mathcal{H}} = P_{\mathcal{H}} U_2 U_1^n\big|_{\mathcal{H}} = RT^n.$$

Finally, $1 \leq \|R\| \leq \|S\| \leq \|U_2\| = 1$ and so $\|R\| = \|S\|$. □

The following equivalent re-formulation of Ando's result is also useful. Given $T_i \in B(\mathcal{H}_i)$, $i = 1, 2$, and $A \in B(\mathcal{H}_1, \mathcal{H}_2)$, we say that A *intertwines* T_1 and T_2 provided that $AT_1 = T_2 A$. Note that this is equivalent to $\left(\begin{smallmatrix} 0 & 0 \\ A & 0 \end{smallmatrix}\right)$ commuting with $\left(\begin{smallmatrix} T_1 & 0 \\ 0 & T_2 \end{smallmatrix}\right)$.

Corollary 5.9 (Intertwining dilation theorem). *Let T_i, $i = 1, 2$, be contraction operators on Hilbert spaces $\mathcal{H}_i$ with minimal unitary dilations $(U_i, \mathcal{K}_i)$. If A intertwines T_1 and T_2, then there exists R intertwining U_1 and U_2 such that $\|A\| = \|R\|$ and*

$$AT_1^n = T_2^n A = P_{\mathcal{H}_2} R U_1^n\big|_{\mathcal{H}_1} = P_{\mathcal{H}_2} U_2^n R\big|_{\mathcal{H}_1}$$

for all $n \geq 0$.

Proof. First note that the minimal unitary dilation of

$$\hat{T} = \begin{pmatrix} T_1 & 0 \\ 0 & T_2 \end{pmatrix}$$

is

$$\hat{U} = \begin{pmatrix} U_1 & 0 \\ 0 & U_2 \end{pmatrix} \qquad \text{acting on } \mathcal{K}_1 \oplus \mathcal{K}_2.$$

Applying commutant lifting to $\hat{A} = \left(\begin{smallmatrix} 0 & 0 \\ A & 0 \end{smallmatrix}\right)$, we obtain $\hat{R} = \left(\begin{smallmatrix} A & B \\ R & D \end{smallmatrix}\right)$ satisfying $\hat{A}\hat{T}^n = P_{\mathcal{H}_1 \oplus \mathcal{H}_2} \hat{R}\hat{U}^n|_{\mathcal{H}_1 \oplus \mathcal{H}_2}$. Equating the (2,1) entries of these operators yields $AT^n = P_{\mathcal{H}_2} RU_1^n|_{\mathcal{H}_1}$ for $n \geq 0$. The other relations follow from the intertwining condition.

Finally, $\|A\| \leq \|R\| \leq \|\hat{R}\| = \|\hat{A}\| = \|A\|$ and so $\|A\| = \|R\|$. □

We now apply these results to characterize the norms of some operator-valued matrices. First, some formalities are in order. Let's assume that $\mathcal{H}$ is a separable Hilbert space and that $f\colon \mathbb{T} \to \mathcal{H}$ is measurable and square integrable. Defining $h_n \in \mathcal{H}$ via $h_n = (1/2\pi i)\int_0^{2\pi} e^{-in\theta} f(e^{i\theta})\, d\theta$ yields a formal Fourier series $\sum_{n=-\infty}^{+\infty} e^{in\theta} h_n$ with

$$\frac{1}{2\pi}\int_0^{2\pi} \|f(e^{i\theta})\|^2\, d\theta = \sum_{n=-\infty}^{+\infty} \|h_n\|^2.$$

This leads to identifications

$$L^2(\mathbb{T};\mathcal{H}) = L^2(\mathbb{T}) \otimes \mathcal{H} = \ell^2_{\mathbb{Z}}(\mathcal{H}).$$

If $A_n \in B(\mathcal{H})$ is a norm-bounded sequence of operators and for $0 < r < 1$ we set

$$B_r(e^{i\theta}) = \sum_{n=-\infty}^{+\infty} A_n r^{|n|} e^{in\theta},$$

then the series converges uniformly to define a continuous operator-valued function $B_r\colon \mathbb{T} \to B(\mathcal{H})$. For $f \in L^2(\mathbb{T};\mathcal{H})$ we have that $\theta \to B_r(e^{i\theta}) f(e^{i\theta}) \in L^2(\mathbb{T};\mathcal{H})$ and that multiplication by B_r, M_{B_r}, is a bounded operator on $L^2(\mathbb{T};\mathcal{H})$ with

$$\|M_{B_r}\| = \sup_\theta \|B_r(e^{i\theta})\| \equiv \|B_r\|_\infty$$

where $B_r(e^{i\theta}) \in B(\mathcal{H})$. With respect to the identification $L^2(\mathbb{T};\mathcal{H}) = \ell^2_{\mathbb{Z}}(\mathcal{H})$, the operator M_{B_r} becomes the doubly infinite operator-valued Toeplitz matrix $(r^{|i-j|}A_{i-j})_{i,j=-\infty}^{+\infty}$. It can be verified that

$$\|(A_{i-j})\| = \sup_{r<1} \left\| \left(r^{|i-j|}A_{i-j}\right) \right\| = \sup_{r<1} \left\| M_{B_r} \right\| = \sup_{r<1} \|B_r\|_\infty.$$

Thus, when $\|(A_{i-j})\|$ is finite, we identify this operator-valued Toeplitz matrix with the operator of multiplication by $B(e^{i\theta}) = \sum_{n=-\infty}^{+\infty} A_n e^{in\theta}$ and write $\|(A_{i-j})\| = \|B\|_\infty$, even though the convergence of this latter series can be a touchy business. Thus, one should always interpret $\|B\|_\infty \equiv \sup_{r<1} \|B_r\|_\infty$.

We shall need the following operator-valued version of Nehari's theorem in Chapter 10.

Theorem 5.10 (Nehari–Page). *Let $\mathcal{H}$ be a separable Hilbert space, and let $A_n \in B(\mathcal{H})$, $n \geq 0$, be a sequence of operators. Then the operator-valued Hankel matrix $(A_{i+j})_{i,j=0}^{\infty}$ is bounded on $\ell^2(\mathcal{H})$ if and only if there exists $A_n \in B(\mathcal{H})$, $n < 0$, such that $\|B\|_\infty \equiv \sup_{r<1} \|B_r\| < +\infty$, where $B(e^{i\theta}) = \sum_{n=-\infty}^{+\infty} A_n e^{in\theta}$. Moreover, in this case there exists a particular choice of $A_n \in B(\mathcal{H})$, $n < 0$, such that*

$$\|(A_{i+j})\| = \|B\|_\infty, \qquad B(e^{i\theta}) = \sum_{n=-\infty}^{+\infty} A_n e^{in\theta}.$$

Proof. We only prove that if $H = (A_{i+j})$ is bounded on $\ell^2(\mathcal{H})$, then there exists $A_n \in B(\mathcal{H})$, $n < 0$, such that $\|H\| = \|B\|_\infty$.

To this end let $S\colon \ell^2(\mathcal{H}) \to \ell^2(\mathcal{H})$ via $S((h_0, h_1, \dots)) = (0, h_0, h_1, \dots)$ with adjoint $S^*((h_0, h_1, \dots)) = (h_1, h_2, \dots)$. These operators can be identified with the operator matrices

$$S = \begin{pmatrix} 0 & \dots & \dots \\ I & 0 & \dots \\ 0 & I & \ddots \\ \vdots & 0 & \ddots \\ \vdots & \vdots & \ddots \end{pmatrix}, \qquad S^* = \begin{pmatrix} 0 & I & 0 & \dots & \dots \\ 0 & 0 & I & 0 & \dots \\ \vdots & \vdots & \ddots & \ddots & \ddots \end{pmatrix}.$$

We have that $S^*H = (A_{i+1+j}) = HS$. It is straightforward to verify that the minimal unitary dilations of S and S^* are the forward and backward shifts $\hat{S}$ and $\hat{S}^*$ on $\ell^2_{\mathbb{Z}}(\mathcal{H})$. By Corollary 5.9 there exists $R = (R_{i,j})\colon \ell^2_{\mathbb{Z}}(\mathcal{H}) \to \ell^2_{\mathbb{Z}}(\mathcal{H})$ with $\|R\| = \|H\|$ such that $\hat{S}^*R = R\hat{S}$. This implies that $R_{i,j} = R_{i+j}$ for some sequence $R_n \in B(\mathcal{H}), n \in \mathbb{Z}$, with $R_n = A_n$ for $n \geq 0$. Now let $W\colon \ell^2_{\mathbb{Z}}(\mathcal{H}) \to \ell^2_{\mathbb{Z}}(\mathcal{H})$ be the unitary defined by

$$W((\dots, h_{-1}, h_0, h_{+1}, \dots)) = (\dots, h_{+1}, h_0, h_{-1}, \dots).$$

Then $RW = (R_{i-j})$ and hence

$$\|H\| = \|R\| = \|RW\| = \left\| \sum_{n=-\infty}^{+\infty} R_n e^{in\theta} \right\|_\infty. \qquad \square$$

When the sequence $\{A_n\}$ is just numbers, then the above theorem has a clearer interpretation. We have $B(e^{i\theta}) = \sum_{n=-\infty}^{+\infty} a_n e^{in\theta}$, and $\|B\|_\infty$ is just the norm of B in $L^\infty(\mathbb{T})$. Thus, the theorem identifies the sequences $\{a_n\}_{n\geq 0}$ which can be "completed" by choosing $\{a_n\}_{n\leq -1}$ to be the Fourier coefficients of an L^∞ function as those for which the corresponding Hankel matrix is bounded.

Set

$$\hat{f}(n) = \frac{1}{2\pi i}\int_0^{2\pi} f(e^{i\theta})e^{-in\theta}\,d\theta,$$

and let $G = \{f \in L^\infty(\mathbb{T}) : \hat{f}(n) = 0 \text{ for all } n \geq 0\}$. Then the map $Q: L^\infty(\mathbb{T}) \to B(\ell^2)$, $Q(f) = (\hat{f}(i+j))_{i,j=0}^{+\infty}$ has kernel G, and by the above theorem $\dot{Q}$: $L^\infty(\mathbb{T})/G \to B(\ell^2)$ is an isometric linear isomorphism onto the space of Hankel matrices.

Alternatively, if we look at the map

$$P: L^\infty(\mathbb{T}) \to B(\ell^2), \qquad P(f) = (\hat{f}(-1-i-j)),$$

then $\ker P = \{f \in L^\infty : \hat{f}(n) = 0 \text{ for all } n \leq -1\} \equiv H^\infty(\mathbb{T})$, and by the above theorem we have that $\dot{P}: L^\infty(\mathbb{T})/H^\infty(\mathbb{T}) \to B(\ell^2)$ is an isometric isomorphism onto the space of Hankel matrices.

This last isomorphism gives rise to an important duality. Recall that the dual of $L^1(\mathbb{T})$ is $L^\infty(\mathbb{T})$. If we let $H^1(\mathbb{T}) = \{f \in L^1(\mathbb{T}) : \hat{f}(n) = 0 \text{ for all } n \leq -1\}$, then the annihilator of $H^1(\mathbb{T})$ is

$$\{f \in L^\infty(\mathbb{T}) : \hat{f}(n) = 0 \text{ for all } n \geq 0\} = e^{i\theta} \cdot H^\infty(\mathbb{T}).$$

Consequently we have isometric isomorphisms between the dual of $H^1(\mathbb{T})$ and $L^\infty(\mathbb{T})/e^{i\theta}H^\infty(\mathbb{T}) \cong L^\infty(\mathbb{T})/H^\infty(\mathbb{T})$, which is in turn isometrically isomorphic to the space of Hankel matrices. Explicitly, this dual pairing is given by

$$\langle A, f\rangle = \sum_{n=0}^{\infty} a_n \hat{f}(n)$$

for $A = (a_{i+j})_{i,j=0}^{\infty}$ a bounded Hankel and $f \in H^1(\mathbb{T})$.

Notes

The fact that von Neumann's inequality holds for two commuting contractions but not three or more is still the source of many surprising results and intriguing questions. Many deep results about analytic functions come from this dichotomy. For example, Agler [3] uses Ando's theorem to deduce an analogue of the classical Nevanlinna–Pick interpolation formula for analytic functions on the bidisk. Because of the failure of a von Neumann inequality for three or more commuting contractions, the analogous formula for the tridisk is known to be false, and the problem of finding the correct analogue of the Nevanlinna–Pick formula for polydisks in three or more variables remains open.

Many results are now known to be equivalent to Ando's theorem, but a truly independent argument for proving any of these results that doesn't ultimately rest on dilating a pair of commuting operators has still not been found. For example, we shall see in Chapter 7 that Ando's theorem is equivalent to the two-variable version of von Neumann's inequality holding for matrices of polynomials. That is, Corollary 5.6 implies Ando's dilation theorem. In Chapter 18, we shall see that Ando's theorem is equivalent to a certain factorization formula holding for matrices of polynomials in two variables. Cole and Wermer [66] prove that Ando's theorem is equivalent to another theorem about factoring polynomials in two variables. Agler's two-variable Nevanlinna–Pick result is also equivalent to Ando's theorem. Yet direct proofs of any of these results that do not rely on Ando's theorem, or one of its close variants, have not been found.

Moreover, because of the failure of the von Neumann inequality for three or more variables, each of the above-cited results is known to fail for three or more variables.

This operator-theoretic approach to results in classical function theory, as represented by Theorem 5.10, Exercise 5.3, and the results cited above, owes much to Sarason [209] and the book of Sz.-Nagy and Foias [231]. This continues to be an active area of research with many engineering applications, particularly to what has come to be known as H^∞-control theory. For an introduction to this area we recommend the text of Foias and Frazho [100].

It is remarkable that we do not know whether or not a constant exists, for each n, such that von Neumann's inequality for n commuting contractions is true up to that constant. This illustrates our lack of understanding of these important topics.

Exercises

5.1 Prove that if $(V_1, \mathcal{K}_1)$ and $(V_2, \mathcal{K}_2)$ are any two minimal isometric dilations of a contraction operator T on $\mathcal{H}$, then there exists a unitary $U\colon \mathcal{K}_1 \to \mathcal{K}_2$ such that $Uh = h$ for all $h \in \mathcal{H}$ and $UV_1U^* = V_2$.

5.2 Let $\mathcal{H}_1, \mathcal{H}_2$ be Hilbert spaces, let $A_n \in B(\mathcal{H}_1, \mathcal{H}_2)$ be a sequence of operators, and let $A = (A_{i+j})\colon \ell^2(\mathcal{H}_1) \to \ell^2(\mathcal{H}_2)$ be the corresponding Hankel operator. Prove the analogue of the Nehari–Page theorem in this setting.

5.3 (Carathéodory's completion theorem) Let $a_0, \ldots, a_n$ be in $\mathbb{C}$. Use commutant lifting to prove that

$$\left\| \begin{pmatrix} a_0 & 0 & \ldots & 0 \\ a_1 & \ddots & \ddots & \vdots \\ \vdots & \ddots & \ddots & 0 \\ a_n & \ldots & a_1 & a_0 \end{pmatrix} \right\|_{M_{n+1}} = \inf \left\| \sum_{j=0}^{n} a_j z^j + \sum_{j=n+1}^{\infty} b_j z^j \right\|_\infty ,$$

where the infimum is over all sequences $\{b_j\}$ such that the resulting power series is bounded on $\mathbb{D}$ and the ∞-norm is the supremum over $\mathbb{D}$. Moreover, there exists a sequence $\{b_j\}$ where the infimum is attained. Thus, a polynomial can be completed to a power series whose supremum over the disk is bounded by 1 by adding higher-order terms if and only if the norm of the corresponding Toeplitz matrix is at most 1. Deduce that the map

$$f \in H^\infty(\mathbb{T}) \to \begin{pmatrix} \hat{f}(0) & 0 & \dots & 0 \\ \hat{f}(1) & \ddots & \ddots & \vdots \\ \vdots & \ddots & \ddots & 0 \\ \hat{f}(n) & \dots & \hat{f}(1) & \hat{f}(0) \end{pmatrix} \in M_{n+1}$$

yields an isometric isomorphism of $H^\infty(\mathbb{T})/e^{i(n+1)\theta}H^\infty(\mathbb{T})$ into M_{n+1}. Generalize to the case where $A_0, \dots, A_n$ are operators on a (separable) Hilbert space. [Hint: The above finite Toeplitz matrix commutes with

$$S_n = \begin{pmatrix} 0 & \dots & \dots & \dots & 0 \\ 1 & \ddots & & & \vdots \\ 0 & \ddots & \ddots & & \vdots \\ \vdots & & \ddots & \ddots & \vdots \\ 0 & \dots & 0 & 1 & 0 \end{pmatrix}.]$$

5.4 Let $\{T_1, \dots, T_n\}$ be contractions on a Hilbert space $\mathcal{H}$ (possibly noncommuting). Prove that there exists a Hilbert space $\mathcal{K}$ containing $\mathcal{H}$ and unitaries $\{U_1, \dots, U_n\}$ on $\mathcal{K}$ such that

$$T_{i_1}^{k_1} \dots T_{i_m}^{k_m} = P_{\mathcal{H}} U_{i_1}^{k_1} \dots U_{i_m}^{k_m}|_{\mathcal{H}}$$

where $m, k_1, \dots, k_m$ are arbitrary nonnegative integers, and $1 \le i_\ell \le n$ for $\ell = 1, \dots, m$.

5.5 (Schaffer [209]) Let T be a contraction on a Hilbert space $\mathcal{H}$, and let $\ell^2_{\mathbb{Z}}(\mathcal{H}) = \sum_{n=-\infty}^{+\infty} \oplus \mathcal{H}$ denote the Hilbert space formed as the direct sum of copies of $\mathcal{H}$ indexed by the integers $\mathbb{Z}$. Define an operator matrix $U = (U_{ij})_{i,j=-\infty}^{+\infty}$ by setting $U_{0,0} = T$, $U_{0,1} = (I - TT^*)^{1/2}$, $U_{-1,0} = (I - T^*T)^{1/2}$, $U_{-1,1} = -T^*$, $U_{n,n+1} = I$ for $n \ge 1$ or $n \le -2$, and $U_{i,j} = 0$ for all other pairs (i, j). Prove that U defines a unitary operator on $\ell^2_{\mathbb{Z}}(\mathcal{H})$ and that if we identify $\mathcal{H}$ with the 0th copy of $\mathcal{H}$ in $\ell^2_{\mathbb{Z}}(\mathcal{H})$, then $T^n = P_{\mathcal{H}} U^n|_{\mathcal{H}}$ for all nonnegative integers n.

5.6 Fix $n \ge 1$, and let $\mathcal{P}_n$ denote the algebra of polynomials in n variables. For each p in $\mathcal{P}_n$ we set $\|p\|_u = \sup\{\|p(T_1, \dots, T_n)\|\}$, where for now the supremum is only taken over all commuting n-tuples of contractions $\{T_1, \ldots, T_n\}$ on the Hilbert space ℓ^2.

(i) Prove that if $\{T_1, \ldots, T_n\}$ are commuting contractions on *any* Hilbert space, then $\|p(T_1, \ldots, T_n)\| \leq \|p\|_u$ and consequently this is the same norm as defined earlier.

(ii) Prove the corresponding result for matrices of polynomials.

(iii) Prove that the universal operator algebra for n-tuples of commuting contractions $(\mathcal{P}_n, \|\cdot\|_{u,k})$ can be represented completely isometrically as an algebra of operators on a separable Hilbert space.

5.7 Prove that $\|\cdot\|_u$ is a norm on $\mathcal{P}_n$ and that the completion of $\mathcal{P}_n$ in this norm is a Banach algebra. *We denote this Banach algebra by $A_u(\mathbb{D}^n)$, whereas we let $A(\mathbb{D}^n)$ denote the Banach algebra of functions that are continuous on the closed polydisk $\mathbb{D}^n$ and analytic on $\mathbb{D}^n$, equipped with the supremum norm, $\|\cdot\|_\infty$. In fact, $A(\mathbb{D}^n) \subseteq C(\mathbb{D}^{n^-})$ is the closure of $\mathcal{P}_n$ in the supremum norm.*

5.8 (i) Prove that $A_u(\mathbb{D}^n) \subseteq A(\mathbb{D}^n)$ as sets.

(ii) Prove that $A_u(\mathbb{D}^n) = A(\mathbb{D}^n)$ if and only if there exists a constant K_n such that $\|p\|_u \leq K_n\|p\|_\infty$ for every p in $\mathcal{P}_n$.

5.9 Let $\mathbb{K} \subseteq B(\ell^2)$ denote the ideal of compact operators on ℓ^2. Let $A(\mathbb{D}^n; \mathbb{K})$ denote the set of continuous functions from the closure of $\mathbb{D}^n$ into $\mathbb{K}$ that are analytic on $\mathbb{D}^n$, and let $\mathcal{P}(\mathbb{D}^n; \mathbb{K}) \subseteq A(\mathbb{D}^n; \mathbb{K})$ denote the functions that are finite sums of the form $\sum z^J K_J$ with K_J in $\mathbb{K}$. If we identify M_k as the subspace of operators in $\mathbb{K}$ that are supported on the first $k \times k$ block, then we may regard $M_k(A(\mathbb{D}^n)) \subseteq M_{k+1}(A(\mathbb{D}^n)) \subseteq A(\mathbb{D}^n; \mathbb{K})$.

(i) Prove that $A(\mathbb{D}^n; \mathbb{K})$ equipped with the supremum norm and pointwise product is a Banach algebra.

(ii) Prove that $\mathcal{P}(\mathbb{D}^n; \mathbb{K})$ is dense in $A(\mathbb{D}^n; \mathbb{K})$.

(iii) Prove that the union over k of $M_k(A(\mathbb{D}^n))$ is dense in $A(\mathbb{D}^n; \mathbb{K})$.

(iv) Give a definition of $A_u(\mathbb{D}^n; \mathbb{K})$ and prove the analogues of (i), (ii), and (iii).

(v) Prove that there exists a constant C_n such that for all matrices of polynomials, $\|(p_{i,j})\|_u \leq C_n\|(p_{i,j})\|_\infty$ if and only if $A_u(\mathbb{D}^n; \mathbb{K}) = A(\mathbb{D}^n; \mathbb{K})$.

Chapter 6

Completely Positive Maps into M_n

In this chapter we characterize completely positive maps into M_n. This characterization allows us to prove an extension theorem for completely positive maps and lends further insight into the properties of positive maps on operator systems. These results are all consequences of a duality between maps into M_n and linear functionals.

Let $\mathcal{M}$ be an operator space, and let $\{e_j\}_{j=1}^n$ be the canonical basis for $\mathbb{C}^n$. If A is in M_n, then let $A_{(i,j)}$ denote the (i, j) entry of A, so that $A_{(i,j)} = \langle Ae_j, e_i \rangle$. If $\phi \colon \mathcal{M} \to M_n$ is a linear map, then we associate to ϕ a linear functional s_ϕ on $M_n(\mathcal{M})$ by the following formula:

$$s_\phi((a_{i,j})) = \frac{1}{n} \sum_{i,j} \phi(a_{i,j})_{(i,j)}.$$

Alternatively, if we let x denote the vector in $\mathbb{C}^{n^2} = \mathbb{C}^n \oplus \cdots \oplus \mathbb{C}^n$ given by $x = e_1 \oplus \cdots \oplus e_n$, then

$$s_\phi((a_{i,j})) = \frac{1}{n} \langle \phi_n((a_{i,j}))x, x \rangle,$$

where the inner product is taken in $\mathbb{C}^{n^2}$.

We leave it to the reader to verify that $\phi \to s_\phi$ defines a linear map from $\mathcal{L}(\mathcal{M}, M_n)$, the vector space of linear maps from $\mathcal{M}$ into M_n, into the vector space $\mathcal{L}(M_n(\mathcal{M}), \mathbb{C})$. If $\mathcal{M}$ contains the unit and $\phi(1) = 1$, then $s_\phi(1) = 1$.

Finally, if $s \colon M_n(\mathcal{M}) \to \mathbb{C}$, then we define $\phi_s \colon \mathcal{M} \to M_n$ via

$$(\phi_s(a))_{(i,j)} = n \cdot s(a \otimes E_{i,j}),$$

where $a \otimes E_{i,j}$ is the element of $M_n(\mathcal{M})$ which has a for its (i, j) entry and is 0 elsewhere. We leave it to the reader to verify that the maps $\phi \to s_\phi$ and $s \to \phi_s$ are mutual inverses.

Theorem 6.1. *Let $\mathcal{A}$ be a C^*-algebra with unit, let $\mathcal{S}$ be an operator system in $\mathcal{A}$, and let $\phi\colon \mathcal{S} \to M_n$. The following are equivalent:*

(i) ϕ is completely positive,
(ii) ϕ is n-positive,
(iii) s_ϕ is positive.

Proof. Obviously, (i) implies (ii). Also, that (ii) implies (iii) is clear by the alternative definition of s_ϕ. So assume that s_ϕ is positive, and we shall prove that ϕ is completely positive.

By Krein's theorem (Exercise 2.10), we may extend s_ϕ from $M_n(\mathcal{S})$ to a positive, linear functional s on $M_n(\mathcal{A})$. Since s extends s_ϕ, the map $\psi\colon \mathcal{A} \to M_n$ associated with s extends ϕ. Clearly, if we can prove that ψ is completely positive, then ϕ will be completely positive.

To verify that ψ is m-positive by Lemma 3.13, it is sufficient to consider a positive element of $M_m(\mathcal{A})$ of the form $(a_i^* a_j)$. Since $\psi_m((a_i^* a_j))$ acts on $\mathbb{C}^{mn}$, to see that it is positive, it is sufficient to take $x = x_1 \oplus \cdots \oplus x_m$, where each $x_j = \sum_k \lambda_{j,k} e_k$ is in $\mathbb{C}^n$, and calculate

$$\begin{aligned}\langle \psi_m((a_i^* a_j))x, x\rangle &= \sum_{i,j} \langle \psi(a_i^* a_j)x_j, x_i\rangle \qquad (*)\\ &= \sum_{i,j,k,\ell} \lambda_{j,k}\bar{\lambda}_{i,\ell}\langle \psi(a_i^* a_j)e_k, e_\ell\rangle\\ &= \sum_{i,j,k,\ell} \lambda_{j,k}\bar{\lambda}_{i,\ell} s(a_i^* a_j \otimes E_{\ell,k}).\end{aligned}$$

Let A_i be the $n \times n$ matrix whose first row is $(\lambda_{i,1}, \ldots, \lambda_{i,n})$ and whose remaining rows are 0. We have that

$$A_i^* A_j = \sum_{k,\ell} \bar{\lambda}_{i,\ell}\lambda_{j,k} E_{\ell,k},$$

and thus $(*)$ becomes

$$\sum_{i,j} s(a_i^* a_j \otimes A_i^* A_j) = s\left(\left(\sum_i a_i \otimes A_i\right)^* \left(\sum_j a_j \otimes A_j\right)\right),$$

which is positive, since s is positive. Thus, ψ is m-positive for all m. □

Theorem 6.2. *Let $\mathcal{A}$ be a C^*-algebra with unit, $\mathcal{S}$ an operator system contained in $\mathcal{A}$, and $\phi\colon \mathcal{S} \to M_n$ completely positive. Then there exists a completely positive map $\psi\colon \mathcal{A} \to M_n$ extending ϕ.*

Proof. Let s_ϕ be the positive, linear functional on $M_n(\mathcal{S})$ associated with ϕ, and extend it to a positive, linear functional s on $M_n(\mathcal{A})$ by Krein's theorem. By Theorem 6.1, the map ψ associated with s is completely positive. Finally, since s extends s_ϕ, it is easy to see that ψ extends ϕ. □

We now restate these results for operator spaces with unit.

Theorem 6.3. *Let $\mathcal{A}$ be a C^*-algebra with unit, let $\mathcal{M}$ be a subspace of $\mathcal{A}$ containing 1, and let $\phi\colon \mathcal{M} \to M_n$ with $\phi(1) = 1$. The following are equivalent:*

(i) ϕ is completely contractive,
(ii) ϕ is n-contractive,
(iii) s_ϕ is contractive.

Proof. Let $\mathcal{S} = \mathcal{M} + \mathcal{M}^*$. Clearly, (i) implies (ii) and (ii) implies (iii). Assuming (iii), since s_ϕ is unital and contractive, we may extend it to a positive, unital map $\tilde{s}_\phi$ on $M_n(\mathcal{M}) + M_n(\mathcal{M})^* = M_n(\mathcal{S})$. By Theorem 6.1, the linear functional $\tilde{s}_\phi$ is associated with a completely positive map on $\mathcal{S}$. This map is readily seen to be $\tilde{\phi}$, where $\tilde{\phi}(x + y^*) = \phi(x) + \phi(y)^*$. Hence, $\tilde{\phi}$ is completely positive, and so ϕ must be completely contractive. □

Theorem 6.4. *Let $\mathcal{A}$ be a C^*-algebra with unit, let $\mathcal{M}$ be a subspace of $\mathcal{A}$ containing 1, and let $\phi\colon \mathcal{M} \to M_n$ be an n-contractive map with $\phi(1) = 1$. Then ϕ extends to a completely positive map on $\mathcal{A}$.*

Proof. In the proof of Theorem 6.3, we saw that $\tilde{\phi}$ is completely positive and hence extends to a completely positive map on $\mathcal{A}$ by Theorem 6.2. □

There is one way in which the above correspondence between linear functionals on $M_n(\mathcal{S})$ and linear maps of $\mathcal{S}$ into M_n is *not* well behaved. Suppose that $s\colon M_n(\mathcal{S}) \to \mathbb{C}$ is positive and unital, so that $\|s\| = 1$. Then s gives rise to a completely positive map $\phi_s\colon \mathcal{S} \to M_n$, but ϕ_s is not necessarily unital. Indeed, since

$$\phi_s(1)_{(i,j)} = ns(1 \otimes E_{i,j}) = ns(E_{i,j}),$$

we have that ϕ_s is unital if and only if

$$s(E_{i,j}) = \begin{cases} 1/n, & i = j, \\ 0, & i \neq j. \end{cases}$$

Because of this fact, ϕ_s is not necessarily a contraction. In fact, it is not hard to construct examples of a unital, positive s such that $\|\phi_s\| = n$ (Exercise 6.1).

To obtain generalizations of the above results to the case where M_n is replaced by $B(\mathcal{H})$ requires some topological preliminaries, which we postpone until the next chapter. We turn instead to some other applications of this correspondence between linear functionals and linear maps.

Let $\mathcal{S}$ be an operator system and $\mathcal{S}^+$ its cone of positive elements. We define

$$\mathcal{S}^+ \otimes M_n^+ = \left\{ \sum_i p_i \otimes Q_i : p_i \in \mathcal{S}^+,\ Q_i \in M_n^+ \right\},$$

where the sum is finite. Note that $\mathcal{S}^+ \otimes M_n^+$ is a cone contained in the cone $M_n(\mathcal{S})^+$ of positive elements of $M_n(\mathcal{S})$.

We have seen above that if $\phi\colon \mathcal{S} \to M_n$, then ϕ is completely positive if and only if its associated linear functional s_ϕ is positive on $M_n(\mathcal{S})^+$. The following points out the relevance of the set defined above.

Lemma 6.5. *Let $\phi\colon \mathcal{S} \to M_n$. Then ϕ is positive if and only if $s_\phi\colon M_n(\mathcal{S}) \to \mathbb{C}$ assumes positive values on $\mathcal{S}^+ \otimes M_n^+$.*

Proof. Let $\phi\colon \mathcal{S} \to M_n$ be positive, let p be in $\mathcal{S}^+$, and let Q be in M_n^+. In order to prove that s_ϕ assumes positive values on $\mathcal{S}^+ \otimes M_n^+$, it will be sufficient to prove that $s_\phi(p \otimes Q)$ is positive. Furthermore, since by Lemma 3.13 Q can be written as a convex sum of matrices of the form $(\bar{\alpha}_i \alpha_j)$, it will suffice to assume that $Q = (\bar{\alpha}_i \alpha_j)$. Thus, $p \otimes Q = (\bar{\alpha}_i \alpha_j p)$ and

$$\begin{aligned} n \cdot s_\phi(p \otimes Q) &= \sum_{i,j} \phi(\bar{\alpha}_i \alpha_j p)_{(i,j)} \\ &= \sum_{i,j} \bar{\alpha}_i \alpha_j \langle \phi(p) e_j, e_i \rangle = \langle \phi(p) x, x \rangle \geq 0, \end{aligned}$$

where $x = \alpha_1 e_1 + \cdots + \alpha_n e_n$.

Conversely, assume that s_ϕ is positive on $\mathcal{S}^+ \otimes M_n^+$, let p be in $\mathcal{S}^+$, and let $x = \alpha_1 e_1 + \cdots + \alpha_n e_n$ be a vector in $\mathbb{C}^n$. We have that

$$\langle \phi(p) x, x \rangle = \sum_{i,j} \langle \phi(\bar{\alpha}_i \alpha_j p) e_j, e_i \rangle = n \cdot s_\phi((\bar{\alpha}_i \alpha_j p)) \geq 0,$$

since $(\bar{\alpha}_i \alpha_j p)$ is in $\mathcal{S}^+ \otimes M_n^+$. Thus, ϕ is positive. □

Theorem 6.6. *Let $\mathcal{S}$ be an operator system. Then the following are equivalent:*

(i) every positive map $\phi\colon \mathcal{S} \to M_n$ is completely positive,

(ii) every unital, positive map $\phi: \mathcal{S} \to M_n$ is completely positive,
(iii) $\mathcal{S}^+ \otimes M_n^+$ is dense in $M_n(\mathcal{S})^+$.

Proof. Clearly, (i) implies (ii). The proof that (ii) implies (i) is left as an exercise (Exercise 6.2). We prove the equivalence of (i) and (iii).

If $\mathcal{S}^+ \otimes M_n^+$ is dense in $M_n(\mathcal{S})^+$ and $\phi: \mathcal{S} \to M_n$ is positive, then, by Lemma 6.5, s_ϕ will be positive on $M_n(\mathcal{S})^+$, and so, by Theorem 6.1, ϕ is completely positive.

Conversely, if $\mathcal{S}^+ \otimes M_n^+$ is not dense in $M_n(\mathcal{S})^+$, then fix p in $M_n(\mathcal{S})^+$, but not in the closure of $\mathcal{S}^+ \otimes M_n^+$. By the Krein–Milman theorem, there will exist a linear functional s on $M_n(\mathcal{S})$ that is positive on $\mathcal{S}^+ \otimes M_n^+$, but such that $s(p) < 0$. The linear map $\phi_s: \mathcal{S} \to M_n$, induced by s, is then positive by Lemma 6.5, but not completely positive. □

Corollary 6.7. *Let $\mathcal{S}$ be an operator system. Then the following are equivalent:*

(i) for every C^*-algebra $\mathcal{B}$, every positive $\phi: \mathcal{S} \to \mathcal{B}$ is completely positive,
(ii) for every n, every positive $\phi: \mathcal{S} \to M_n$ is completely positive,
(iii) $\mathcal{S}^+ \otimes M_n^+$ is dense in $M_n(\mathcal{S})^+$ for all n.

Proof. Clearly, (ii) and (iii) are equivalent and (i) implies (ii). To see that (ii) implies (i), it is sufficient to consider $\mathcal{B} = B(\mathcal{H})$. Given $(a_{i,j})$ in $M_n(\mathcal{S})^+$, to check that $\phi_n((a_{i,j}))$ is positive, it is enough to choose $x_1, \dots, x_n$ in $\mathcal{H}$ and check that

$$\sum_{i,j} \langle \phi(a_{i,j}) x_j, x_i \rangle \geq 0.$$

Let $\mathcal{F}$ be the finite-dimensional subspace spanned by these n vectors, and let $\psi: \mathcal{S} \to B(\mathcal{F})$ be the compression of ϕ to $\mathcal{F}$. Identifying $B(\mathcal{F})$ with M_k, where $k = \dim(\mathcal{F})$, we have that ψ is completely positive by (ii) and hence

$$0 \leq \sum_{i,j} \langle \psi(a_{i,j}) x_j, x_i \rangle = \sum_{i,j} \langle \phi(a_{i,j}) x_j, x_i \rangle,$$

as desired. □

As well as being related to complete positivity, the above cone $\mathcal{S}^+ \otimes M_n^+$ determines the norm behavior of positive maps. Let $\mathcal{S}$ be an operator system, and let x be in $\mathcal{S}$. We say that $\mathcal{S}$ has a *partition of unity for* x, and say that x *is partitionable with respect to* $\mathcal{S}$, provided that for every $\varepsilon > 0$, there exist positive elements $p_1, \dots, p_n$ in $\mathcal{S}$ with $\sum_i p_i \leq 1$, and scalars $\lambda_1, \dots, \lambda_n$ with

$|\lambda_i| \le \|x\|$, such that

$$\left\| x - \sum_i \lambda_i p_i \right\| < \varepsilon.$$

We say that $\mathcal{S}$ has a *partition of unity* for a subset $\mathcal{M}$ of $\mathcal{S}$ provided that every element of $\mathcal{M}$ is partitionable with respect to $\mathcal{S}$.

Lemma 6.8. *Let $\mathcal{S}$ be an operator system and let x be in $\mathcal{S}$, with $\|x\| \le 1$. Then the following are equivalent:*

(i) x is partitionable with respect to $\mathcal{S}$,
(ii) every positive map ϕ with domain $\mathcal{S}$ satisfies $\|\phi(x)\| \le \|\phi(1)\|$,
(iii) $\left[\begin{smallmatrix} 1 & x \\ x^* & 1 \end{smallmatrix}\right]$ is in the closure of $\mathcal{S}^+ \otimes M_2^+$.

Proof. The proof that (i) implies (ii) is identical to the proof of Theorem 2.4.

To see that (ii) implies (iii), assume that (iii) is not met. Let $s\colon M_2(\mathcal{S}) \to \mathbb{C}$ be a linear functional such that

$$s\left(\begin{bmatrix} 1 & x \\ x^* & 1 \end{bmatrix}\right) < 0,$$

while s is positive on $\mathcal{S}^+ \otimes M_2^+$, and let $\phi\colon \mathcal{S} \to M_2$ be the linear map associated with s. By Lemma 6.5, ϕ is positive. Since

$$\left\langle \begin{bmatrix} \phi(1) & \phi(x) \\ \phi(x)^* & \phi(1) \end{bmatrix} \begin{bmatrix} e_1 \\ e_2 \end{bmatrix}, \begin{bmatrix} e_1 \\ e_2 \end{bmatrix} \right\rangle = 2s\left(\begin{bmatrix} 1 & x \\ x^* & 1 \end{bmatrix}\right) < 0,$$

we have that

$$\phi_2\left(\begin{bmatrix} 1 & x \\ x^* & 1 \end{bmatrix}\right)$$

is not positive. If we construct a unital, positive map ϕ' from ϕ as prescribed by Exercise 6.2(i), then by Exercise 6.2(ii) we have that

$$\phi_2'\left(\begin{bmatrix} 1 & x \\ x^* & 1 \end{bmatrix}\right) = \begin{bmatrix} 1 & \phi'(x) \\ \phi'(x)^* & 1 \end{bmatrix}$$

is not positive. Thus, by Lemma 3.1, $\|\phi'(x)\| > 1 = \|\phi'(1)\|$, and so (ii) does not hold.

Now suppose that (iii) is true, let $\varepsilon > 0$, and let p_i in $\mathcal{S}^+$ and Q_i in M_2^+, $i = 1, \ldots, n$, be such that

$$\left\| \begin{bmatrix} 1 & x \\ x^* & 1 \end{bmatrix} - \sum_i p_i \otimes Q_i \right\| < \varepsilon. \qquad (*)$$

We have that

$$Q_i = \begin{bmatrix} r_i & \lambda_i \\ \bar{\lambda}_i & t_i \end{bmatrix}, \qquad \text{with} \quad r_i \geq 0, \quad t_i \geq 0 \quad \text{for all } i.$$

We set $s_i = (r_i + t_i)/2$. Note that if $s_i = 0$, then $\lambda_i = 0$. The equation $(*)$ implies the following inequalities;

$$1 - \varepsilon < \sum_i r_i p_i < 1 + \varepsilon,$$

$$1 - \varepsilon < \sum_i t_i p_i < 1 + \varepsilon,$$

$$\left\| x - \sum_i \lambda_i p_i \right\| < \varepsilon.$$

Thus,

$$1 - \varepsilon < \sum_i s_i p_i < 1 + \varepsilon.$$

Setting $\lambda_i / s_i = 0$ when $s_i = 0$, we have

$$\left\| x - \sum_i (\lambda_i / s_i) s_i p_i \right\| < \varepsilon,$$

with $|\lambda_i / s_i| \leq 1$. These last two equations are clearly enough to guarantee that x is partitionable. □

Theorem 6.9. *Let $\mathcal{S}$ be an operator system, $\mathcal{M}$ a subset of $\mathcal{S}$. Then every positive map on $\mathcal{S}$ has norm $\|\phi(1)\|$ when restricted to $\mathcal{M}$ if and only if $\mathcal{S}$ has a partition of unity for $\mathcal{M}$.*

Proof. The proof is an immediate application of the above lemma. □

Corollary 6.10. *Let $\mathcal{B}$ be a C^*-algebra with unit, let $\mathcal{A}$ be a subalgebra of $\mathcal{B}$ containing the unit, and let $\mathcal{S} = \mathcal{A} + \mathcal{A}^*$. Then $\mathcal{S}$ has a partition of unity for $\mathcal{A}$.*

Proof. By Corollary 2.8, every positive map on $\mathcal{S}$ satisfies $\|\phi(a)\| \leq \|\phi(1)\| \cdot \|a\|$ for a in $\mathcal{A}$. Thus, by Theorem 6.9, $\mathcal{S}$ has a partition of unity for $\mathcal{A}$. □

We now study two examples to illustrate the use of the above ideas. The second arises from the theory of hypo-Dirichlet algebras, which we shall examine in more detail in Chapter 11. A uniform algebra $\mathcal{A}$ on a compact Hausdorff space X is called *hypo-Dirichlet* provided that the closure of $\mathcal{A} + \bar{\mathcal{A}}$ is a finite-codimension subspace of $C(X)$.

Our first example is just a reexamination of an earlier one. Let $\mathcal{S}$ and ϕ be the operator system and positive map of Example 2.2. Since $\phi(1) = 1$, while $\|\phi(z)\| = 2$ with $\|z\| = 1$, we see that z is not partitionable with respect to $\mathcal{S}$. It is interesting to attempt to show this directly.

The second example is somewhat longer.

Example 6.11. Let $0 < R_1 < R_2$, and let $\mathbb{A}$ denote the annulus in $\mathbb{C}$ with inner radius R_1 and outer radius R_2, i.e.,

$$\mathbb{A} = \{z \in \mathbb{C}\colon R_1 \leq |z| \leq R_2\}.$$

We let $\mathcal{R}(\mathbb{A})$, as before, denote the algebra of rational functions on $\mathbb{A}$, and let $\mathcal{S}$ denote the closure of $\mathcal{R}(\mathbb{A}) + \overline{\mathcal{R}(\mathbb{A})}$ in $C(\partial\mathbb{A})$. By a theorem of Walsh [239], $\mathcal{S}$ is a codimension 1 subspace of $C(\partial\mathbb{A})$ and hence $\mathcal{R}(\mathbb{A})$ is a hypo-Dirichlet algebra.

We let $\Gamma_j = \{z\colon |z| = R_j\}$, $j = 1, 2$, denote the two boundary circles of $\mathbb{A}$, and define positive linear functionals s_j, $j = 1, 2$, on $C(\partial\mathbb{A})$ by

$$s_j(f) = \frac{1}{2\pi} \int_0^{2\pi} f(R_j e^{it})\, dt.$$

If $f(z) = \sum_{k=-n}^{+n} a_k z^k$ is a finite Laurent polynomial, then

$$s_1(f) = s_2(f) = a_0.$$

Since the finite Laurent polynomials are dense in $\mathcal{R}(\mathbb{A})$ and since s_1 and s_2 are self-adjoint functionals, we have that $s_1(f) = s_2(f)$ for all f in $\mathcal{S}$. Since $\mathcal{S}$ is of codimension 1, we have that

$$\mathcal{S} = \{f \in C(\partial A)\colon s_1(f) = s_2(f)\}.$$

We claim that $\mathcal{S}$ has no partition of unity for $\mathcal{S}$. Before proving this claim, let us explore some of the consequences of the claim.

Set

$$p(z) = \begin{cases} 0, & z \in \Gamma_1, \\ 1, & z \in \Gamma_2, \end{cases}$$

so that $s_1(p) = 0 \neq 1 = s_2(p)$. From this, we see that $p \notin \mathcal{S}$, and so the span of $\mathcal{S}$ and p is all of $C(\partial A)$.

We leave it to the reader to verify that the claim that $\mathcal{S}$ has no partition of unity for $\mathcal{S}$ yields a positive, unital map $\phi\colon \mathcal{S} \to M_2$ with the following properties (Exercise 6.4):

(i) ϕ is not contractive.
(ii) ϕ is not completely positive.
(iii) ϕ has no positive extension to $C(\partial\mathbb{A})$.
(iv) If $\mathcal{L}_1 = \{f \in \mathcal{S}\colon f \leq p\}$, $\mathcal{L}_2 = \{f \in \mathcal{S}\colon f \geq p\}$, then there is no matrix P in M_2 satisfying

$$\phi(f_1) \leq P \leq \phi(f_2)$$

for all f_1 in $\mathcal{L}_1$, f_2 in $\mathcal{L}_2$.
(v) If $\psi = \phi|_{\mathcal{R}(\mathbb{A})}$, then ψ is a unital contraction, but $\tilde{\psi} = \phi$ is not contractive.
(vi) ψ has no contractive extension to $\mathcal{S}$, and so none to $C(\partial\mathbb{A})$ either.
(vii) ψ is not 2-contractive.

Part (vi), in particular, shows that a direct generalization of the Hahn–Banach extension theorem to operator-valued mappings fails even when the domain is commutative.

To verify the claim, it will suffice to construct an element of $\mathcal{S}$ that is not partitionable. Fix a small positive number $\delta > 0$ and set $j = 1, 2$,

$$X_j = \{R_j e^{it}\colon \delta \leq t \leq \pi - \delta\},$$
$$Y_j = \{R_j e^{it}\colon \pi + \delta \leq t \leq 2\pi - \delta\}.$$

Define a continuous function f on $\partial\mathbb{A}$ by setting $f(X_1) = 1$, $f(Y_1) = -1$, $f(X_2) = i$, $f(Y_2) = -i$, and extending f linearly on the remaining arcs, so that $\|f\| = 1$. It is easy to check that $s_1(f) = s_2(f)$, so that f is in $\mathcal{S}$.

Suppose that we were given scalars $\lambda_1, \ldots, \lambda_n$, $|\lambda_j| \leq 1$, $j = 1, \ldots, n$, and positive functions $p_1, \ldots, p_n$ in $\mathcal{S}$, summing to less than 1, such that f is not merely approximated by the sum, but in fact

$$f = \sum_i \lambda_i p_i.$$

Since $f(X_1) = 1$ and $|\lambda_j| \leq 1$, we must have that some subset of the λ_j's is exactly equal to 1, which, after reordering, we may take to be $\lambda_1, \ldots, \lambda_k$. It is not difficult to see that the above equations imply that

$$p_1(x) + \cdots + p_k(x) = 1$$

for all x in X_1. Similarly, after reordering the remaining λ_j's, we find a set $\lambda_{k+1}, \ldots, \lambda_m$ that are all equal to -1, and

$$p_{k+1}(x) + \cdots + p_m(x) = 1,$$

for all x in Y_1. But this implies that

$$s_1(p_1 + \cdots + p_m) \geq 1 - 4\delta,$$

and since s_1 and s_2 agree on $\mathcal{S}$,

$$s_2(p_{m+1} + \cdots + p_n) \leq 4\delta.$$

We have that $\text{Im}(f) = \text{Im}(\lambda_{m+1})p_{m+1} + \cdots + \text{Im}(\lambda_n)p_n$, so

$$1 - 4\delta \leq s_1(|\text{Im}(f)|) \leq s_2(p_{m+1} + \cdots + p_n) \leq 4\delta,$$

a clear contradiction for $\delta < 1/8$.

Thus, we see that no sum of the above form could actually equal f. An analogous, but somewhat more detailed, argument shows that for the same δ and ε sufficiently small, no sum of the above form can approximate f to within ε. We leave it to the reader to verify this claim.

Notes

The correspondence between linear maps from a space into M_n and linear functionals on the tensor product of that space with M_n is a recurring theme in this subject (see for example, [46], [47], [134], [223]). Our presentation owes a great deal to [6] and [218].

The study of $\mathcal{S}^+ \otimes M_n^+$ is adapted from [46], where it is used to study positive maps between matrix algebras. In contrast to Proposition 4.7, a number of unanswered questions and surprising examples arise in the study of positive maps between matrix algebras ([46], [222], and [244]). For example, positive maps between M_n and M_k are characterized by linear functionals on $M_n(M_k)$, which are positive on $M_n^+ \otimes M_k^+$. A decent characterization of the matrices in $M_n(M_k) = M_{nk}$ that belong to this cone is still not available. See [46] and [245] for an introduction to this topic. Example 6.11 is adapted from a similar example in [77]. The partition-of-unity techniques originate there also.

Exercises

6.1 Let $\mathcal{A}$ be any unital C^*-algebra. Give an example of a linear functional $s\colon M_n(\mathcal{A}) \to \mathbb{C}$ such that s is unital and positive, but such that the associated linear map, $\phi_s\colon \mathcal{A} \to M_n$, has norm n. (Hint: First consider the case $\mathcal{A} = \mathbb{C}$.)

6.2 Let $\phi\colon \mathcal{S} \to M_n$ be positive and set $\phi(1) = P$. Let Q be the projection onto the range of P, and let R be positive with $(I - Q)R = 0$, $RPR = Q$. Let $\psi\colon \mathcal{S} \to M_n$ be any positive, unital map, and set $\phi'(a) = R\phi(a)R + (I - Q)\psi(a)(I - Q)$.

(i) Show that ϕ' is a unital, positive map.
(ii) Show that if $(a_{i,j})$ is positive in $M_k(\mathcal{S})$, but $\phi_k((a_{i,j}))$ is not positive, then $\phi'_k((a_{i,j}))$ is not positive either.
(iii) Deduce the equivalence of (i) and (ii) in Theorem 6.6.

6.3 Assume that the equivalent conditions of Theorem 6.9 are not met. Show that then there always exists a unital, positive map $\phi\colon \mathcal{S} \to M_2$ that is not contractive.

6.4 Verify the existence of a $\phi\colon \mathcal{S} \to M_2$ with properties (i)–(vii) of Example 6.11.

6.5 Let $\mathcal{S}$ be an operator system. Prove that the following are equivalent:
(i) For every C^*-algebra $\mathcal{B}$, every positive $\phi\colon \mathcal{S} \to \mathcal{B}$ is n-positive.
(ii) $\mathcal{S}^+ \otimes M_n^+$ is dense in $M_n(\mathcal{S})^+$.

6.6 Use Corollary 6.7 to give an alternate proof of the fact that every positive map with domain $C(X)$ is completely positive.

Chapter 7
Arveson's Extension Theorems

In this chapter we extend the results of Chapter 6 from finite-dimensional ranges M_n to maps with range $B(\mathcal{H})$. We then develop the immediate applications of the extension theorems to dilation theory. We begin with some observations of a general functional-analytic nature.

Let X and Y be Banach spaces, let Y^* denote the dual of Y, and let $B(X, Y^*)$ denote the bounded linear transformations of X into Y^*. We wish to construct a Banach space such that $B(X, Y^*)$ is isometrically isomorphic to its dual. This will allow us to endow $B(X, Y^*)$ with a weak* topology.

Fix vectors x in X and y in Y, and define a linear functional $x \otimes y$ on $B(X, Y^*)$ by $x \otimes y(L) = L(x)(y)$. Since $|x \otimes y(L)| \leq \|L\| \cdot \|x\| \cdot \|y\|$, we see that $x \otimes y$ is in $B(X, Y^*)^*$ with $\|x \otimes y\| \leq \|x\|\|y\|$. In fact, $\|x \otimes y\| = \|x\|\|y\|$ (Exercise 7.1).

It is not difficult to check that the above definition is bilinear, i.e., $x \otimes (y_1 + y_2) = x \otimes y_1 + x \otimes y_2$, $(x_1 + x_2) \otimes y = x_1 \otimes y + x_2 \otimes y$, and $(\lambda x) \otimes y = x \otimes (\lambda y) = \lambda(x \otimes y)$ for $\lambda \in \mathbb{C}$. We let Z denote the closed linear span in $B(X, Y^*)^*$ of these elementary tensors. Actually, Z can be identified as the completion of $X \otimes Y$ with respect to a cross-norm (Exercise 7.1), but we shall not need that fact here.

Lemma 7.1. *$B(X, Y^*)$ is isometrically isomorphic to Z^* with the duality given by*

$$\langle L, x \otimes y \rangle = x \otimes y(L).$$

Proof. It is straightforward to verify that the above pairing defines an isometric isomorphism of $B(X, Y^*)$ into Z^*. To see that it is onto, fix $f \in Z^*$, and for each x, define a linear map, $f_x \colon Y \to \mathbb{C}$, via $f_x(y) = f(x \otimes y)$. Since $|f_x(y)| \leq \|f\|\|x\|\|y\|$, we have that $f_x \in Y^*$. It is straightforward to verify that if we set $L(x) = f_x$, then $L: X \to Y^*$ is linear, and that L is bounded with $\|L\| \leq \|f\|$.

Thus, $L \in B(X, Y^*)$, and clearly under our correspondence $L \to f$, which completes the proof. □

We call the weak* topology that is induced on $B(X, Y^*)$ by this identification the *BW topology* (for bounded weak). The following lemma explains the name.

Lemma 7.2. *Let $\{L_\lambda\}$ be a bounded net in $B(X, Y^*)$. Then L_λ converges to L in the BW topology if and only if $L_\lambda(x)$ converges weakly to $L(x)$ for all x in X.*

Proof. If L_λ converges to L in the BW topology, then

$$L_\lambda(x)(y) = \langle L_\lambda, (x \otimes y)\rangle \to \langle L, (x \otimes y)\rangle = L(x)(y)$$

for all y in Y. Thus, $L_\lambda(x)$ converges weakly to $L(x)$ for all x.

Conversely, if $L_\lambda(x)$ converges weakly to $L(x)$ for all x, then $\langle L_\lambda, (x \otimes y)\rangle$ converges to $\langle L, (x \otimes y)\rangle$ for all x and y and hence on the linear span of the elementary tensors. But since the net is bounded, this implies that it converges on the closed linear span. □

If $\mathcal{H}$ is a Hilbert space, then $B(\mathcal{H}) = B(\mathcal{H}, \mathcal{H})$ is the dual of a Banach space by Lemma 7.1. This Banach space can also be identified with the space of ultraweakly continuous linear functionals or with the trace-class operators (TC) on $\mathcal{H}$ with trace norm, $\|T\|_1 = tr(|T|)$, where $tr(\cdot)$ denotes the trace [68, Theorem I.4.5]. We prefer to focus on the duality with the trace class operators.

Under this duality, an operator $A \in B(\mathcal{H})$ is identified with the linear functional $\mathrm{Tr}(AT)$ for $T \in \mathrm{TC}$. If h, k are in $\mathcal{H}$, let $R_{h,k}$ denote the elementary rank-one operator on $\mathcal{H}$ given by $R_{h,k}(x) = \langle x, k\rangle h$. The linear span of these operators is dense in TC in the trace norm (Exercise 7.2), and for $A \in B(\mathcal{H})$,

$$\mathrm{Tr}(AR_{h,k}) = \langle Ah, k\rangle.$$

Proposition 7.3. *Let X be a Banach space, and let $\mathcal{H}$ be a Hilbert space. Then a bounded net $\{L_\lambda\}$ in $B(X, B(\mathcal{H}))$ converges in the BW topology to L if and only if $\langle L_\lambda(x)h, k\rangle$ converges to $\langle L(x)h, k\rangle$ for all h, k in $\mathcal{H}$ and x in X.*

Proof. We have that $\{L_\lambda\}$ converges in the BW topology to L if and only if $\mathrm{Tr}(L_\lambda(x)T) \to \mathrm{Tr}(L(x)T)$ for all $T \in \mathrm{TC}$ and $x \in X$. But again, since the net is bounded, it is enough to consider $T = R_{h,k}$. □

Let $\mathcal{A}$ be a C^*-algebra, $\mathcal{S}$ an operator system, $\mathcal{M}$ a subspace. We make the following definitions:

$$B_r(\mathcal{M}, \mathcal{H}) = \{L \in B(\mathcal{M}, B(\mathcal{H})) \colon \|L\| \le r\},$$

$$\mathrm{CB}_r(\mathcal{M}, \mathcal{H}) = \{L \in B(\mathcal{M}, B(\mathcal{H})) \colon \|L\|_{\mathrm{cb}} \le r\},$$

$$\mathrm{CP}_r(\mathcal{S}, \mathcal{H}) = \{L \in B(\mathcal{S}, B(\mathcal{H})) \colon L \text{ is completely positive}, \|L\| \le r\},$$

$$\mathrm{CP}(\mathcal{S}, \mathcal{H}; P) = \{L \in B(\mathcal{S}, B(\mathcal{H})) \colon L \text{ is completely positive}, L(1) = P\}.$$

Theorem 7.4. *Let $\mathcal{A}$ be a C^*-algebra, let $\mathcal{S}$ be a closed operator system, and let $\mathcal{M}$ be a closed subspace. Then each of the four above sets is compact in the BW topology.*

Proof. Since the BW topology is a weak* topology, the set $B_r(\mathcal{M}, \mathcal{H})$ is compact by the Banach–Alaoglu theorem. Since the remaining sets are subsets of this set, it is enough to show that they are closed.

We argue for $\mathrm{CB}_r(\mathcal{M}, \mathcal{H})$; the rest are similar. Let $\{L_\lambda\}$ be a net in $\mathrm{CB}_r(\mathcal{M}, \mathcal{H})$, and suppose $\{L_\lambda\}$ converges to L. If $(a_{i,j})$ is in $M_n(\mathcal{M})$, and $x = x_1 \oplus \cdots \oplus x_n$, $y = y_1 \oplus \cdots \oplus y_n$ are in $\mathcal{H} \oplus \cdots \oplus \mathcal{H}$, then $\langle (L(a_{i,j}))x, y\rangle = \lim_\lambda \langle (L_\lambda(a_{i,j}))x, y\rangle$. Hence, $\|(L(a_{i,j}))\| \le r \cdot \|(a_{i,j})\|$ for all n, and so $\|L\|_{\mathrm{cb}} \le r$. □

We're now in a position to prove the main result of the chapter.

Theorem 7.5 (Arveson's extension theorem). *Let $\mathcal{A}$ be a C^*-algebra, $\mathcal{S}$ an operator system contained in $\mathcal{A}$, and $\phi \colon \mathcal{S} \to B(\mathcal{H})$ a completely positive map. Then there exists a completely positive map, $\psi \colon \mathcal{A} \to B(\mathcal{H})$, extending ϕ.*

Proof. Let $\mathcal{F}$ be a finite-dimensional subspace of $\mathcal{H}$, and let $\phi_{\mathcal{F}} \colon \mathcal{S} \to B(\mathcal{F})$ be the compression of ϕ to $\mathcal{F}$, i.e., $\phi_{\mathcal{F}}(a) = P_{\mathcal{F}}\phi(a)|_{\mathcal{F}}$, where $P_{\mathcal{F}}$ is the projection onto $\mathcal{F}$. Since $B(\mathcal{F})$ is isomorphic to M_n for some n, by Theorem 6.2 there exists a completely positive map $\psi_{\mathcal{F}} \colon \mathcal{A} \to B(\mathcal{F})$ extending $\phi_{\mathcal{F}}$. Let $\psi'_{\mathcal{F}} \colon \mathcal{A} \to B(\mathcal{H})$ be defined by setting $\psi'_{\mathcal{F}}(a)$ equal to $\psi_{\mathcal{F}}(a)$ on $\mathcal{F}$ and extending it to be 0 on $\mathcal{F}^\perp$.

The set of finite-dimensional subspaces of $\mathcal{H}$ is a directed set under inclusion, and so $\{\psi'_{\mathcal{F}}\}$ is a net in $\mathrm{CP}_r(\mathcal{A}, \mathcal{H})$ where $r = \|\phi\|$. Since this latter set is compact, we may choose a subnet which converges to some element $\psi \in \mathrm{CP}_r(\mathcal{A}, \mathcal{H})$.

We claim that ψ is the desired extension. Indeed, if $a \in \mathcal{S}$ and x, y are in $\mathcal{H}$. Let $\mathcal{F}$ be the space spanned by x and y. Then for any $\mathcal{F}_1 \supseteq \mathcal{F}$, $\langle \phi(a)x, y\rangle = \langle \psi'_{\mathcal{F}_1}(a)x, y\rangle$, and since the set of such $\mathcal{F}_1$ is cofinal, we have that $\langle \phi(a)x, y\rangle = \langle \psi(a)x, y\rangle$.

This completes the proof of the theorem. □

Corollary 7.6 (Arveson). *Let $\mathcal{A}$ be a C^*-algebra, $\mathcal{M}$ a subspace with $1 \in \mathcal{M}$, and $\phi\colon \mathcal{M} \to B(\mathcal{H})$ a unital, complete contraction. Then there exists a completely positive map $\psi\colon \mathcal{A} \to B(H)$ extending ϕ.*

We've seen earlier that positive maps need not have positive extensions (Example 2.13) and that unital, contractive maps need not have contractive extensions [Example 6.11(vi)] even when the range is finite-dimensional. These facts make the above results all the more striking.

A C^*-algebra $\mathcal{B}$ is called *injective* if for every C^*-algebra $\mathcal{A}$ and operator system $\mathcal{S}$ contained in $\mathcal{A}$, every completely positive map $\phi\colon \mathcal{S} \to \mathcal{B}$ can be extended to a completely positive map on all of $\mathcal{A}$. Thus, Theorem 7.5 is the assertion that $B(\mathcal{H})$ is injective. Exercise 7.5 gives an elementary characterization of injective C^*-algebras.

Corollary 7.6 is the basis for a general dilation theory. Let $\mathcal{B}$ be a unital C^*-algebra, and let $\mathcal{A}$ be a subalgebra (not necessarily $*$-closed) with $1 \in A$. We shall call $\mathcal{A}$ an *operator algebra*. A unital homomorphism $\rho\colon \mathcal{A} \to \mathcal{B}(\mathcal{H})$ is said to have a *$\mathcal{B}$-dilation* if there exists a Hilbert space $\mathcal{K}$ containing $\mathcal{H}$ and a unital $*$-homomorphism $\pi\colon \mathcal{B} \to B(\mathcal{K})$ such that

$$\rho(a) = P_{\mathcal{H}}\pi(a)|_{\mathcal{H}} \quad \text{for all } a \text{ in } \mathcal{A}.$$

This definition is motivated in part by the theory of normal ∂X-dilations. Recall that if $T \in B(\mathcal{H})$, then a compact set X is a spectral set for X provided that the homomorphism $\rho\colon \mathcal{R}(X) \to B(\mathcal{H})$ given by $\rho(r) = r(T)$ is contractive. It is clear that T has a normal ∂X-dilation if and only if ρ has a $C(\partial X)$-dilation.

Corollary 7.7 (Arveson). *Let $\mathcal{A}$ be an operator algebra contained in the C^*-algebra $\mathcal{B}$, let $\rho\colon \mathcal{A} \to B(\mathcal{H})$ be a unital homomorphism, and let $\tilde{\rho}\colon \mathcal{A} + \mathcal{A}^* \to B(\mathcal{H})$ be the positive extension of ρ. Then the following are equivalent:*

(i) ρ has a $\mathcal{B}$-dilation,
(ii) ρ is completely contractive,
(iii) $\tilde{\rho}$ is completely positive.

Moreover, in this case there exists a $\mathcal{B}$-dilation $\pi\colon \mathcal{B} \to B(\mathcal{K})$ such that $\pi(\mathcal{B})\mathcal{H}$ is dense in $\mathcal{K}$.

Proof. We have seen the equivalence of (ii) and (iii) in Chapter 3. If ρ has a $\mathcal{B}$-dilation, then the map $\phi\colon \mathcal{B} \to B(\mathcal{H})$ defined by

$$\phi(b) = P_{\mathcal{H}}\pi(b)|_{\mathcal{H}}$$

is completely positive and extends ρ, so $\tilde{\rho}$ is completely positive.

Conversely, if $\tilde{\rho}$ is completely positive, then we may extend $\tilde{\rho}$ to a completely positive map $\phi: \mathcal{B} \to \mathcal{B}(\mathcal{H})$. The Stinespring representation of ϕ gives rise to the $\mathcal{B}$-dilation of ρ.

A minimal Stinespring representation of ϕ will have the property that $\pi(\mathcal{B})\mathcal{H}$ is dense in $\mathcal{K}$. □

A $\mathcal{B}$-dilation with the property that $\pi(\mathcal{B})\mathcal{H}$ is dense in $\mathcal{K}$ will be called a *minimal* $\mathcal{B}$-dilation of ρ. These need not be unique (Exercise 7.3).

Let $T \in B(\mathcal{H})$, and let X be a spectral set for T. If the homomorphism $\rho: \mathcal{R}(X) \to B(\mathcal{H})$ is completely contractive, then we shall call X a *complete spectral set* for T.

Corollary 7.8. *Let $T \in B(\mathcal{H})$, and let X be a spectral set for T. Then the following are equivalent:*

(i) T has a normal ∂X-dilation,
(ii) X is a complete spectral set,
(iii) $\tilde{\rho}$ is completely positive,
(iv) $\tilde{\rho}$ has a positive extension to $C(\partial X)$.

Moreover, in this case there is a normal ∂X-dilation N for T such that the smallest closed, reducing subspace for N containing $\mathcal{H}$ is $\mathcal{K}$.

Proof. The equivalence of (i)–(iii) is just Corollary 7.7. If $\tilde{\rho}$ is completely positive, then it has a (completely) positive extension to $C(\partial X)$ by Arveson's extension theorem. Conversely, if $\tilde{\rho}$ does have a positive extension to $C(\partial X)$, then that extension is automatically completely positive by Theorem 3.11, and hence its restriction to $\mathcal{S}$, $\tilde{\rho}$, is completely positive.

If $\mathcal{K}$ is the smallest closed, reducing subspace for N containing $\mathcal{H}$, then the representation $\pi: C(\partial X) \to B(\mathcal{K})$ given by $\pi(z) = N$ has no closed, reducing subspaces containing $\mathcal{H}$. But this is equivalent to the requirement that $\pi(C(\partial X))\mathcal{H}$ be dense in $\mathcal{K}$, since this latter space is clearly reducing. □

A normal ∂X-dilation of T with no closed, reducing subspace containing $\mathcal{H}$ is called a *minimal* normal ∂X-dilation of T. Unlike the case of Sz.-Nagy's minimal unitary dilation of a contraction (Theorem 4.3), the minimal normal ∂X-dilation of an operator need not be unique up to unitary equivalence. Exercise 7.3 illustrates the difficulty.

A unital homomorphism of an operator algebra $\mathcal{A}$ into $B(\mathcal{H})$ will be called a *representation* of $\mathcal{A}$. If $\mathcal{A}$ is a C^*-algebra, then every contractive representation of $\mathcal{A}$ is automatically a $*$-representation. Corollary 7.7 gives a useful characterization of the completely contractive representations, for it shows that these representations can be studied via the $*$-representation theory of $\mathcal{B}$. Consequently, it is important to know when representations are completely contractive.

At present we have no natural source of contractive representations that are not completely contractive. In general, such examples are hard to construct, since in some sense they are the pathological cases that are difficult to fit to a general theory. They can occur, however, even when the algebra is finite-dimensional and the representation is on a finite-dimensional space. An example of a subalgebra of M_3 and a representation of it on $\mathbb{C}^3$ that is not completely contractive is given in [6].

An example of considerable importance to the theory of dilations of contractions was given by Parrott [156].

Parrott's Example. Let U and V be contractions in $B(\mathcal{H})$ such that U is unitary and U and V don't commute. We define commuting contractions on $B(\mathcal{H} \oplus \mathcal{H})$ by setting

$$T_1 = \begin{bmatrix} 0 & 0 \\ I & 0 \end{bmatrix}, \qquad T_2 = \begin{bmatrix} 0 & 0 \\ U & 0 \end{bmatrix}, \qquad T_3 = \begin{bmatrix} 0 & 0 \\ V & 0 \end{bmatrix}.$$

Let $\mathcal{P}(\mathbb{D}^3)$ be the algebra of polynomials in three variables z_1, z_2, z_3, regarded as a subalgebra of $C(\mathbb{T}^3)$, where $\mathbb{T}^3$ is the 3-torus. In Chapter 5 we saw an example of three commuting contractions such that the induced homomorphism of $\mathcal{P}(\mathbb{D}^3)$ was not contractive. We claim that the homomorphism $\rho\colon \mathcal{P}(\mathbb{D}^3) \to B(\mathcal{H} \oplus \mathcal{H})$ defined by $\rho(z_i) = T_i$, $i = 1, 2, 3$, is contractive but not completely contractive.

To see that ρ is contractive, let $p(z_1, z_2, z_3)$ be an arbitrary element of $\mathcal{P}(\mathbb{D}^3)$, and write

$$p(z_1, z_2, z_3) = a_0 + a_1 z_1 + a_2 z_2 + a_3 z_3 + q(z_1, z_2, z_3),$$

where $q(z_1, z_2, z_3)$ contains all the higher-order terms of p. We have that

$$\rho(p) = \begin{bmatrix} a_0 & 0 \\ a_1 I + a_2 U + a_3 V & a_0 \end{bmatrix},$$

since $T_i \cdot T_j = 0$ for all i and j.

Let $x = x_1 \oplus x_2$ and $y = y_1 \oplus y_2$ be arbitrary unit vectors in $\mathcal{H} \oplus \mathcal{H}$, and calculate

$$\begin{aligned}
|\langle \rho(p)x, y\rangle| &= |a_0\langle x_1, y_1\rangle + \langle (a_1 I + a_2 U + a_3 V)x_1, y_2\rangle + a_0\langle x_2, y_2\rangle| \\
&\leq |a_0|\|x_1\|\|y_1\| + (|a_1| + |a_2| + |a_3|)\|x_1\|\|y_2\| + |a_0|\|x_2\|\|y_2\| \\
&= \left\langle \begin{bmatrix} |a_0| & 0 \\ |a_1| + |a_2| + |a_3| & |a_0| \end{bmatrix} \begin{bmatrix} \|x_1\| \\ \|x_2\| \end{bmatrix}, \begin{bmatrix} \|y_1\| \\ \|y_2\| \end{bmatrix} \right\rangle.
\end{aligned}$$

Thus, we have that

$$\|\rho(p)\| \leq \left\| \begin{bmatrix} |a_0| & 0 \\ |a_1| + |a_2| + |a_3| & |a_0| \end{bmatrix} \right\|,$$

where the latter matrix is an element of M_2.

But by Exercise 2.11,

$$\left\| \begin{bmatrix} |a_0| & 0 \\ |a_1| + |a_2| + |a_3| & |a_0| \end{bmatrix} \right\| \leq \inf_r \{\| |a_0| + (|a_1| + |a_2| + |a_3|)z + r(z)\|\},$$

where $r(z)$ is an arbitrary polynomial whose lowest-order term is at least of degree 2, and the latter norm is the supremum norm over the unit circle. Let $\lambda_0, \lambda_1, \lambda_2, \lambda_3$ be numbers of modulus 1 such that $\lambda_0 a_0, \lambda_1 a_1, \lambda_2 a_2, \lambda_3 a_3$ are positive; then

$$\begin{aligned}
\|\rho(p)\| &\leq \left\| \begin{bmatrix} |a_0| & 0 \\ |a_1| + |a_2| + |a_3| & |a_0| \end{bmatrix} \right\| \\
&\leq \| |a_0| + (|a_1| + |a_2| + |a_3|)z + \lambda_0 q(\bar{\lambda}_0\lambda_1 z, \bar{\lambda}_0\lambda_2 z, \bar{\lambda}_0\lambda_3 z)\| \\
&= \|p(\bar{\lambda}_0\lambda_1 z, \bar{\lambda}_0\lambda_2 z, \bar{\lambda}_0\lambda_3 z)\| \\
&\leq \|p(z_1, z_2, z_3)\|,
\end{aligned}$$

where the third and fourth norms are the supremum over $\mathbb{T}$ and the last norm is the supremum over $\mathbb{T}^3$. Thus, ρ is contractive.

Now assume that ρ is completely contractive. Consider an element of $M_n(\mathcal{P}(\mathbb{D}^3))$ of the form $(a_{i,j}z_1 + b_{i,j}z_2 + c_{i,j}z_3)$. Its image under ρ_n is

$$\left[\begin{array}{cccc|c}
0 & 0 & \cdots & 0 & 0 \\
\vdots & \vdots & & & \vdots \\
a_{11}I + b_{11}U + c_{11}V & 0 & \cdots & a_{1n}I + b_{1n}U + c_{1n}V & 0 \\
\hdashline
\vdots & \vdots & \ddots & \vdots & \vdots \\
\hdashline
0 & 0 & \cdots & 0 & 0 \\
\vdots & \vdots & & \vdots & \vdots \\
a_{n1}I + b_{n1}U + c_{n1}V & 0 & \cdots & a_{nn}I + b_{nn}U + c_{nn}V & 0
\end{array}\right],$$

which is an operator acting on the direct sum of n copies of $\mathcal{H} \oplus \mathcal{H}$.

Think of this space as $(\mathcal{H}_{11} \oplus \mathcal{H}_{21}) \oplus \cdots \oplus (\mathcal{H}_{1n} \oplus \mathcal{H}_{2n})$, where each $\mathcal{H}_{i,j} = \mathcal{H}$. If we reorder the spaces by writing

$$(\mathcal{H}_{11} \oplus \cdots \oplus \mathcal{H}_{1n}) \oplus (\mathcal{H}_{21} \oplus \cdots \oplus \mathcal{H}_{2n}),$$

then the matrix of this same operator will be

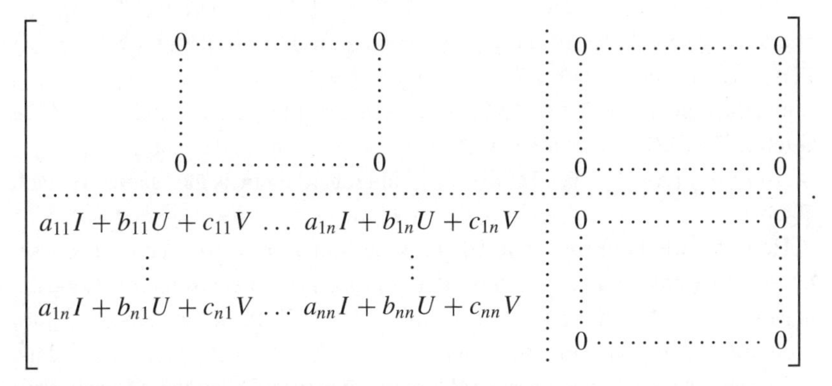

Since ρ is assumed to be completely contractive, the operator matrix in the (2,1) block of the above matrix has norm bounded by

$$\begin{aligned} &\|\rho_n((a_{i,j}z_1 + b_{i,j}z_2 + c_{i,j}z_3))\| \\ &\qquad \le \sup\{\|(a_{i,j}z_1 + b_{i,j}z_2 + c_{i,j}z_3)\| : |z_1| = |z_2| = |z_3| = 1\}. \end{aligned}$$

But this latter supremum is equal to

$$\sup\{\|(a_{i,j} + b_{i,j}w_1 + c_{i,j}w_2)\| : |w_1| = |w_2| = 1\}.$$

Let $\mathcal{M}$ be the subspace of $C(\mathbb{T}^2)$ spanned by 1, w_1, and w_2, where w_1 and w_2 are the coordinate functions.

The above discussion shows that if ρ is completely contractive, then the map $\psi \colon \mathcal{M} \to B(\mathcal{H})$ defined by $\psi(1) = I$, $\psi(w_1) = U$, and $\psi(w_2) = V$ is completely contractive. By Arveson's extension theorem, we may extend ψ to a completely positive map on $C(\mathbb{T}^2)$ and then obtain a Stinespring representation $(\pi, S, \mathcal{K})$ with S an isometry and $\pi \colon C(\mathbb{T}^2) \to B(\mathcal{K})$ a $*$-homomorphism.

Identifying $S\mathcal{H}$ and $\mathcal{H}$, we have that

$$\psi(\cdot) = P_{\mathcal{H}}\pi(\cdot)|_{\mathcal{H}}.$$

But since $\pi(w_1)$ and $U = \psi(w_1)$ are both unitary, this implies that $\mathcal{H}$ reduces $\pi(w_1)$. Finally, since $\pi(w_1)$ and $\pi(w_2)$ commute, and $\mathcal{H}$ reduces $\pi(w_1)$, this implies that $\psi(w_1)$ commutes with $\psi(w_2)$ – that is, that U commutes with V, a contradiction. Therefore, ρ is not completely contractive.

If one chooses U and V to both be unitaries, then it is possible to show that ρ is not even 4-contractive (Exercise 7.9).

Another remark to be made is that there exist U and V in M_2 satisfying the above conditions, and hence there are three commuting contractions in M_4 for which ρ is a contraction but not a complete contraction.

Parrott's example also shows that we must be extremely careful about what we mean by the expression "operator algebra." Consider any $*$-isomorphism $\pi\colon C(\mathbb{T}^3) \to B(\mathcal{K})$ and let $A_i = \pi(z_i) \oplus T_i$, $i = 1, 2, 3$, where the T_i's are the operators given above. For any polynomial p in three variables we have that $p(A_1, A_2, A_3) = \pi(p) \oplus p(T_1, T_2, T_3)$ and hence $\|p(A_1, A_2, A_3)\| = \|p\|_\infty$. Thus, the algebra $\mathcal{A} \subseteq B(\mathcal{K} \oplus \mathcal{H} \oplus H)$ generated by A_1, A_2, and A_3 and the algebra $\mathcal{P}(\mathbb{D}^3) \subseteq C(\mathbb{T}^3)$ are isometrically isomorphic via the map $\gamma\colon \mathcal{P}(\mathbb{D}^3) \to \mathcal{A}$, $\gamma(p) = p(A_1, A_2, A_3)$. Hence, we could regard them as the "same" Banach algebra.

However, we should not regard them as the "same" *operator* algebra, because, by the above calculations, we know that γ is not a complete isometry. Thus, the norms on $M_n(\mathcal{A})$ differ from the norms on $M_n(\mathcal{P}(\mathbb{D}^3))$. In fact, since ρ is not 4-contractive, we have that $\gamma_4\colon M_4(\mathcal{P}(\mathbb{D}^3)) \to M_4(\mathcal{A})$ is not an isometry. Thus, $\mathcal{P}(\mathbb{D}^3)$ and $\mathcal{A}$ are distinct operator algebras, since in this book we are concerned with their matrix norm structures.

This subtlety does not arise with C^*-algebras. The discussion in Chapter 1 shows that if $\gamma\colon \mathcal{A} \to \mathcal{B}$ is a $*$-isomorphism between two C^*-algebras, then $\gamma_n\colon M_n(\mathcal{A}) \to M_n(\mathcal{B})$ is also a $*$-isomorphism for all n. Hence every isometric isomorphism between C^*-algebras is automatically a completely isometric isomorphism.

Now that we've seen an example of a commutative operator algebra for which every contractive representation is not necessarily completely contractive, we would like to present an example in the opposite extreme. That is, some highly noncommutative operator algebras for which every contractive representation is completely contractive. In general, it is hard to test if a homomorphism is contractive. Thus, the following theorem is quite nice.

Proposition 7.9 (McAsey–Muhly). *Let $\mathcal{A}$ be the algebra of upper triangular matrices in M_n, i.e.,*

$$\mathcal{A} = \{(a_{i,j})\colon a_{i,j} = 0 \text{ for } i > j\},$$

and let $\{E_{i,j}\}$ be the standard matrix units for M_n. Then every representation ρ of $\mathcal{A}$ with $\|\rho(E_{ij})\| \leq 1$ for $i \leq j$ is completely contractive.

Proof. Let $\rho\colon \mathcal{A} \to \mathcal{B}(\mathcal{H})$ be a representation. We define a map $\phi\colon M_n \to B(\mathcal{H})$ by $\phi(E_{ij}) = \rho(E_{ij})$, $i \leq j$, and $\phi(E_{ij}) = \rho(E_{ji})^*$, $i > j$. We must prove that ϕ is completely positive.

Since ρ is unital and a homomorphism, $\{\rho(E_{i,i})\}$ will be projections that sum to the identity. Since $\|\rho(E_{ii})\| \leq 1$, they will be orthogonal projections. Let $\mathcal{H}_i$ be the space that $\rho(E_{i,i})$ projects onto, so $\mathcal{H} = \mathcal{H}_1 \oplus \cdots \oplus \mathcal{H}_n$.

Since $\rho(E_{i,j}) = \rho(E_{i,i})\rho(E_{i,j})\rho(E_{j,j})$, there will exist operators $T_{i,j}\colon \mathcal{H}_j \to \mathcal{H}_i$ such that the operator matrix of $\rho(E_{i,j})$ relative to the above decomposition of $\mathcal{H}$ is $T_{i,j}$ in the (i, j)th position and 0 elsewhere. Relative to this decomposition we have that $\rho((a_{ij})) = (a_{ij}T_{ij})$. Set $T_i = T_{i,i+1}$, and note that for $i < j$,

$$T_{i,j} = T_i \cdot T_{i+1} \cdots T_j.$$

By Theorem 3.14, to prove that ϕ is completely positive, it is sufficient to prove that $(\phi(E_{i,j}))$ is a positive operator on the direct sum of n copies of $\mathcal{H}, \mathcal{H}^1 \oplus \cdots \oplus \mathcal{H}^n$. Decomposing each $\mathcal{H}^i$ as $\mathcal{H}_1^i \oplus \cdots \oplus \mathcal{H}_n^i$, we have that $(\phi(E_{i,j}))$ is represented as an $n^2 \times n^2$ operator matrix with each $n \times n$ block having only one nonzero entry.

Reorder the subspaces so that $\mathcal{H}_1^1, \mathcal{H}_2^2, \ldots, \mathcal{H}_n^n$ are listed first and the remaining spaces occur in any order. The operator matrix for $(\phi(E_{i,j}))$ with respect to this reordering is $(T_{i,j})$ in the first $n \times n$ entries and 0 elsewhere.

Thus, it will be sufficient to prove that $(T_{i,j})$ is positive. Set

$$R = \begin{bmatrix} 0 & T_1 & 0 & \dots & 0 \\ 0 & 0 & T_2 & \ddots & \vdots \\ \vdots & & \ddots & \ddots & 0 \\ \vdots & & & \ddots & T_{n-1} \\ 0 & \dots & \dots & \dots & 0 \end{bmatrix};$$

then $\|R\| \leq 1$, $R^{n+1} = 0$, and $(T_{i,j}) = (I - R)^{-1} + (I - R^*)^{-1} - I$. Consequently, as in the proof of Theorem 2.6, $(T_{i,j})$ is positive. □

If we examine the Stinespring representation of ρ, then it can be seen to give a factorization of the operator $(T_{i,j})$. Indeed, if $\rho(A) = V^*\pi(A)V$ for some $*$-homomorphism $\pi\colon M_n \to B(\mathcal{K})$ and isometry $V\colon \mathcal{H} \to \mathcal{K}$, then we know that up to a unitary equivalence we have that $\mathcal{K} = \mathcal{L}^{(n)}$ for some Hilbert space $\mathcal{L}$ and $\pi(E_{i,j})$ is the operator matrix that is $I_{\mathcal{L}}$ in the (i, j) entry and 0 elsewhere. Writing $V\colon \mathcal{H} \to \mathcal{L}^{(n)}$ as $Vh = (V_1h, \ldots, V_nh)$ where $V_i\colon \mathcal{H} \to \mathcal{L}$, we have that $\rho(E_{i,j}) = V_i^*V_j$.

Since so many positive maps on M_n are not completely positive, it is somewhat surprising that every map induced by a representation of $\mathcal{A}$ is completely positive.

Notes

The BW topology was introduced in [6], where the proofs of the extension theorems also appear, as well as some material on $\mathcal{B}$-dilations.

Parrott [156] originally proved that the set of operators in his example did not have a set of commuting unitary dilations, i.e., that $\rho(z_1, z_2, z_3) \rightarrow \rho(T_1, T_2, T_3)$ does not have a $C(\mathbb{T}^3)$ dilation, by examining spatial relations among the unitaries. By Corollary 7.7, this is equivalent to the fact that Parrott's example yields a contractive homomorphism that is not completely contractive. We've avoided Parrott's spatial proof in an attempt to cast a different perspective. It is interesting to note that when V is also unitary, a direct proof of the noncomplete contractivity of the homomorphism simplifies (Exercise 7.9), as does the spatial proof [231]. We do not know if the homomorphism is not 2-contractive for arbitrary V.

Let us reinterpret the material on commuting contractions in light of Corollary 7.7. Ando's theorem [5] shows that every pair of commuting contractions induces a completely contractive homomorphism of the bidisk algebra $\mathcal{P}(\mathbb{D}^2)$. So, in particular, every contractive homomorphism of that algebra is completely contractive. Parrott's example shows that $\mathcal{P}(\mathbb{D}^3)$ differs in this respect; contractive homomorphisms need not be completely contractive. The internal properties of the algebra $\mathcal{P}(\mathbb{D}^3)$ that lead to this difference do not seem to be understood. Neither do the internal properties of the bidisk algebra. For example, it is not known if every unital contraction (including nonhomomorphisms) on $\mathcal{P}(\mathbb{D}^2)$ is completely contractive. This is the case for the disk algebra, since it is a Dirichlet algebra. We shall see some additional explanation of the difference between the two- and three-variable situations in Chapter 18, but our understanding of this situation is still not satisfactory.

The results of Crabbe and Davies [69] and Varopoulos [236] examined in Chapter 5, that not every set of $n \geq 3$ commuting contractions need satisfy von Neumann's inequality, reflect a different property of the polydisk algebras than Parrott's example. In Chapter 18, we will formalize this difference a bit more.

Gaspar and Racz [104] have shown that every set of cyclically commuting contractions has a set of cyclically commuting, unitary dilations. If given a C^*-algebraic interpretation, their result should add some additional insights.

Choi and Effros [50] prove that for a von Neumann algebra $\mathcal{A}$, semidiscreteness (Exercise 7.14) is equivalent to the existence of a net of finite-rank, completely positive maps $R_\lambda \colon \mathcal{A} \rightarrow \mathcal{A}$ such that $R_\lambda(x)$ converges to x σ-weakly for all x. Connes [67] proves that semidiscreteness and injectivity are equivalent for von Neumann algebras.

Proposition 7.9 is adapted from McAsey and Muhly [143].

Exercises

7.1 Let X and Y be Banach spaces, and for $f \in X^*$, $g \in Y^*$ define $L_{f,g} \in B(X, Y^*)$ by $L_{f,g}(x) = f(x)g$. By considering these operators, prove that the map $j: X \otimes Y \to B(X, Y^*)^*$ defined in this chapter has the following properties:

(i) j is linear,

(ii) $\|j(x \otimes y)\| = \|x\|\|y\|$,

(iii) j is one-to-one.

Since j is one-to-one, the identification of $X \otimes Y$ with $j(X \otimes Y)$ endows $X \otimes Y$ with a norm (rather than just a seminorm). Conclude that Z of Lemma 7.1 can be identified with the completion of $X \otimes Y$ with respect to this norm. *A norm on the tensor product of two normed spaces that satisfies* $\|x \otimes y\| = \|x\|\|y\|$ *is called a cross-norm.*

7.2 Let $R_{h,k}$ be the operator defined by $R_{h,k}(x) = \langle x, k \rangle h$. Show that $R_{h,k}$ is trace-class, that $\text{Tr}(AR_{h,k}) = \langle Ah, k \rangle$, and that the linear span of $\{R_{h,k}: h, k \in \mathcal{H}\}$ is dense in TC in the trace norm.

7.3 Let $\mathcal{A}$ be an operator algebra contained in the C^*-algebra $\mathcal{B}$, let $\rho: \mathcal{A} \to B(\mathcal{H})$ be a completely contractive, unital homomorphism, and let $\pi_i: \mathcal{B} \to B(\mathcal{K}_i)$, $i = 1, 2$, define minimal $\mathcal{B}$-dilations of ρ. Define completely positive maps $\phi_i: \mathcal{B} \to \mathcal{B}(\mathcal{H})$ by $\phi_i(b) = P_{\mathcal{H}}\pi_i(b)|_{\mathcal{H}}$, $i = 1, 2$.

(i) Show that there exists a unitary $U: \mathcal{K}_1 \to \mathcal{K}_2$ with $Uh = h$ for h in $\mathcal{H}$, and $U^*\pi_2(b)U = \pi_1(b)$ if and only if $\phi_1 = \phi_2$. Such dilations are called *unitarily equivalent*.

(ii) Show that there is a one-to-one correspondence between unitarily equivalent, minimal $\mathcal{B}$-dilations of ρ and completely positive extensions of ρ to $\mathcal{B}$.

(iii) Show that the set of completely positive extensions of ρ is a compact, convex set in the BW toplogy on CP($\mathcal{B}$, $\mathcal{H}$).

7.4 (Extension of bimodule maps) Let $\mathcal{A}, \mathcal{C}$ be C^*-algebras, let $\mathcal{S}$ be an operator system, and suppose that $\mathcal{C} \subseteq \mathcal{S} \subseteq \mathcal{A}$. If $\mathcal{C} \subseteq B(\mathcal{H})$, then $\phi: \mathcal{S} \to B(\mathcal{H})$ is a $\mathcal{C}$-bimodule map provided $\phi(c_1ac_2) = c_1\phi(a)c_2$. Prove that if $\phi: \mathcal{S} \to B(\mathcal{H})$ is a completely positive $\mathcal{C}$-bimodule map, then every completely positive extension of ϕ to $\mathcal{A}$ is also a $\mathcal{C}$-bimodule map.

7.5 Let $\mathcal{B} \subseteq B(\mathcal{H})$ be a unital C^*-algebra. Prove that $\mathcal{B}$ is injective if and only if there exists a completely positive map $\phi: B(\mathcal{H}) \to \mathcal{B}$ such that $\phi(b) = b$ for all b in $\mathcal{B}$. Show that ϕ is necessarily a $\mathcal{B}$-bimodule map. A map with the above properties is called a *completely positive conditional expectation.*

7.6 (Sarason) Let $\mathcal{A} \subseteq B(\mathcal{K})$ be an algebra, and let $\mathcal{H}$ be a subspace of $\mathcal{K}$ such that $A \to P_{\mathcal{H}}A|_{\mathcal{H}}$ is a homomorphism on $\mathcal{A}$. Prove that the subspaces $\mathcal{H}_1 =$

span$\{Ah\colon A \in \mathcal{A}, h \in \mathcal{H}\}^-$ and $\mathcal{H}_2 = \mathcal{H}_1 \cap \mathcal{H}^\perp$ of $\mathcal{K}$ both are invariant for every element of $\mathcal{A}$, and that $\mathcal{H}_1 \oplus \mathcal{H} = \mathcal{H}_2$. Conversely, show that if $\mathcal{H}_1$ and $\mathcal{H}_2$ are invariant for $\mathcal{A}$ and $\mathcal{H}_1 \oplus \mathcal{H} = \mathcal{H}_2$, then compression onto $\mathcal{H}$ is a homomorphism on $\mathcal{A}$. Such a subspace is said to be *semiinvariant* for $\mathcal{A}$.

7.7 Let $Q = (q_{i,j})$ be the element of M_n given by $q_{i,i+1} = 1, q_{i,j} = 0$ for all other (i, j), and let $N \in B(\mathcal{H})$ satisfy $\|N\| \leq 1, N^n = 0$. Prove that $Q \to N$ defines a completely contractive representation of the algebra generated by Q.

7.8 Prove that $\mathcal{B}$ is injective if and only if $M_n(\mathcal{B})$ is injective for some n.

7.9 In Parrott's example, show that if U and V are both unitaries that don't commute, then ρ is not 4-contractive. (Hint: Consider $\left[\begin{smallmatrix} U^* & V \\ V^* & -U \end{smallmatrix}\right]$.)

7.10 (Parrott) Prove that there exist three commuting operators, T_1, T_2, and T_3, with $\|T_i\| < 1$ such that von Neumann's inequality holds, but that do not have a commuting unitary dilation.

7.11 Prove that for every $n \geq 3$, there exist n commuting operators such that $\|T_i\| < 1, i = 1, \ldots, n$, and von Neumann's inequality holds, but that do not have commuting unitary dilations.

7.12 Let $\mathcal{A} \subseteq B(\mathcal{H})$ be a unital C^*-algebra. If $\mathcal{A}$ is closed in the weak operator topology, then $\mathcal{A}$ is called a *von Neumann algebra*. Prove that every von Neumann algebra has a predual. The induced weak* topology on $\mathcal{A}$ is called the *σ-weak topology*.

7.13 Let $\mathcal{A}$ be a von Neumann algebra. If there exists a net of σ-weakly continuous, completely positive maps, $\phi_\lambda\colon \mathcal{A} \to M_{n_\lambda}$, $\psi_\lambda\colon M_{n_\lambda} \to \mathcal{A}$, with $\phi_\lambda(1) = 1$, $\psi_\lambda(1) = 1$, such that $\psi_\lambda \circ \phi_\lambda(a) \to a$, σ-weakly for all a in $\mathcal{A}$, then $\mathcal{A}$ is called *semidiscrete*. Prove that every semidiscrete von Neumann algebra is injective.

Chapter 8
Completely Bounded Maps

In this chapter, we extend many of the results obtained in previous chapters concerning completely positive maps to the completely bounded maps. Our main technique is to realize completely bounded maps as the off-diagonal corners of completely positive maps.

Let us consider for a moment $M_m(M_n(\mathcal{A}))$. For a C^*-algebra $\mathcal{A}$, a typical element of this algebra is of the form $A = (A_{i,j})_{i,j=1}^m$, where each $A_{i,j}$ is in $M_n(\mathcal{A})$. Thus, $A_{i,j} = (a_{i,j,k,\ell})_{k,\ell=1}^n$, with $a_{i,j,k,\ell}$ in $\mathcal{A}$. Setting $B_{k,\ell} = (a_{i,j,k,\ell})_{i,j=1}^m$, we obtain an element of $M_m(\mathcal{A})$, and thus $B = (B_{k,\ell})_{k,\ell=1}^n$ is in $M_n(M_m(\mathcal{A}))$. Now, $M_m(M_n(\mathcal{A}))$ and $M_n(M_m(\mathcal{A}))$ are both isomorphic to $M_{nm}(\mathcal{A})$ by just deleting the extra parentheses. With these identifications, A and B are unitarily equivalent elements of $M_{nm}(\mathcal{A})$; in fact, the unitary is just a permutation matrix.

To see this, note that if we regard A as an element of $M_{mn}(\mathcal{A})$, say $A = (c_{s,t})_{s,t=1}^{mn}$, then $c_{s,t} = a_{i,j,k,\ell}$, where $s = n(i-1)+k$, $t = n(j-1)+\ell$; while if we regard B as an element of $M_{mn}(\mathcal{A})$, say $B = (d_{s,t})_{s,t=1}^{mn}$, then $d_{s,t} = a_{i,j,k,\ell}$, where $s = m(k-1)+i$, $t = m(\ell-1)+j$.

Since the above operation for passing from $M_m(M_n(\mathcal{A}))$ to $M_n(M_m(\mathcal{A}))$ is just a permutation, it is a $*$-isomorphism. We shall refer to this $*$-isomorphism as the *canonical shuffle*. It is important to note that since the canonical shuffle is a $*$-isomorphism, it preserves norm and positivity.

Note that we've encountered this permutation earlier in Parrott's example, in the proof of Proposition 7.9, and in our discussion of the Kronecker product in Chapter 3. It is also useful to understand this canonical shuffle in the tensor notation. Let $\{E_{i,j}\}_{i,j=1}^m$ and $\{F_{k,\ell}\}_{k,\ell=1}^n$ denote the standard matrix units for M_m and M_n, respectively. Our element A of $M_m(M_n(\mathcal{A})) \cong (\mathcal{A} \otimes M_n) \otimes M_m$ is just $A = \sum_{i,j=1}^m A_{i,j} \otimes E_{i,j}$, where each $A_{i,j}$ is in $\mathcal{A} \otimes M_n$ and has the form $A_{i,j} = \sum_{k,\ell=1}^n a_{i,j,k,\ell} \otimes F_{k,\ell}$. Thus, $A = \sum_{i,j=1}^m \sum_{k,\ell=1}^n a_{i,j,k,\ell} \otimes F_{k,\ell} \otimes E_{i,j}$. On the other hand, $B_{k,\ell} = \sum_{i,j=1}^m a_{i,j,k,\ell} \otimes E_{i,j}$, so that $B = \sum_{k,\ell=1}^n B_{k,\ell} \otimes$

$F_{k,\ell} = \sum_{k,\ell=1}^{n} \sum_{i,j=1}^{m} a_{i,j,k,\ell} \otimes E_{i,j} \otimes F_{k,\ell}$, which lies in $(\mathcal{A} \otimes M_m) \otimes M_n \cong M_n(M_m(\mathcal{A}))$. When we regard A and B as elements of $M_{mn}(\mathcal{A})$, then we see that B is the image of A under the following string of isomorphisms: $M_{mn}(\mathcal{A}) \cong M_m(M_n(\mathcal{A})) \cong (\mathcal{A} \otimes M_n) \otimes M_m \cong \mathcal{A} \otimes (M_n \otimes M_m) \cong \mathcal{A} \otimes (M_m \otimes M_n) \cong (\mathcal{A} \otimes M_m) \otimes M_n \cong M_n(M_m(\mathcal{A})) \cong M_{mn}(\mathcal{A})$. Most of these isomorphisms are so canonical that it is customary to ignore them. Setting $\mathcal{A} = \mathbb{C}$, it is interesting to note that when we identify $M_{mn} = M_m(M_n) = M_n \otimes M_m$ and $M_{mn} = M_n(M_m) = M_m \otimes M_n$, then it is only the isomorphism between $M_m \otimes M_n$ and $M_n \otimes M_m$ that leads to the permutation in M_{mn}.

The following lemma is central to this chapter and introduces the off-diagonal technique.

Lemma 8.1. *Let $\mathcal{A}$, $\mathcal{B}$ be C^*-algebras with unit 1, let $\mathcal{M}$ be an operator space in $\mathcal{A}$, and let ϕ: $\mathcal{M} \to \mathcal{B}$. Define an operator system $\mathcal{S}_{\mathcal{M}} \subseteq M_2(\mathcal{A})$ by*

$$\mathcal{S}_{\mathcal{M}} = \left\{ \begin{bmatrix} \lambda 1 & a \\ b^* & \mu 1 \end{bmatrix} : \lambda, \mu \in \mathbb{C}, a, b, \in \mathcal{M} \right\},$$

and Φ: $\mathcal{S}_{\mathcal{M}} \to M_2(\mathcal{B})$ via

$$\Phi\left[\begin{bmatrix} \lambda 1 & a \\ b^* & \mu 1 \end{bmatrix}\right] = \begin{bmatrix} \lambda 1 & \phi(a) \\ \phi(b)^* & \mu 1 \end{bmatrix}.$$

If ϕ is completely contractive, then Φ is completely positive.

Proof. Let $(S_{i,j})$ be in $M_n(\mathcal{S}_{\mathcal{M}})$, say

$$S_{i,j} = \begin{bmatrix} \lambda_{i,j} & a_{i,j} \\ b_{i,j}^* & \mu_{i,j} \end{bmatrix}.$$

Since $M_n(\mathcal{S}_{\mathcal{M}})$ is a subspace of $M_n(M_2(\mathcal{A}))$, if we perform the canonical shuffle, then $(S_{i,j})$ becomes an element of $M_2(M_n(\mathcal{A}))$. Indeed, if we set $H = (\lambda_{i,j})$, $A = (a_{i,j})$, $B = (b_{j,i})$, $K = (\mu_{i,j})$, then the image of $(S_{i,j})$ under the canonical shuffle is

$$\begin{bmatrix} H & A \\ B^* & K \end{bmatrix}. \tag{8.1}$$

Similarly, after the canonical shuffle, $\Phi_n((S_{i,j}))$ becomes

$$\begin{bmatrix} H & \phi_n(A) \\ \phi_n(B)^* & K \end{bmatrix}. \tag{8.2}$$

Thus, to prove that Φ is completely positive, it is sufficient to prove that for all n, if (8.1) is positive, then (8.2) is positive.

Now, if (8.1) is positive, then $A = B$ and H and K must be positive. Fix $\varepsilon > 0$, and set $H_\varepsilon = H + \varepsilon I$, $K_\varepsilon = K + \varepsilon I$, so that H_ε and K_ε are positive and invertible. We have that

$$\begin{bmatrix} I & H_\varepsilon^{-1/2} A K_\varepsilon^{-1/2} \\ K_\varepsilon^{-1/2} A^* H_\varepsilon^{-1/2} & I \end{bmatrix} = \begin{bmatrix} H_\varepsilon^{-1/2} & 0 \\ 0 & K_\varepsilon^{-1/2} \end{bmatrix} \times \begin{bmatrix} H_\varepsilon & A \\ A^* & K_\varepsilon \end{bmatrix} \begin{bmatrix} H_\varepsilon^{-1/2} & 0 \\ 0 & K_\varepsilon^{-1/2} \end{bmatrix}$$

is positive, and consequently by Lemma 3.1,

$$\| H_\varepsilon^{-1/2} A K_\varepsilon^{-1/2} \| \le 1.$$

Also, $\phi_n(H_\varepsilon^{-1/2} A K_\varepsilon^{-1/2}) = H_\varepsilon^{-1/2} \phi_n(A) K_\varepsilon^{-1/2}$ (Exercise 8.1), and so

$$\begin{bmatrix} H_\varepsilon & \phi_n(A) \\ \phi_n(A)^* & K_\varepsilon \end{bmatrix} = \begin{bmatrix} H_\varepsilon^{1/2} & 0 \\ 0 & K_\varepsilon^{1/2} \end{bmatrix} \times \begin{bmatrix} I & \phi_n\left(H^{-1/2} A K^{-1/2}\right) \\ \phi_n\left(H^{-1/2} A K^{-1/2}\right)^* & I \end{bmatrix} \begin{bmatrix} H_\varepsilon^{-1/2} & 0 \\ 0 & K_\varepsilon^{1/2} \end{bmatrix}.$$

But, since ϕ is completely contractive, $\|\phi_n(H^{-1/2} A K^{-1/2})\| \le 1$, and so by another application of Lemma 3.1, the middle matrix on the right of the above equation is positive. Thus, the left hand side is positive for all ε, and so (8.1) is positive.

Consequently, Φ is completely positive, which completes the proof. □

We can now prove a Hahn–Banach-type extension theorem for completely bounded maps.

Theorem 8.2 (Wittstock's extension theorem). *Let $\mathcal{A}$ be a unital C^*-algebra, $\mathcal{M}$ a subspace of $\mathcal{A}$, and let $\phi\colon \mathcal{M} \to B(\mathcal{H})$ be completely bounded. Then there exists a completely bounded map $\psi\colon \mathcal{A} \to B(\mathcal{H})$ that extends ϕ, with $\|\phi\|_{\mathrm{cb}} = \|\psi\|_{\mathrm{cb}}$.*

Proof. We may assume without loss of generality that $\|\phi\|_{\mathrm{cb}} = 1$. Let $\mathcal{S}_\mathcal{M}$ and Φ be as in Lemma 8.1. Since Φ is completely positive, by Arveson's extension theorem there exists a completely positive $\Psi\colon M_2(\mathcal{A}) \to M_2(B(\mathcal{H})) = B(\mathcal{H} \oplus \mathcal{H})$ extending Φ.

Define ψ via

$$\Psi\left(\begin{bmatrix} 0 & a \\ 0 & 0 \end{bmatrix}\right) = \begin{bmatrix} * & \psi(a) \\ * & * \end{bmatrix}.$$

Clearly, ψ is linear, and since Ψ extends Φ, ψ extends ϕ. Also, since Ψ is unital,

$$\|\psi(a)\| \leq \left\| \Psi\left(\begin{bmatrix} 0 & a \\ 0 & 0 \end{bmatrix}\right) \right\| \leq \|\Psi\| \cdot \|a\| \leq \|a\|,$$

and so ψ is contractive.

To see that ψ is completely contractive, let $A = (a_{i,j})$ be in $M_n(\mathcal{A})$; then

$$\Psi_n\left(\left[\begin{bmatrix} 0 & a_{i,j} \\ 0 & 0 \end{bmatrix}\right]_{i,j=1}^n\right) = \left[\begin{bmatrix} * & \psi(a_{i,j}) \\ * & * \end{bmatrix}\right]_{i,j=1}^n.$$

After performing the canonical shuffle, the right hand side becomes

$$\begin{bmatrix} * & \psi_n(A) \\ * & * \end{bmatrix},$$

where the asterisk now indicates an $n \times n$ rather than a 1×1 entry. Thus, by the same inequality as above,

$$\|\psi_n(A)\| \leq \left\| \left[\begin{bmatrix} 0 & a_{i,j} \\ 0 & 0 \end{bmatrix}\right]_{i,j=1}^n \right\|.$$

But after a canonical shuffle,

$$\left[\begin{bmatrix} 0 & a_{i,j} \\ 0 & 0 \end{bmatrix}\right]_{i,j=1}^n \quad \text{becomes} \quad \begin{bmatrix} 0 & A \\ 0 & 0 \end{bmatrix}.$$

Thus, $\|\psi_n(A)\| \leq \|A\|$, which is what we needed to prove. □

If $\mathcal{M}$ is an operator space in the C^*-algebra $\mathcal{A}$, $\mathcal{B}$ is another C^*-algebra, and $\phi\colon \mathcal{M} \to \mathcal{B}$ is a linear map, we set $\mathcal{M}^* = \{a\colon a^* \in \mathcal{M}\}$ and define a linear map

$$\phi^*\colon \mathcal{M}^* \to \mathcal{B}, \qquad \text{via} \quad \phi^*(a) = \phi(a^*)^*.$$

When $\mathcal{M} = \mathcal{M}^*$, we set

$$\operatorname{Re}\phi = (\phi + \phi^*)/2, \qquad \operatorname{Im}\phi = (\phi - \phi^*)/2i,$$

so that $\operatorname{Re}\phi$, $\operatorname{Im}\phi$ are self-adjoint, linear maps with $\phi = \operatorname{Re}\phi + i \operatorname{Im}\phi$. We note that some care is needed, since, in general, $(\operatorname{Re}\phi)(a) \neq \operatorname{Re}(\phi(a))$, but they are equal when $a = a^*$.

The above lemma also yields the following decomposition theorem for completely bounded maps.

Theorem 8.3. *Let $\mathcal{A}$ be a C^*-algebra with unit, and let $\phi\colon \mathcal{A} \to B(\mathcal{H})$ be completely bounded. Then there exist completely positive maps $\phi_i\colon \mathcal{A} \to B(\mathcal{H})$ with $\|\phi_i\|_{\text{cb}} = \|\phi\|_{\text{cb}}$, $i = 1, 2$, such that the map $\Phi\colon M_2(\mathcal{A}) \to B(\mathcal{H} \oplus \mathcal{H})$ given by*

$$\Phi\left(\begin{bmatrix} a & b \\ c & d \end{bmatrix}\right) = \begin{bmatrix} \phi_1(a) & \phi(b) \\ \phi^*(c) & \phi_2(d) \end{bmatrix}$$

is completely positive. Moreover, if $\|\phi\|_{\text{cb}} \le 1$, then we may take $\phi_1(1) = \phi_2(1) = I_{\mathcal{H}}$.

Proof. Clearly, we may assume that $\|\phi\|_{\text{cb}} = 1$. Applying Lemma 8.1 with $\mathcal{M} = \mathcal{A}$, we obtain a completely positive map $\Phi\colon \mathcal{S}_{\mathcal{A}} \to B(\mathcal{H} \oplus \mathcal{H})$ where $\mathcal{S}_{\mathcal{A}} \subseteq M_2(\mathcal{A})$.

By Arveson's extension theorem, Φ extends to a completely positive map on all of $M_2(\mathcal{A})$, which we still denote by Φ. Since $\left[\begin{smallmatrix} 0 & b \\ c & 0 \end{smallmatrix}\right]$ is in $\mathcal{S}_{\mathcal{A}}$, by the definition of Φ,

$$\Phi\left(\begin{bmatrix} 0 & b \\ c & 0 \end{bmatrix}\right) = \begin{bmatrix} 0 & \phi(b) \\ \phi(c^*)^* & 0 \end{bmatrix} = \begin{bmatrix} 0 & \phi(b) \\ \phi^*(c) & 0 \end{bmatrix}.$$

Now, let p be positive, $p \le 1$; then since

$$\begin{bmatrix} p & 0 \\ 0 & 0 \end{bmatrix} \le \begin{bmatrix} 1 & 0 \\ 0 & 0 \end{bmatrix},$$

we have

$$\begin{aligned} \begin{bmatrix} 0 & 0 \\ 0 & 0 \end{bmatrix} &\le \Phi\left(\begin{bmatrix} p & 0 \\ 0 & 0 \end{bmatrix}\right) \\ &\le \Phi\left(\begin{bmatrix} 1 & 0 \\ 0 & 0 \end{bmatrix}\right) = \begin{bmatrix} 1 & 0 \\ 0 & 0 \end{bmatrix}. \end{aligned}$$

A straightforward calculation shows that these inequalities taken together imply that

$$\Phi\left(\begin{bmatrix} p & 0 \\ 0 & 0 \end{bmatrix}\right) = \begin{bmatrix} * & 0 \\ 0 & 0 \end{bmatrix}.$$

Since $\mathcal{A}$ is the span of its positive elements, there must exist a linear map $\phi_1\colon \mathcal{A} \to B(\mathcal{H})$ such that

$$\Phi\left[\begin{bmatrix} a & 0 \\ 0 & 0 \end{bmatrix}\right] = \begin{bmatrix} \phi_1(a) & 0 \\ 0 & 0 \end{bmatrix}.$$

By an argument similar to the proof in Theorem 8.2, that ψ is completely contractive, one obtains that ϕ_1 is completely positive.

By analogous arguments, one obtains a completely positive map $\phi_2\colon \mathcal{A} \to B(\mathcal{H})$ satisfying

$$\Phi\left[\begin{bmatrix} 0 & 0 \\ 0 & d \end{bmatrix}\right] = \begin{bmatrix} 0 & 0 \\ 0 & \phi_2(d) \end{bmatrix}.$$

Thus, we see that any completely positive map Φ obtained via extension from $\mathcal{S}_{\mathcal{A}}$ to $M_2(\mathcal{A})$ has the desired form,

$$\Phi\left[\begin{bmatrix} a & b \\ c & d \end{bmatrix}\right] = \Phi\left[\begin{bmatrix} 0 & b \\ c & 0 \end{bmatrix}\right] + \Phi\left[\begin{bmatrix} a & 0 \\ 0 & d \end{bmatrix}\right] = \begin{bmatrix} \phi_1(a) & \phi(b) \\ \phi^*(c) & \phi_2(d) \end{bmatrix},$$

for some ϕ_1 and ϕ_2.

Since $\phi_1(1) = \phi_2(1) = 1$, we have that $\|\phi_1\|_{\text{cb}} = \|\phi_2\|_{\text{cb}} = \|\phi\|_{\text{cb}}$. This completes the proof of the theorem. □

An alternative way to prove the above theorem is to use the fact that Φ fixes the C^*-subalgebra, $\mathbb{C} \oplus \mathbb{C}$, consisting of scalar diagonal matrices, and consequently by Corollary 3.19 is necessarily a bimodule map over this algebra. An extension of this idea is included in Exercise 8.6.

The above decomposition leads readily to a generalization of Stinespring's representation.

Theorem 8.4. *Let $\mathcal{A}$ be a C^*-algebra with unit, and let $\phi\colon \mathcal{A} \to B(\mathcal{H})$ be a completely bounded map. Then there exists a Hilbert space $\mathcal{K}$, a $*$-homomorphism $\pi\colon \mathcal{A} \to B(\mathcal{K})$, and bounded operators $V_i\colon \mathcal{H} \to \mathcal{K}$, $i = 1, 2$, with $\|\phi\|_{\text{cb}} = \|V_1\| \cdot \|V_2\|$ such that*

$$\phi(a) = V_1^* \pi(a) V_2$$

for all $a \in \mathcal{A}$. Moreover, if $\|\phi\|_{\text{cb}} = 1$, then V_1 and V_2 may be taken to be isometries.

Proof. Clearly, we may assume that $\|\phi\|_{\text{cb}} = 1$. Let ϕ_1, ϕ_2, and Φ be as in Theorem 8.3. Let $(\pi_1, V, \mathcal{K}_1)$ be a minimal Stinespring representation for Φ, and note that since Φ is unital, V may be taken to be an isometry and π_1 to be unital.

Since $M_2(\mathcal{A})$ contains a copy of M_2, the Hilbert space $\mathcal{K}_1$ may be decomposed as $\mathcal{K}_1 = \mathcal{K} \oplus \mathcal{K}$ in such a way that the $*$-homomorphism $\pi_1\colon M_2(\mathcal{A}) \to B(\mathcal{K} \oplus \mathcal{K})$ has the form

$$\pi_1\left(\begin{bmatrix} a & b \\ c & d \end{bmatrix}\right) = \begin{bmatrix} \pi(a) & \pi(b) \\ \pi(c) & \pi(d) \end{bmatrix},$$

where $\pi\colon \mathcal{A} \to B(\mathcal{K})$ is a unital $*$-homomorphism (Exercise 8.3).

Thus, we have that $V\colon \mathcal{H} \oplus \mathcal{H} \to \mathcal{K} \oplus \mathcal{K}$ is an isometry, and

$$\begin{bmatrix} \phi_1(a) & \phi(b) \\ \phi^*(c) & \phi_2(d) \end{bmatrix} = V^* \begin{bmatrix} \pi(a) & \pi(b) \\ \pi(c) & \pi(d) \end{bmatrix} V.$$

For h in $\mathcal{H}$,

$$\begin{aligned} \begin{bmatrix} h \\ 0 \end{bmatrix} = \begin{bmatrix} \phi_1(1) & 0 \\ 0 & 0 \end{bmatrix} \begin{bmatrix} h \\ 0 \end{bmatrix} &= V^* \begin{bmatrix} \pi(1) & 0 \\ 0 & 0 \end{bmatrix} V \begin{bmatrix} h \\ 0 \end{bmatrix} \\ &= V^* \begin{bmatrix} 1_{\mathcal{K}} & 0 \\ 0 & 0 \end{bmatrix} V \begin{bmatrix} h \\ 0 \end{bmatrix}, \end{aligned}$$

and since V is an isometry, we must have that

$$V \begin{bmatrix} h \\ 0 \end{bmatrix} = \begin{bmatrix} * \\ 0 \end{bmatrix}.$$

Thus, there is a linear map $V_1\colon \mathcal{H} \to \mathcal{K}$ such that

$$V \begin{bmatrix} h \\ 0 \end{bmatrix} = \begin{bmatrix} V_1 h \\ 0 \end{bmatrix},$$

and V_1 must also be an isometry. Similarly, there exists $V_2\colon \mathcal{H} \to \mathcal{K}$ such that

$$V \begin{bmatrix} 0 \\ h \end{bmatrix} = \begin{bmatrix} 0 \\ V_2 h \end{bmatrix}.$$

Consequently,

$$\begin{aligned} \begin{bmatrix} \phi_1(a) & \phi(b) \\ \phi^*(c) & \phi_2(d) \end{bmatrix} &= V^* \begin{bmatrix} \pi(a) & \pi(b) \\ \pi(c) & \pi(d) \end{bmatrix} V \\ &= \begin{bmatrix} V_1^*\pi(a)V_1 & V_1^*\pi(b)V_2 \\ V_2^*\pi(c)V_1 & V_2^*\pi(d)V_2 \end{bmatrix}, \end{aligned}$$

which completes the proof of the theorem. □

Unlike the Stinespring representation of a completely positive map, there are no extra conditions known to impose on the above representation of a completely bounded map that will make it unique up to unitary equivalence. Given Φ, one can always take its minimal (and hence unique) Stinespring representation, but unfortunately Φ is not uniquely determined by ϕ. Recall that ϕ uniquely determines a completely positive map on the operator system

$$S = \left\{ \begin{pmatrix} \lambda & a \\ b^* & \mu \end{pmatrix} : \lambda, \mu \in \mathbb{C}, a, b \in \mathcal{A} \right\} \subseteq M_2(\mathcal{A})$$

and that Φ is obtained by extending this completely positive map from S to $M_2(\mathcal{A})$. Thus, in a certain sense the nonuniqueness of the representation is parameterized by the set of possible completely positive extensions. It is possible to impose conditions on the above representation such that it is unique up to conjugation by a (generally unbounded) closed, densely defined similarity [171].

Theorem 8.5 (Wittstock's decomposition theorem). *Let $\mathcal{A}$ be a C^*-algebra with unit, and let $\phi\colon \mathcal{A} \to B(\mathcal{H})$ be completely bounded. Then there exists a completely positive map $\psi\colon \mathcal{A} \to B(\mathcal{H})$ with $\|\psi\|_{\text{cb}} \leq \|\phi\|_{\text{cb}}$ such that $\psi \pm$ Re(ϕ) and $\psi \pm$ Im(ϕ) are all completely positive. In particular, the completely bounded maps are the linear span of the completely positive maps.*

Proof. Let $\phi(a) = V_1^*\pi(a)V_2$ be as in Theorem 8.4 with $\|V_1\| = \|V_2\| = \|\phi\|_{\text{cb}}^{1/2}$, and set $\psi(a) = (V_1^*\pi(a)V_1 + V_2^*\pi(a)V_2)/2$, so that ψ is completely positive and $\|\psi\|_{\text{cb}} = \|\psi(1)\| \leq \|\phi\|_{\text{cb}}$. Notice that $\phi^*(a) = V_2^*\pi(a)V_1$, so that

$$2\psi(a) \pm 2 \operatorname{Re} \phi(a) = (V_1 \pm V_2)^*\pi(a)(V_1 \pm V_2),$$
$$2\psi(a) \pm 2 \operatorname{Im} \phi(a) = (V_1 \mp iV_2)^*\pi(a)(V_1 \mp iV_2),$$

and each of these four maps is completely positive.

For the last statement, note that

$$2\phi = ((\psi + \operatorname{Re} \phi) - (\psi - \operatorname{Re} \phi)) + i((\psi + \operatorname{Im} \phi) - (\psi - \operatorname{Im} \phi))$$

is a decomposition of ϕ into the span of four completely positive maps. □

Operator-Valued Measures

Let X be a compact Hausdorff space, let E be a bounded, regular, operator-valued measure on X, and let $\phi\colon C(X) \to B(\mathcal{H})$ be the bounded, linear map associated with E by integration as described in Chapter 4. We call E completely bounded when ϕ is completely bounded.

By Wittstock's decomposition theorem, E is completely bounded if and only if it can be expressed as a linear combination of positive operator-valued measures.

When $\mathcal{H}$ is one-dimensional, that is, when E is a bounded, regular, complex-valued measure on X, then the associated bounded, linear map $\phi: C(X) \to \mathbb{C}$ is automatically completely bounded by Proposition 3.8. Wittstock's decomposition becomes the statement that every complex measure is the span of four positive measures. If we let ψ be defined by integration against the total variation measure $|E|$ associated with E, then $\psi \pm \operatorname{Re} \phi$ and $\psi \pm \operatorname{Im} \phi$ are completely positive maps, and $\|\psi\|_{\text{cb}} = \|\psi(1)\| = |E|(X) = \|\phi\| = \|\phi\|_{\text{cb}}$.

Thus, we see that in some sense, the ψ of Wittstock's decomposition can be thought of as a total variation or an "absolute value" of the completely bounded map ϕ. This analogy is pursued in [139] and [171].

In a closely related development, Loebl [139] defines a self-adjoint map $\phi: C(X) \to B(\mathcal{H})$ to have *finite total variation* if

$$\sup \left\{ \sum \| |\phi(f_i)| \| : f_i \text{ is a partition of unity} \right\}$$

is finite, and proves that every such map decomposes as $\phi = \phi_1 - \phi_2$ with ϕ_1 and ϕ_2 (completely) positive. In particular, this shows that such maps are completely bounded. However, Hadwin [109] has shown that there are completely bounded, self-adjoint maps that are not of finite total variation.

Unfortunately, there is no known analytic characterization of the completely bounded, operator-valued measures. To illustrate the difficulties, let us suppose for simplicity that E is self-adjoint and completely bounded.

From Wittstock's decomposition, we obtain a positive, operator-valued measure F such that $F(B) \pm E(B) \geq 0$ for all Borel sets B. If E and F were scalar-valued, this inequality would imply that $F(B) \geq |E(B)|$, but for operators, this is far from the case. In fact, it is possible for E to be completely bounded while

$$\sup \left\{ \left\| \sum |E(B_i)| \right\| : B_i \text{ disjoint, Borel} \right\} = +\infty.$$

This phenomenon was first described by Hadwin [109], and we reproduce his example here.

Let $X = \{x_n\}_{n=1}^{\infty}$ be a countable, compact Hausdorff space, and let A_n, $n = 1, 2, \dots$, be a sequence of self-adjoint operators on $\mathcal{H}$. If these operators are summable in the weak operator topology, then setting $E(\{x_n\}) = A_n$ defines a self-adjoint, operator-valued measure on X.

Now, let $\mathcal{H}$ be a separable Hilbert space with orthonormal basis, $e_1, e_2, \ldots,$ and define A_n via

$$A_n f = \frac{1}{n}(\langle f, e_1\rangle e_n + \langle f, e_n\rangle e_1),$$

so that

$$|A_n| f = \frac{1}{n}(\langle f, e_1\rangle e_1 + \langle f, e_n\rangle e_n).$$

It is easily checked that the A_n's are summable in the weak operator topology, but that the $|A_n|$'s are not, since $\langle |A_n| e_1, e_1\rangle$ is the harmonic series. Thus, setting $F(\{x_n\}) = |A_n|$ will not define an operator-valued measure. However, setting

$$B_n f = \frac{1}{n^2}\langle f, e_1\rangle e_1 + \langle f, e_n\rangle e_n$$

yields a sequence of positive operators that are summable in the weak operator topology and satisfy

$$B_n \pm A_n \geq 0.$$

Thus, setting $F(\{x_n\}) = B_n$ yields a positive operator-valued measure such that $F \pm E$ are positive.

It is easy to see that the linear map $\phi\colon C(X) \to B(\mathcal{H})$ induced by E is completely bounded, but not of finite total variation.

Note that the decomposition of $A_n = (B_n + A_n)/2 - (B_n - A_n)/2$ into the difference of two positive matrices is not the usual decomposition, $A_n = A_n^+ - A_n^-$, for if it were, then necessarily, we should have $B_n = A_n^+ + A_n^- = |A_n|$.

In fact, since $\{|A_n|\}$ is not summable, either $\{A_n^+\}$ or $\{A_n^-\}$ must not be summable. This happens in spite of the fact that we can decompose each A_n into a difference of positive operators, namely $(B_n \pm A_n)/2$, with both of these sequences of positive operators summable.

Bimodule Maps

At the moment we have many results about completely bounded maps, but few tools for determining when bounded maps are completely bounded. For bimodule maps there is a nice theory.

Let $\mathcal{C} \subseteq \mathcal{B}$ be C^*-algebras. We shall call $\mathcal{C}$ *matrically norming for* $\mathcal{B}$ provided for every (B_{ij}) in $M_n(\mathcal{B})$ and every n we have that

$$\|(B_{ij})\| = \sup\left\{\left\|\sum_{i,j=1}^{n} C_i B_{ij} D_j\right\| : C_i, D_j \in \mathcal{C}\right\}$$

with $\|\sum_{i=1}^{n} C_i C_i^*\| \leq 1$ and $\|\sum_{j=1}^{n} D_j^* D_j\| \leq 1$. This condition looks strange until one realizes that

$$\sum_{i,j} C_i B_{ij} D_j = (C_1, \ldots, C_n)(B_{ij}) \begin{pmatrix} D_1 \\ \vdots \\ D_n \end{pmatrix}$$

and the above inequalities are simply the requirement that the row of C's and column of D's each have norm less than or equal to 1. Thus

$$\left\| \sum_{i,j} C_i B_{ij} D_j \right\| \leq \|(C_1, \ldots, C_n)\| \|(B_{ij})\| \left\| \begin{pmatrix} D_1 \\ \vdots \\ D_n \end{pmatrix} \right\| \leq \|(B_{ij})\|,$$

and $\mathcal{C}$ is matrically norming for $\mathcal{B}$ precisely when we can achieve $\|(B_{ij})\|$ as the supremum of such products. Note that $\mathcal{C}$ plays the same role as the complex numbers play for computing the norms of scalar matrices.

For an easy example of this phenomenon, consider $\mathcal{B} = C(X)$, and let $\mathcal{C}$ be the constant functions, which can be identified with $\mathbb{C}$. By compactness we have that for $(f_{ij}) \in M_n(C(X))$ there exists x_0 with $\|(f_{ij})\| = \|(f_{ij}(x_0))\|$. One can then choose unit vectors $h = (c_1, \ldots, c_n)$, $k = (d_1, \ldots, d_n)$ such that $\|(f_{ij}(x_0))\| = |\sum_{i,j} c_i f_{ij}(x_0) d_j|$ and hence $\|(f_{ij})\| = \|\sum_{i,j} c_i f_{ij} d_j\|$.

Proposition 8.6. *Let $\mathcal{A}$, $\mathcal{B}$, and $\mathcal{C}$ be C^*-algebras with $\mathcal{C}$ a subalgebra of both $\mathcal{A}$ and $\mathcal{B}$. If $\mathcal{C}$ is matrically norming for $\mathcal{B}$ and $\phi \colon \mathcal{A} \to \mathcal{B}$ is a $\mathcal{C}$-bimodule map, then $\|\phi\| = \|\phi\|_{\text{cb}}$.*

Proof. For any (A_{ij}) in $M_n(\mathcal{A})$ we have that

$$\|(\phi(A_{ij}))\| = \sup\left\{ \left\| \sum C_i \phi(A_{ij}) D_j \right\| : C_i, D_j \in \mathcal{C}, \left\| \sum C_i C_i^* \right\| \leq 1 \right.$$

and $\|\sum D_j^* D_j\| \leq 1\}$. But

$$\left\| \sum C_i \phi(A_{ij}) D_j \right\| = \left\| \phi\left(\sum C_i A_{ij} D_j \right) \right\| \leq \|\phi\| \cdot \|(A_{ij})\|$$

and the result follows. □

Since every linear map is a $\mathbb{C}$-bimodule map, the above result gives another proof of the fact that for maps into $\mathcal{B} = C(X)$ the norm and cb norm are the same.

There is another instance in the theory of completely bounded maps where the usual decomposition of a self-adjoint operator A into A^+ and A^- is not the "best" decomposition into a difference of positive operators. This occurs in the theory of Schur products.

Schur Products Revisited

Let A be in M_n, and let $S_A\colon M_n \to M_n$ be given by $S_A(B) = A * B$. Note that if $D = \operatorname{diag}(d_i)$ is a diagonal matrix, then $S_A(DB) = (a_{ij}d_ib_{ij}) = DS_A(B)$ and similarly $S_A(BD) = S_A(B)D$. Thus, Schur product maps are bimodule maps over the C^*-algebra $\mathcal{D}_n$ of diagonal matrices. In fact, by Exercise 4.4, if $\phi\colon M_n \to M_n$ is a $\mathcal{D}_n$-bimodule map, then $\phi = S_A$ for some matrix A. Using the representation theorem for completely bounded maps and Proposition 8.6, we can now characterize the norm of Schur product maps.

Theorem 8.7. *Let $A = (a_{ij})$ be in M_n. Then the following are equivalent:*

(i) $\|S_A\| \leq 1$,
(ii) $\|S_A\|_{\text{cb}} \leq 1$,
(iii) there exist $2n$ vectors $x_1, \ldots, x_n, y_1, \ldots, y_n$, in the unit ball of some Hilbert space such that $a_{ij} = \langle x_j, y_i \rangle$.

Proof. Clearly (ii) implies (i). To see that (i) implies (ii), we will prove that $\mathcal{D}_n$ is matrically norming for M_n. Suppose that we can prove that for every $h_1, \ldots, h_k$ in $\mathbb{C}^n$ with $\|h_1\|^2 + \cdots + \|h_k\|^2 \leq 1$ there exists a unit vector h and $D_1, \ldots, D_k$ such that $h_j = D_j h$ for all j and $\| \sum D_j^* D_j \| \leq 1$. Then given $B = (B_{ij})$ in $M_k(M_n)$, there will exist $h_1, \ldots, h_k, \tilde{h}_1, \ldots, \tilde{h}_k$ with $\|h_1\|^2 + \cdots + \|h_k\|^2 = \|\tilde{h}_1\|^2 + \cdots + \|\tilde{h}_k\|^2 = 1$ such that

$$\|B\| = \sum_{i,j} \langle B_{ij}\tilde{h}_j, h_i \rangle.$$

For these vectors we choose $D_1, \ldots, D_k, \tilde{D}_1, \ldots, \tilde{D}_k, h$, and $\tilde{h}$ as above and find that $\|B\| = \sum_{i,j} \langle B_{ij}\tilde{D}_j\tilde{h}, D_i h \rangle$. Consequently, $\|B\| = \| \sum_{i,j} D_i^* B_{ij} \tilde{D}_j \|$, and we will have that $\mathcal{D}_n$ is matrically norming for M_n.

To establish the claim, let $h_1, \ldots, h_k$ be as above, and let e denote the vector all of whose entries are 1's. Set $H_j = \operatorname{diag}(h_j)$, so that $h_j = H_j e$ and $\operatorname{Tr}(H_j^* H_j) = \|h_j\|^2$. Let $P = \sum H_j^* H_j$; set $h = P^{1/2}e$ and $D_j = H_j P^{-1/2}$. Then $\sum D_j^* D_j = P^{-1/2}(\sum H_j^* H_j)P^{-1/2} = I$, $D_j h = (H_j P^{-1/2})(P^{1/2}e) = h_j$, and $\|h\|^2 = \langle P^{1/2}e, P^{1/2}e \rangle = \langle Pe, e \rangle = \operatorname{Tr}(P) = \sum_j \|h_j\|^2 \leq 1$, and we have shown the claim. Thus, (i) implies (ii), and $\|S_A\| = \|S_A\|_{\text{cb}}$ for every Schur product map.

To prove that (ii) implies (iii), use Theorem 8.4 to write $S_A(B) = V_1^*\pi(B)V_2$ where $V_i\colon \mathbb{C}^n \to \mathcal{H}$, $i = 1, 2$, are isometries and $\pi\colon M_n \to B(\mathcal{H})$ is a

$*$-homomorphism. Let $e_1, \dots, e_n$ be the canonical basis for $\mathbb{C}^n$, set $E_j = E_{1j}$ and define $2n$ vectors in $\mathcal{H}$ via

$$x_j = \pi(E_j)V_2 e_j, \qquad j = 1, \dots, n,$$
$$y_i = \pi(E_i)V_1 e_1, \qquad i = 1, \dots, n.$$

We have

$$\begin{aligned} \|x_j\|^2 &= \langle V_2^* \pi(E_j^* \cdot E_j) V_2 e_j, e_j \rangle \\ &= \langle V_2^* \pi(E_{j,j}) V_2 e_j, e_j \rangle \\ &\leq 1, \qquad j = 1, \dots, n, \end{aligned}$$

and similarly that

$$\|y_i\|^2 \leq 1, \qquad i = 1, \dots, n.$$

Let $a_{i,j}$ denote the (i, j)th entry of A; then

$$\begin{aligned} \langle x_j, y_i \rangle &= \langle V_1^* \pi(E_i^* E_j) V_2 e_j, e_i \rangle \\ &= \langle V_1^* \pi(E_{i,j}) V_2 e_j, e_i \rangle \\ &= \langle S_A(E_{i,j}) e_j, e_i \rangle \\ &= a_{i,j}. \end{aligned}$$

Thus, if $\|S_A\|_{\text{cb}} \leq 1$, then there is a Hilbert space and $2n$ vectors in its unit ball such that

$$A = (\langle x_j, y_i \rangle).$$

Thus, (ii) implies (iii).

Finally, we prove that (iii) implies (i). Assume that $A = (\langle x_j, y_i \rangle)$, where $\|x_j\| \leq 1$ and $\|y_i\| \leq 1$, $1 \leq i, j \leq n$, are vectors in a Hilbert space $\mathcal{H}$. Given $B \in M_n$, $h = (\alpha_1, \dots, \alpha_n)$, and $k = (\beta_1, \dots, \beta_n)$ in $\mathbb{C}^n$, we must prove that $|\langle S_A(B)h, k \rangle| \leq \|B\| \|h\| \|k\|$. We have

$$\begin{aligned} |\langle S_A(B)h, k \rangle| &= \left| \sum_{i,j} \langle x_j, y_i \rangle_{\mathcal{H}} b_{i,j} \alpha_j \bar{\beta}_i \right| \\ &= \left| \left\langle (b_{i,j} I_{\mathcal{H}}) \begin{pmatrix} \alpha_1 x_1 \\ \vdots \\ \alpha_n x_n \end{pmatrix}, \begin{pmatrix} \beta_1 y_1 \\ \vdots \\ \beta_n y_n \end{pmatrix} \right\rangle_{\mathcal{H}^n} \right| \\ &\leq \|(b_{ij} I_{\mathcal{H}})\| \cdot \left(\sum \|\alpha_j x_j\|^2 \right)^{1/2} \left(\sum \|\beta_i y_i\|^2 \right)^{1/2} \\ &\leq \|B\| \cdot \|h\| \cdot \|k\|, \end{aligned}$$

since $\|\alpha_j x_j\| \leq |\alpha_j|$ and $\|\beta_i y_i\| \leq |\beta_i|$ for all i and j. Thus, $\|S_A\| \leq 1$, and the proof of the theorem is complete. □

We are now in a position to characterize completely bounded Schur product maps on $B(\ell^2)$. We leave the analogous result for completely positive maps to the exercises.

Corollary 8.8. *Let $A = (a_{ij})_{i,j=1}^{\infty}$ be an infinite matrix. Then the following are equivalent:*

(i) $S_A\colon B(\ell^2) \to B(\ell^2)$ is bounded with $\|S_A\| \le 1$,
(ii) $S_A\colon B(\ell^2) \to B(\ell^2)$ is completely bounded with $\|S_A\|_{\text{cb}} \le 1$,
(iii) there exists a Hilbert space $\mathcal{H}$ and vectors $\{x_j\}_{j=1}^{\infty}$ and $\{y_i\}_{i=1}^{\infty}$ in $\mathcal{H}$ with $\|x_j\| \le 1$ and $\|y_i\| \le 1$ for all i and j, such that $a_{ij} = \langle x_j, y_i \rangle$.

Proof. Given $B = (b_{ij})_{i,j=1}^{\infty}$, let $B_n = (b_{i,j})_{i,j=1}^{n}$. It is easily checked that B is a bounded operator on ℓ^2 if and only if $\sup_n \|B_n\|$ is finite and that in this case $\|B\| = \sup_n \|B_n\|$.

Consequently, S_A is bounded if and only if $\sup_n \|S_{A_n}\|$ is finite and $\|S_A\| = \sup_n \|S_{A_n}\|$.

Similarly, one can see that $\|S_A\|_{\text{cb}} = \sup_n \|S_{A_n}\|_{\text{cb}}$ and so $\|S_A\| = \|S_A\|_{\text{cb}}$, by applying Theorem 8.7. The proofs that (ii) implies (iii) and (iii) implies (i) are as in Theorem 8.7. □

From the above two results we see that for Schur product maps,

$$\|S_A\| = \|S_A\|_{\text{cb}} = \inf_{X,Y} \sup_{i,j} \{\|x_j\| \|y_i\|\},$$

where the infimum is over all sets of vectors $X = \{x_j\}$, $Y = \{y_i\}$ satisfying $\langle x_j, y_i \rangle = a_{ij}$. Moreover, the infimum is attained.

The following criterion for an infinite matrix to give rise to a bounded Schur product map is often useful.

Theorem 8.9. *Let $A = (a_{ij})_{i,j=1}^{\infty}$ and set $b_{ij} = a_{ij} - a_{i+1,j} - a_{i,j+1} + a_{i+1,j+1}$. If the limits $\lim_j a_{ij} = s_i$, $\lim_i a_{ij} = t_j$, $\lim_i s_i$, $\lim_j t_j$ all exist and $\{b_{ij}\}$ is absolutely summable, then $S_A\colon B(\ell^2) \to B(\ell^2)$ is completely bounded.*

Proof. First note that

$$\begin{aligned}\sum_{k=j}^{\infty} \sum_{\ell=i}^{\infty} b_{\ell,k} &= \sum_{k=j}^{\infty} a_{i,k} - t_k - a_{i,k+1} + t_{k+1} \\ &= a_{i,j} - s_i - t_j + \lim_k t_k,\end{aligned}$$

while

$$\sum_{\ell=i}^{\infty}\sum_{k=j}^{\infty} b_{\ell,k} = a_{ij} - s_i - t_j + \lim_{\ell} s_\ell.$$

Hence the absolute summability of $\{b_{ij}\}$ implies that $\lim_i s_i = \lim_j t_j = w$.

Choose complex numbers $\{c_{ij}\}$, $\{d_{ij}\}$ such that $|c_{ij}| = |d_{ij}|$ and $c_{ij}\bar{d}_{ij} = b_{ij}$. Fix a Hilbert space and a set of orthonormal vectors $\{e_{ij}\}_{i,j=1}^{\infty} \cup \{e, f, g\}$. Define

$$x_j = -we + t_j f + g + \sum_{k=j}^{\infty}\sum_{\ell=1}^{\infty} c_{k\ell}e_{k\ell},$$

$$y_i = e + f + s_i g + \sum_{k=1}^{\infty}\sum_{\ell=i}^{\infty} d_{k\ell}e_{k\ell}.$$

Then it is easily checked that $\{x_j\}$ and $\{y_i\}$ are norm-bounded sets with

$$\langle x_j, y_i\rangle = -w + t_j + s_i + \sum_{k=j}^{\infty}\sum_{\ell=i}^{\infty} b_{k\ell} = a_{ij},$$

and the proof is complete. □

It is interesting to see how the decomposition in Theorem 8.3 relates to Schur product maps. If $A = (\langle x_j, y_i\rangle)$ with $\|x_j\| \leq 1$, $\|y_i\| \leq 1$ for all i, j, then let $P_1 = (\langle x_j, x_i\rangle)$, $P_2 = (\langle y_j, y_i\rangle)$, and note that

$$\begin{bmatrix} P_1 & A \\ A & P_2 \end{bmatrix} \geq 0.$$

By Exercise 8.7, we have that S_{P_1} and S_{P_2} are completely positive and this is the decomposition of Theorem 8.3 for S_A.

It is interesting to see how Wittstock's decomposition of S_A for A self-adjoint relates to the usual decomposition of a self-adjoint matrix as a difference of positive matrices. Suppose that $A = A^*$, so that the usual decomposition of A into a difference of positive matrices is given by $A = \frac{1}{2}(|A| + A) - \frac{1}{2}(|A| - A) = A^+ - A^-$. From this it follows that $S_{|A|} \pm S_A \geq 0$, that $S_A = S_{A^+} - S_{A^-}$ is a decomposition of S_A as the difference of two completely positive maps and that $\|S_A\|_{cb} \leq \|S_{|A|}\|_{cb}$. If equality holds in this last inequality, then $S_{|A|}$ could play the role of ψ in the Wittstock decomposition theorem for $\phi = S_A$. However, we shall see [Exercise 8.7(vii)] that, in general, $\|S_A\|_{cb} < \|S_{|A|}\|_{cb}$.

Thus, the usual decomposition of a self-adjoint matrix into a difference of positive matrices, while minimal in some senses, is not minimal for the Wittstock decomposition of S_A. This makes it somewhat surprising that when $A = A^*$, then there always is a positive matrix P with $P \pm A \geq 0$ and $\|S_P\|_{cb} = \|S_A\|_{cb}$ [Exercise 8.7(v)].

Let $\mathcal{M}$ be an operator space, let $\phi\colon \mathcal{M} \to \mathcal{B}$, and let Φ and $\mathcal{S}_{\mathcal{M}}$ be as in Lemma 8.1. It is easy to see that if $\|\phi_k\| \leq 1$, then the proof of Lemma 8.1 shows that Φ is k-positive. Thus, if $\phi\colon \mathcal{M} \to M_n$ and $\|\phi_{2n}\| \leq 1$, then $\Phi\colon \mathcal{S}_{\mathcal{M}} \to M_2(M_n) = M_{2n}$ will be $2n$-positive, and consequently completely positive. But, if Φ is completely positive, then ϕ must be completely contractive. Hence, $\|\phi_{2n}\| = \|\phi\|_{cb}$ for maps into M_n. Because of the special nature of the subspace $\mathcal{S}_{\mathcal{M}}$, it turns out that if Φ is n-positive, then it is completely positive, and consequently, $\|\phi_n\| = \|\phi\|_{cb}$. To obtain this more delicate result, we need a preliminary lemma.

Lemma 8.10. *Let $\mathcal{A}$ be a C^*-algebra, and let $P = (P_{i,j})_{i,j=1}^{2n}$ be a positive element in $M_{2n}(M_2(\mathcal{A}))$ where*

$$P_{i,j} = \begin{bmatrix} a_{i,j} & b_{i,j} \\ c_{i,j} & d_{i,j} \end{bmatrix}, \qquad i, j = 1, \ldots, 2n.$$

Then

$$\left[\begin{bmatrix} a_{i,j} & b_{i,j+n} \\ c_{i+n,j} & d_{i+n,j+n} \end{bmatrix}\right]_{i,j=1}^{n}$$

is positive in $M_n(M_2(\mathcal{A}))$.

Proof. Set $A = (a_{i,j})$, $B = (b_{i,j})$, $C = (c_{i,j})$, and $D = (d_{i,j})$ for $i, j = 1, \ldots, 2n$. After the canonical shuffle, P becomes $[\begin{smallmatrix} A & B \\ C & D \end{smallmatrix}]$, and so this latter matrix is positive. Now partition each of the $2n \times 2n$ matrices into 2×2 matrices consisting of $n \times n$ blocks in the natural fashion. That is,

$$A = \begin{bmatrix} A_{11} & A_{12} \\ A_{21} & A_{22} \end{bmatrix},$$

where $A_{11} = (a_{i,j})_{i,j=1}^n$, $A_{12} = (a_{i,j+n})_{i,j=1}^n$, $A_{21} = (a_{i+n,j})_{i,j=1}^n$, and $A_{22} = (A_{i+n,j+n})_{i,j=1}^n$ with the same definitions for B, C, and D. Thus,

$$\begin{bmatrix} A & B \\ C & D \end{bmatrix} = \begin{bmatrix} A_{11} & A_{12} & B_{11} & B_{12} \\ A_{21} & A_{22} & B_{21} & B_{22} \\ C_{11} & C_{12} & D_{11} & D_{12} \\ C_{21} & C_{22} & D_{21} & B_{22} \end{bmatrix}$$

is positive.

A moment's reflection shows that if such a matrix is positive, then the four corners will form a positive matrix, that is,

$$\begin{bmatrix} A_{11} & B_{12} \\ C_{21} & D_{22} \end{bmatrix}$$

will be positive.

The result follows by performing the canonical shuffle on this last matrix. □

Proposition 8.11 (Smith). *Let $\mathcal{M}$ be an operator space, and let $\phi\colon \mathcal{M} \to M_n$ be a linear map. Then $\|\phi_n\| = \|\phi\|_{\text{cb}}$.*

Proof. It is sufficient to assume that $\|\phi_n\| = 1$ and prove that $\Phi\colon \mathcal{S}_{\mathcal{M}} \to M_{2n}$ is completely positive, where Φ and $\mathcal{S}_{\mathcal{M}}$ are defined as in Lemma 8.1. By Theorem 5.1, to prove Φ is completely positive, it is sufficient to prove that the associated linear functional $S_\Phi\colon M_{2n}(\mathcal{S}_{\mathcal{M}}) \to \mathbb{C}$ is positive.

Thus, let $P = (P_{i,j})_{i,j=1}^{2n}$ be positive in $M_{2n}(\mathcal{S}_{\mathcal{M}})$, with

$$P_{i,j} = \begin{bmatrix} \lambda_{i,j} & a_{i,j} \\ b_{i,j}^* & \mu_{i,j} \end{bmatrix},$$

and calculate

$$\begin{aligned} 2n \cdot S_\Phi(P) &= \sum_{i,j=1}^{n} \Phi(P_{i,j})_{(i,j)} + \sum_{i,j=1}^{n} \Phi(P_{i,j+n})_{(i,j+n)} \\ &\quad + \sum_{i,j=1}^{n} \Phi(P_{i+n,j})_{(i+n,j)} + \sum_{i,j=1}^{n} \Phi(P_{i+n,j+n})_{(i+n,j+n)} \\ &= \sum_{i=1}^{n} \lambda_{i,i} + \sum_{i,j=1}^{n} \phi(a_{i,j+n})_{(i,j+n)} \\ &\quad + \sum_{i,j=1}^{n} \phi(b_{i,j})^*_{(i+n,j)} + \sum_{i=1}^{n} \mu_{i+n,i+n} \\ &= \sum_{i,j=1}^{n} \phi(A_{i,j})_{(i,j)}, \end{aligned}$$

where

$$A_{i,j} = \begin{bmatrix} \lambda_{i,j} & a_{i,j+n} \\ b_{i+n,j}^* & \mu_{i+n,j+n} \end{bmatrix}.$$

By the above lemma, $A = (A_{i,j})$ is a positive element of $M_n(\mathcal{S}_{\mathcal{M}})$, and if we set $x = e_1 \oplus \cdots \oplus e_n$, then the last sum above is easily recognized as $\langle \Phi_n(A)x, x \rangle$.

But this last expression is positive, because the assumption that $\|\phi_n\| = 1$ is enough to guarantee that Φ is n-positive, as observed in the remarks proceeding Lemma 8.10. □

The counterpart of Theorem 3.14 for completely bounded maps is not true. If $\phi\colon M_n \to \mathcal{B}$, then, in general, $\|\phi_n\| \neq \|\phi\|_{\text{cb}}$, and in fact, there does not exist any integer $m = m(n)$ such that $\|\phi_m\| = \|\phi\|_{\text{cb}}$ for all ϕ. This follows from work of Haagerup [108].

Combining Proposition 8.11 with Exercise 3.10(ii), we see that for maps into M_n, $\|\phi\|_{\text{cb}} \leq n\|\phi\|$, which gives another proof of Exercise 3.11.

Notes

Wittstock obtained his decomposition theorem for completely bounded maps in [242] and proved the extension theorem for completely bounded maps in [243]. His proofs used his theory of matricial sublinear functionals, which is a generalization of the Hahn–Banach theorem to set-valued mappings into $B(\mathcal{H})$.

Haagerup [107] obtained these same results by exploiting the correspondence between maps from $\mathcal{M}$ into M_n and linear functionals on $M_n(\mathcal{M})$, as was done for completely positive maps in Chapter 6. The difficult part of this approach in the completely bounded case is that the norm of the associated linear functional and the cb norm of the mapping into M_n are not very closely related. Thus, for example, to obtain an extension of a map from $\mathcal{M}$ into M_n to $\mathcal{A}$ into M_n, where $\mathcal{M} \subseteq \mathcal{A}$, with the same cb norm, one must extend the associated linear functional on $M_n(\mathcal{M})$ to $M_n(\mathcal{A})$, but in a very particular fashion. This technical stumbling block was the main obstruction to the theory of completely bounded maps developing simultaneously with the theory of completely positive maps. To overcome this difficulty, Haagerup used a technique that eventually led to the definition of the Haagerup tensor norm, which is studied in Chapter 17.

The proofs that we have presented here using the off-diagonal technique appeared in [158] and [159].

The generalization of Stinespring's representation theorem to completely bounded maps is referred to as the factorization theorem for completely bounded maps in some other texts. This result first appeared in [107] and [159]. Although this result can be deduced from Wittstock's decomposition and extension theorem, it is a bit tricky to get the statement about the equality of the norms via that approach. Even in the abelian case and with $\mathbb{C}$ for the range, it is difficult to express the norm of a measure in terms of inequalities involving its real and imaginary parts. Arguably the only way to achieve this in the noncommutative case is via the off-diagonal trick.

The results in [107], [159], and [243] were obtained independently and within about a six-month time period. It is certain that Wittstock obtained his proofs first, and in any case his early work was the inspiration for [107] and [159].

Proposition 8.11 was obtained by Smith [214].

Haagerup [107] was the first to prove that for Schur product maps, $\|S_A\| = \|S_A\|_{\text{cb}}$. Smith [217] later gave a simpler proof that contained the idea of a matricially norming subalgebra. This concept has subsequently been developed further by Pop, Sinclair, and Smith [195] and has many applications to computing cohomology.

Theorem 8.9 is a slight variation of a result of Bennett [14].

The generalization of the off-diagonalization technique to $\mathcal{C}$-bimodule maps (Exercise 8.6) is in Suen [225]. Wittstock had proved the decomposition and extension theorems in this more general setting earlier ([242] and [243]).

Exercise 8.8 is proven in Haagerup [108], where a partial converse of the Wittstock decomposition theorem is obtained. Namely, if a von Neumann algebra has the property that the span of the completely positive maps from it into itself is the completely bounded maps, then it is injective. An example is given by Huruya [125] of a C^*-algebra that is not injective, but is such that every completely bounded map into it is in the span of the completely positive maps. However, this decomposition does not meet the norm inequality in the statement of the Wittstock decomposition property. It is known that a C^*-algebra that has the Wittstock decomposition property, including the norm inequality, is injective. See [220] and [193] for more on this topic.

Smith [214] proved that the C^*-algebra $C([0, 1])$ has the property that not every completely bounded map from it to itself is in the span of the completely positive maps.

Exercises

8.1 Show that $\text{Re}(\phi_n) = (\text{Re}\,\phi)_n$, and that $(\phi^*)_n = (\phi_n)^*$.

8.2 Let $\phi\colon \mathcal{M} \to B$, let H, K be in M_n, and let A be in $M_n(\mathcal{M})$. Prove that $\phi_n(H \cdot A \cdot K) = H \cdot \phi_n(A) \cdot K$. Thus, $\phi_n\colon M_n(\mathcal{M}) \to M_n(\mathcal{B})$ is an M_n-bimodule map.

8.3 Verify the claim of Theorem 8.4.

8.4 Show that if ϕ is completely bounded, and $\phi(a) = V_1^*\pi(a)V_2$ is the representation of Theorem 8.4 with $\|V_1\| = \|V_2\|$, then setting $\phi_i(a) = V_i^*\pi(a)V_i$ yields the map Φ of Theorem 8.3.

8.5 Prove that the conclusions of Theorems 8.2, 8.3, and 8.5 still hold when the range is changed from $B(\mathcal{H})$ to an arbitrary injective C^*-algebra.

8.6 Let $\mathcal{A}$, $\mathcal{B}$, and $\mathcal{C}$ be C^*-algebras with unit, with $\mathcal{C}$ contained in both $\mathcal{A}$ and $\mathcal{B}$, and $1_{\mathcal{A}} = 1_{\mathcal{C}}$, $1_{\mathcal{B}} = 1_{\mathcal{C}}$. Let $\mathcal{M} \subseteq \mathcal{A}$ be a subspace such that $c_1 \cdot \mathcal{M} \cdot c_2 \subseteq \mathcal{M}$ for all c_1, c_2 in $\mathcal{C}$, and set

$$\mathcal{S} = \left\{ \begin{bmatrix} c_1 & a \\ b^* & c_2 \end{bmatrix} : a, b \in \mathcal{M},\ c_1, c_2 \in \mathcal{C} \right\}.$$

(i) Prove that if $\phi\colon \mathcal{M} \to \mathcal{B}$ is a completely contractive $\mathcal{C}$-bimodule map, then $\Phi\colon \mathcal{S} \to M_2(\mathcal{B})$ defined by

$$\Phi\left(\begin{bmatrix} c_1 & a \\ b^* & c_2 \end{bmatrix} \right) = \begin{bmatrix} c_1 & \phi(a) \\ \phi(b)^* & c_2 \end{bmatrix}$$

is completely positive.

(ii) Prove that if $\mathcal{B}$ is injective, then the conclusions of Theorems 8.2, 8.3, and 8.5 still hold with the additional requirement that the maps be $\mathcal{C}$-bimodule maps. (Hint: Use Theorem 3.18.)

8.7 Let $A = (a_{ij})_{i,j=1}^{\infty}$. Prove that the following are equivalent:

(i) $S_A\colon B(\ell^2) \to B(\ell^2)$ is positive,

(ii) $S_A\colon B(\ell^2) \to B(\ell^2)$ is completely positive,

(iii) there exists a Hilbert space $\mathcal{H}$ and a bounded sequence of vectors $\{x_i\}$ in $\mathcal{H}$ such that $a_{ij} = \langle x_j, x_i \rangle$.

8.8 Let $S_A\colon M_n \to M_n$ be the Schur product map $S_A(B) = A * B$. Let $A = H + iK$ be the decomposition of A into its real and imaginary parts.

(i) Show that $\mathrm{Re}(S_A) = S_H$.

(ii) Prove that there exist positive matrices P_1 and P_2 such that $\phi_i = S_{P_i}$, $i = 1, 2$, satisfy the conclusions of Theorem 8.3 for $\phi = S_A$. (Hint: Recall Exercise 4.4.)

(iii) Prove that P_1 and P_2 are such a pair of positive matrices if and only if $\|S_{P_1}\|_{\mathrm{cb}} = \|S_{P_2}\|_{\mathrm{cb}} = \|S_A\|_{\mathrm{cb}}$ and $\left[\begin{smallmatrix} P_1 & A \\ A^* & P_2 \end{smallmatrix}\right]$ is positive.

(iv) Let $\|S_A\|_{\mathrm{cb}} \le 1$, and write $A = (\langle x_j, y_i \rangle)$ as in Theorem 8.7. Show that $P_1 = (\langle y_j, y_i \rangle)$, $P_2 = (\langle x_j, x_i \rangle)$ are such a pair of positive matrices.

(v) Let $d(B)$ denote the maximum diagonal element of a matrix. Conclude that

$$\|S_A\|_{\mathrm{cb}} = \inf \left\{ d\begin{bmatrix} P_1 & A \\ A^* & P_2 \end{bmatrix} : \begin{bmatrix} P_1 & A \\ A^* & P_2 \end{bmatrix} \ge 0 \right\}.$$

(vi) Show that

$$\begin{bmatrix} |A^*| & A \\ A^* & |A| \end{bmatrix} \geq 0.$$

(vii) Give an example where $\|S_A\|_{cb} < \max\{d(|A^*|), d(|A|)\}$, with $A = A^*$.

8.9 (Haagerup) Let $\phi_1, \phi_2, \phi\colon \mathcal{A} \to \mathcal{B}$. Prove that $\Phi\colon M_2(\mathcal{A}) \to M_2(\mathcal{B})$, defined by

$$\Phi\begin{bmatrix} a & b \\ c & d \end{bmatrix} = \begin{bmatrix} \phi_1(a) & \phi(b) \\ \phi^*(c) & \phi_2(d) \end{bmatrix},$$

is completely positive if and only if $\Psi\colon \mathcal{A} \to M_2(\mathcal{B})$, defined by

$$\Psi(a) = \begin{bmatrix} \phi_1(a) & \phi(a) \\ \phi^*(a) & \phi_2(a) \end{bmatrix},$$

is completely positive.

8.10 Let $\Phi\colon M_2(M_n) \to M_2(\mathcal{B})$ be given by

$$\Phi\begin{bmatrix} a & b \\ c & d \end{bmatrix} = \begin{bmatrix} \phi^+(a) & \phi(b) \\ \phi^*(a) & \phi^-(d) \end{bmatrix},$$

and let $E = (E_{i,j})$ be in $M_n(M_n)$, where $E_{i,j}$ are the canonical matrix units. Prove that Φ is completely positive if and only if

$$\begin{bmatrix} \phi_n^+(E) & \phi(E) \\ \phi_n^*(E) & \phi_n^-(E) \end{bmatrix}$$

is positive.

8.11 Let $\phi\colon M_n \to \mathcal{B}$, set $B = (\phi(E_{i,j}))$ in $M_n(\mathcal{B})$, and let $|B^*| = (p_{i,j})$, $|B| = (q_{i,j})$. Define linear maps $\phi_1, \phi_2\colon M_n \to \mathcal{B}$ by $\phi_1(E_{i,j}) = p_{i,j}$, $\phi_2(E_{i,j}) = q_{i,j}$.

(i) Prove that $\Phi\colon M_2(M_n) \to M_2(\mathcal{B})$, defined by

$$\Phi\left[\begin{bmatrix} a & b \\ c & d \end{bmatrix}\right] = \begin{bmatrix} \phi_1(a) & \phi(b) \\ \phi^*(c) & \phi_2(d) \end{bmatrix},$$

is completely positive.

(ii) Prove that $\|\phi\|_{cb}^2 \leq \|p_{11} + \cdots + p_{nn}\|^2 \cdot \|q_{11} + \cdots + q_{nn}\|^2$, and that this estimate is sharp.

(iii) Let $B^*B = (c_{i,j})$. Prove that $\|c_{11} + \cdots + c_{nn}\| \leq n \cdot \|\phi\|_{\text{cb}}^2$, and that this estimate is sharp.

8.12 Prove directly that if $\phi: C(X) \to B(\mathcal{H})$ has finite total variation, then ϕ is completely bounded.

8.13 Let E be a completely bounded, operator-valued measure. Prove that E has finite total 2-variation, i.e., that $\sup\{\sum \||E(B_i)|^2\|: B_i$ disjoint, Borel$\} < +\infty$.

8.14 (Sakai) Let $\mathcal{A}$ be a C^*-algebra, and let $f: \mathcal{A} \to \mathbb{C}$ be a bounded, self-adjoint linear functional with $\|f\| \leq 1$. Show that $f = f_1 - f_2$, with f_1 and f_2 positive linear functionals and $\|f_1 + f_2\| = 1$.

8.15 Let $A = (a_{ij})_{i,j=1}^{\infty}$, and assume that S_A is bounded. Prove that if $\lim_i \lim_j a_{ij}$ and $\lim_j \lim_i a_{ij}$ both exist, then they must be equal. (Hint: Write $a_{ij} = \langle x_j, y_i \rangle$, and use the weak* compactness of the ball in Hilbert space.)

8.16 Prove that if for every $B = (b_{ij})$ on $B(\ell^2)$ the matrix

$$C = (c_{ij}) \qquad \text{where} \quad c_{ij} = \begin{cases} 0, & i > j, \\ b_{ij}, & i \leq j, \end{cases}$$

is in $B(\ell^2)$, then S_A is bounded, where

$$A = (a_{ij}), \qquad a_{ij} = \begin{cases} 0, & i > j \\ 1, & i \leq j \end{cases}.$$

Use Exercise 8.15 to show that S_A is not bounded, and deduce that there exists B in $B(\ell^2)$ for which C is unbounded.

8.17 Fix a subset $P \subseteq \{1, \ldots, n\} \times \{1, \ldots, n\}$, and define a subspace $\mathcal{M}_P \subseteq M_n$ via $B \in \mathcal{M}_P$ if and only if $B = (b_{ij})$ with $b_{ij} = 0$ whenever $(i, j) \notin P$. Show that $\mathcal{M} \subseteq M_n$ is a $\mathcal{D}_n$-bimodule if and only if $\mathcal{M} = \mathcal{M}_P$ for some P.

8.18 Let $P, \mathcal{M}_P$ be as in Exercise 8.17, and assume we are given complex numbers (a_{ij}) for every $(i, j) \in P$. Such a matrix is called *partially defined*. Show that there is a well-defined $\mathcal{D}_n$-bimodule map $\phi: \mathcal{M}_p \to \mathcal{M}_p$ given by $\phi((b_{ij})) = (a_{ij}b_{ij})$ and that every $\mathcal{D}_n$-bimodule map arises this way. Prove that $\|\phi\| = \|\phi\|_{\text{cb}} = \inf_{X,Y} \sup_{(i,j)\in P}\{\|x_j\|\|y_i\|\}$, where the infimum is over every set $X = \{x_j\}$, $Y = \{y_i\}$ satisfying $\langle x_j, y_i \rangle = a_{ij}$ for all $(i, j) \in P$. Given a partially defined matrix $A = (a_{ij})$ as above, a choice of the remaining undefined entries is called a *completion* of A.

8.19 Show that $\mathcal{M}_P$ is an operator system if and only if $(i, i) \in P$ for $1 \leq i \leq n$ and if $(i, j) \in P$ then $(j, i) \in P$. Let $\mathcal{M}_P$ be an operator system, and let $\phi: \mathcal{M}_P \to \mathcal{M}_P$ and $A = (a_{ij})$ be as in Exercise 8.18. Prove that the following are equivalent:

(i) ϕ is positive,

(ii) ϕ is completely positive,

(iii) A has a positive completion.

8.20 Say E has *finite total p-variation* if $\sup\{\sum \||E(B_i)|^p\|: B_i$ disjoint, Borel$\} < +\infty$. Does E have finite total 1-variation if and only if its associated map, $\phi: C(X) \to B(\mathcal{H})$, has finite total variation? If E is completely bounded, must E have finite total p-variation for all $p > 1$?

Chapter 9
Completely Bounded Homomorphisms

In Chapter 7 we saw that if $\mathcal{B}$ is a C^*-algebra with unit, and $\mathcal{A}$ is a subalgebra of $\mathcal{B}$ containing $1_{\mathcal{B}}$, then the unital, completely contractive homomorphisms of $\mathcal{A}$ into $B(\mathcal{H})$ are precisely the homomorphisms with $\mathcal{B}$-dilations. In this chapter, we prove that the unital homomorphisms of $\mathcal{A}$, which are similar to homomorphisms with $\mathcal{B}$-dilations, are precisely the completely bounded homomorphisms.

If $\mathcal{A}$ is a C^*-algebra, then every unital, contractive homomorphism is a positive map and hence a $*$-homomorphism. Thus, for unital maps of C^*-algebras, the sets of contractive homomorphisms, completely contractive homomorphisms, and $*$-homomorphisms coincide. In this case, the above result says that a unital homomorphism of a C^*-algebra is similar to a $*$-homomorphism if and only if it is a completely bounded homomorphism.

Let's begin with a simple observation. Suppose that S is a similarity, ρ is a homomorphism of some operator algebra $\mathcal{A}$, and $\pi(\cdot) = S^{-1}\rho(\cdot)S$ is a completely contractive homomorphism. Letting S_n denote the direct sum of n copies of S, we have that $S_n^{-1} = (S^{-1})_n$, $\|S_n\| = \|S\|$, and $\rho_n(\cdot) = S_n\pi_n(\cdot)S_n^{-1}$. Thus, ρ is completely bounded with $\|\rho\|_{\text{cb}} \leq \|S^{-1}\| \cdot \|S\|$. So any homomorphism that is similar to a completely contractive homomorphism is necessarily completely bounded with

$$\|\rho\|_{\text{cb}} \leq \inf\{\|S^{-1}\| \cdot \|S\| \colon S^{-1}\rho(\cdot)S \text{ is completely contractive}\}.$$

The next result proves that not only is the converse of the above statement true, but the above infimum is achieved and gives the cb norm of ρ.

Theorem 9.1. *Let $\mathcal{A}$ be an operator algebra, and let $\rho\colon \mathcal{A} \to B(\mathcal{H})$ be a unital, completely bounded homomorphism. Then there exists an invertible operator S with $\|S\| \cdot \|S^{-1}\| = \|\rho\|_{\text{cb}}$ such that $S^{-1}\rho(\cdot)S$ is a completely contractive* *homomorphism. Moreover,*

$$\|\rho\|_{\text{cb}} = \inf\{\|R^{-1}\| \cdot \|R\| \colon R^{-1}\rho(\cdot)R \textit{ is completely contractive}\}.$$

Proof. The remarks preceding the statement of the theorem yield the last equality.

Let $\mathcal{A}$ be contained in a C^*-algebra $\mathcal{B}$. By the extension and representation theorems for completely bounded maps, we know that there exists a Hilbert space $\mathcal{K}$, a $*$-homomorphism $\pi\colon \mathcal{B} \to B(\mathcal{K})$, and two bounded operators $V_i\colon \mathcal{H} \to \mathcal{K}$, $i = 1, 2$, with $\|\rho\|_{\text{cb}} = \|V_1\| \cdot \|V_2\|$, such that $\rho(a) = V_1^*\pi(a)V_2$ for all a in $\mathcal{A}$.

For h in $\mathcal{H}$, we define

$$|h| = \inf\left\{\left\|\sum \pi(a_i)V_2h_i\right\| \colon \sum \rho(a_i)h_i = h,\ a_i \in \mathcal{A},\ h_i \in \mathcal{H}\right\},$$

where the infimum is taken over all finite sums. It is easy to see that $|\cdot|$ defines a seminorm on $\mathcal{H}$.

If $h = \sum \rho(a_i)h_i$, then $\|h\| = \|\sum \rho(a_i)h_i\| = \|\sum V_1^*\pi(a_i)V_2h_i\| \leq \|V_1^*\| \cdot \|\sum \pi(a_i)V_2h_i\|$, and thus $\|h\| \leq \|V_1^*\| \cdot |h|$. Similarly, the equation $\rho(1)h = h$ yields $|h| \leq \|V_2\| \cdot \|h\|$.

Thus, $|\cdot|$ is equivalent to the original norm on $\mathcal{H}$. Hence $|\cdot|$ is also a norm on $\mathcal{H}$, and $(\mathcal{H}, |\cdot|)$ is complete.

We claim that $(\mathcal{H}, |\cdot|)$ is also a Hilbert space. By the Jordan–von Neumann theorem, it is enough to verify that $|\cdot|$ satisfies the parallelogram law

$$|h + k|^2 + |h - k|^2 = 2(|h|^2 + |k|^2)$$

for all h and k in $\mathcal{H}$.

Let $h = \sum \rho(a_i)h_i$, $k = \sum \rho(b_i)k_i$, then $h \pm k = \sum \rho(a_i)h_i \pm \sum \rho(b_i)k_i$, and so

$$\begin{aligned}|h + k|^2 + |h - k|^2 &\leq \left\|\sum \pi(a_i)V_2h_i + \sum \pi(b_i)V_2k_i\right\|^2 \\ &\quad + \left\|\sum \pi(a_i)V_2h_i - \sum \pi(b_i)V_2k_i\right\|^2 \\ &= 2\left\|\sum \pi(a_i)V_2h_i\right\|^2 + 2\left\|\sum \pi(b_i)V_2k_i\right\|^2.\end{aligned}$$

Taking the infimum over all such sums yields

$$|h + k|^2 + |h - k|^2 \leq 2|h|^2 + 2|k|^2.$$

The other inequality follows by substituting $h + k$ and $h - k$ for h and k, respectively. Hence, $(\mathcal{H}, |\cdot|)$ is also a Hilbert space.

Let $S\colon (\mathcal{H}, |\cdot|) \to (\mathcal{H}, \|\cdot\|)$ be the identity map. Then S is bounded and invertible, $\|S^{-1}\| \cdot \|S\| \le \|V_1^*\| \cdot \|V_2\| = \|\rho\|_{\mathrm{cb}}$, and $S^{-1}\rho(\cdot)S$ is just the homomorphism ρ, but with respect to the $|\cdot|$-norm. Thus, to complete the proof of the theorem, it is sufficient to prove that ρ is completely contractive with respect to this new norm. (To see this, let $U\colon (\mathcal{H}, \|\cdot\|) \to (\mathcal{H}, |\cdot|)$ be a unitary, and set $R = US$.)

It is not difficult to see that $\rho(\cdot)$ is contractive with respect to the $|\cdot|$-norm, for if $a \in \mathcal{A}$, $h \in \mathcal{H}$, and $h = \sum \rho(a_i)h_i$, then

$$|\rho(a)h| = \left|\sum \rho(aa_i)h_i\right| \le \left\|\sum \pi(aa_i)V_2h_i\right\| \le \|a\| \cdot \left\|\sum \pi(a_i)V_2h_i\right\|,$$

and so $|\rho(a)h| \le \|a\| \cdot |h|$.

To see that $\rho(\cdot)$ is completely contractive in the $|\cdot|$-norm, fix an integer n, and let $\hat{\mathcal{H}} = \mathcal{H} \oplus \cdots \oplus \mathcal{H}$ (n copies). Let $|\cdot|_n$ denote the Hilbert space norm on $\hat{\mathcal{H}}$ induced by $|\cdot|$, that is,

$$|\hat{h}|_n^2 = |h_1|^2 + \cdots + |h_n|^2$$

for $\hat{h} = (h_1, \ldots, h_n)$ in $\hat{\mathcal{H}}$. We must prove that if $A = (a_{i,j})$ is in $M_n(\mathcal{A})$, then

$$|\rho_n(A)\hat{h}|_n \le \|A\| \cdot |\hat{h}_n|.$$

Consider $\hat{\mathcal{H}}$ with its old norm,

$$\|h\|_n^2 = \|h_1\|^2 + \cdots + \|h_n\|^2.$$

Since ρ is completely bounded, $\rho_n\colon M_n(\mathcal{A}) \to B(\hat{\mathcal{H}}, \|\cdot\|_n)$ is a completely bounded homomorphism, and so, by the first part of the proof, we can endow $\hat{\mathcal{H}}$ with yet another norm $|\cdot|'$ such that ρ_n is contractive in the $|\cdot|'$-norm. By the first part of the proof, to define $|\cdot|'$, we need a Stinespring representation of ρ_n. To this end, let $\hat{\mathcal{K}} = \mathcal{K} \oplus \cdots \oplus \mathcal{K}$ (n copies), $\hat{V}_i = V_i \oplus \cdots \oplus V_i\colon \hat{\mathcal{H}} \to \hat{\mathcal{K}}$, $i = 1, 2$, and $\pi_n\colon M_n(\mathcal{A}) \to B(\hat{\mathcal{K}})$, so that $\rho_n(\cdot) = \hat{V}_1^*\pi_n(\cdot)\hat{V}_2$. If we define

$$|\hat{h}|' = \inf\left\{\left\|\sum \rho_n(A_i)\hat{V}_2\hat{h}_i\right\|_n : \sum \rho_n(A_i)\hat{h}_i = \hat{h}\right\},$$

where the infimum is taken over all finite sums, then $\rho_n(\cdot)$ will be contractive in $|\cdot|'$. The proof of the theorem is now completed by showing that $|\cdot|' = |\cdot|_n$, which we leave to the reader (Exercise 7.1). □

Corollary 9.2. (Haagerup). *Let $\mathcal{A}$ be a C^*-algebra with unit, and let ρ: $\mathcal{A} \to B(\mathcal{H})$ be a bounded, unital homomorphism. Then ρ is similar to a $*$-homomorphism if and only if ρ is completely bounded. Moreover, if ρ is completely bounded, then there exists a similarity S with $S^{-1}\rho(\cdot)S$ a $*$-homomorphism and $\|S^{-1}\| \cdot \|S\| = \|\rho\|_{\mathrm{cb}}$.*

The above result sheds considerable light on a conjecture of Kadison [128]. *Kadison's conjecture* is that every bounded homomorphism of a C^*-algebra into $B(\mathcal{H})$ is similar to a $*$-homomorphism. Haagerup's result shows that Kadison's conjecture is equivalent to determining whether or not bounded homomorphisms are necessarily completely bounded. At the present time the similarity question is still open, although there are a number of deep partial results (see [12], [35], [55], and [106]). In Chapter 19 we will return to this question and study the recent contributions of Pisier to this problem.

There are several important cases where bounded homomorphisms of a C^*-algebra are completely bounded. These are closely tied to the theory of group representations. Let G be a locally compact, topological group, and let $L^\infty(G)$ denote the C^*-algebra of bounded, measurable functions on G. Given g in G, we define the (right) translation operators $R_g\colon L^\infty(G) \to L^\infty(G)$ by $(R_g f)(g') = f(g'g)$. A state $m\colon L^\infty(G)^* \to \mathbb{C}$ is called a (right) *invariant mean* on G if $m(R_g f) = m(f)$ for all g in G. The group G is called *amenable* if there exists an invariant mean on G.

Note that if G is compact, then it is amenable, since we may define

$$m(f) = \int_G f(g)\,dg,$$

where dg denotes Haar measure.

Another important class of groups that are amenable are the commutative groups. To see this, note that the adjoint maps, $R_g^*\colon L^\infty(G)^* \to L^\infty(G)^*$ are weak*-continuous and map states to states. Since the space of states is weak*-compact and convex, and since the maps R_g^* form a commutative group of continuous maps on this space, by the Markov–Kakutani fixed point theorem [78] there will exist a fixed point. This fixed point is easily seen to be an invariant mean.

The importance of the existence of an invariant mean is best illustrated in the following result.

Theorem 9.3 (Dixmier). *Let G be an amenable group, and let $\rho\colon G \to B(\mathcal{H})$ be a strongly continuous homomorphism with $\rho(e) = 1$, such that $\|\rho\| = \sup\{\|\rho(g)\|\colon g \in G\}$ is finite. Then there exists an invertible S in $B(\mathcal{H})$ with $\|S\| \cdot \|S^{-1}\| \le \|\rho\|^2$ such that $S^{-1}\rho(g)S$ is a unitary representation of G.*

Proof. Let m denote an invariant mean on G, and note that for each pair of vectors x, y in $\mathcal{H}$, the function $f_{x,y}(g) = \langle \rho(g)x, \rho(g)y\rangle$ is a bounded, continuous function on G. Set $\langle x, y\rangle_1 = m(f_{x,y})$, and note that since the map $(x, y) \to f_{x,y}$ is sesquilinear, $\langle , \rangle_1$ defines a sesquilinear form on $\mathcal{H}$. Also, since m is a positive linear functional, and $f_{x,x}$ is a positive function, this new sesquilinear form is a semidefinite inner product on $\mathcal{H}$.

We shall show that it is bounded above and below by the original inner product. To see this, set $M = \|\rho\|$ and note that since $\|\rho(g)\| \leq M$ and $\|\rho(g)^{-1}\| \leq M$, we have that

$$1/M^2 \leq \rho(g)^*\rho(g) \leq M^2.$$

Hence, $(1/M^2)\langle x, x\rangle \leq f_{x,x}(g) \leq M^2\langle x, x\rangle$, and applying m to this inequality yields

$$\frac{1}{M^2}\langle x, x\rangle \leq \langle x, x\rangle_1 \leq M^2\langle x, x\rangle.$$

Thus, $\langle x, y\rangle_1$ is an equivalent inner product on $\mathcal{H}$.

If we let $(\mathcal{H}, |\cdot|_1)$ denote our Hilbert space with the norm induced by this new inner product, then just as in the proof of Theorem 9.1, the map $S\colon (\mathcal{H}, \|\cdot\|) \to (\mathcal{H}, |\cdot|_1)$ is bounded and invertible, with $\|S\| \cdot \|S^{-1}\| \leq M^2 = \|\rho\|^2$.

Finally, note that with respect to this new inner product,

$$\langle \rho(g)x, \rho(g)y\rangle_1 = m(R_g f_{x,y}) = m(f_{x,y}) = \langle x, y\rangle_1,$$

and so $\rho(g)$ is a unitary on $(\mathcal{H}, |\cdot|_1)$. □

Corollary 9.4. (Sz.-Nagy). *Let T be an invertible operator on a Hilbert space such that $\|T^n\| \leq M$ for all integers n. Then there exists an invertible operator S, with $\|S^{-1}\| \cdot \|S\| \leq M^2$, such that $S^{-1}TS$ is a unitary operator.*

Proof. Let $\mathbb{Z}$ denote the group of integers, define $\rho(n) = T^n$, and apply Theorem 9.3. □

In a number of cases, similarity results for homomorphisms of C^*-algebras can be obtained from Theorem 9.3 by restricting attention to the group of unitaries in the C^*-algebra. The following result demonstrates this principle.

Lemma 9.5. *Let $\mathcal{A}$ be a C^*-algebra. Then every element in $\mathcal{A}$ is a linear combination of at most four unitaries.*

Proof. Let $h = h^*$ be a self-adjoint element of $\mathcal{A}$, with $\|h\| \leq 1$. Then $u = h + i\sqrt{1 - h^2}$ is easily seen to be unitary, and $h = u + u^*$. This shows that every self-adjoint element is in the span of two unitaries. Using the Cartesian decomposition, we obtain the result. □

Lemma 9.6. *Let $\mathcal{A}$ and $\mathcal{B}$ be C^*-algebras, and let $\rho\colon \mathcal{A} \to \mathcal{B}$ be a homomorphism with $\rho(1) = 1$. If ρ maps unitaries to unitaries, then ρ is a $*$-homomorphism.*

Proof. Let u be unitary in $\mathcal{A}$; then $\rho(u^*) = \rho(u)^{-1} = \rho(u^{-1}) = \rho(u)^*$, and so ρ is self-adjoint on the unitary elements of $\mathcal{A}$. By Lemma 9.5, ρ is self-adjoint on $\mathcal{A}$. □

Theorem 9.7. *Let $\mathcal{A}$ be a commutative, unital C^*-algebra. If $\rho\colon \mathcal{A} \to B(\mathcal{H})$ is a bounded homomorphism, then ρ is completely bounded and $\|\rho\|_{\mathrm{cb}} \leq \|\rho\|^2$.*

Proof. Let G denote the commutative group of unitary elements of $\mathcal{A}$. By Theorem 9.3, there is a similarity S, with $\|S\| \cdot \|S^{-1}\| \leq \|\rho\|^2$, such that $S^{-1}\rho(u)S$ is unitary for all u in G. By Lemma 9.6, $S^{-1}\rho(\cdot)S$ is a $*$-homomorphism. □

Corollary 9.8. *Let T be an operator on a Hilbert space. Then T is similar to a self-adjoint operator if and only if for some interval, $[a, b]$, there is a constant K such that*

$$\|p(T)\| \leq K \cdot \sup\{|p(t)|\colon t \in [a, b]\}$$

for all polynomials p with real coefficients.

By Theorem 9.3 and Lemma 9.6, if $\mathcal{A}$ is a C^*-algebra, G is an amenable group contained in the group of unitaries of $\mathcal{A}$, and $\mathcal{B}$ is the C^*-subalgebra generated by G, then for any bounded homomorphism $\rho\colon \mathcal{A} \to B(\mathcal{H})$ there is an invertible S, with $\|S\|\|S^{-1}\| \leq \|\rho\|^2$ such that $S^{-1}\rho(\cdot)S$ is a $*$-homomorphism on $\mathcal{B}$.

An affirmative answer to Kadison's conjecture for C^*-algebras is known to imply an affirmative answer to two other questions concerning *derivations* and *invariant operator ranges.* Consequently, these problems have an analogous status, they have affirmative answers if and only if certain bounded maps are completely bounded, but at the present time, both questions are still open. We discuss the derivation question in this chapter. For a discussion of invariant operator ranges and their connections with completely bounded maps, see [158].

Let $\mathcal{A}$ be a C^*-algebra and $\pi\colon \mathcal{A} \to B(\mathcal{H})$ a unital $*$-homomorphism. A linear map $\delta\colon \mathcal{A} \to B(\mathcal{H})$ is called a *derivation* if $\delta(AB) = \pi(A)\delta(B) + \delta(A)\pi(B)$. It is known that every derivation is automatically bounded [199]. If $X \in B(\mathcal{H})$, then setting $\delta(A) = \pi(A)X - X\pi(A)$ defines a derivation, and such a derivation is called *inner*. The derivation question asks whether every derivation into $B(\mathcal{H})$ is necessarily inner.

Given a derivation δ, if we define $\rho\colon \mathcal{A} \to B(\mathcal{H} \oplus \mathcal{H})$ by

$$\rho(A) = \begin{bmatrix} \pi(A) & \delta(A) \\ 0 & \pi(A) \end{bmatrix},$$

then ρ will be a homomorphism of $\mathcal{A}$. In fact, it is not difficult to see that ρ is a homomorphism if and only if δ is a derivation.

Proposition 9.9. *The derivation δ is inner if and only if ρ is similar to a $*$-homomorphism.*

Proof. First, suppose that δ is inner, so that $\delta(A) = \pi(A)X - X\pi(A)$. Setting

$$S = \begin{bmatrix} 1 & X \\ 0 & 1 \end{bmatrix}, \qquad S^{-1} = \begin{bmatrix} 1 & -X \\ 0 & 1 \end{bmatrix},$$

we have that

$$S\rho(A)S^{-1} = \begin{bmatrix} \pi(A) & 0 \\ 0 & \pi(A) \end{bmatrix},$$

which is a $*$-homomorphism.

Conversely, if S is a similarity such that $\gamma(A) = S^{-1}\rho(A)S$ is a $*$-homomorphism, then set $X = SS^*$. We have that

$$\rho(A)X = S\gamma(A)S^* = (S\gamma(A^*)S^*)^* = (\rho(A^*)SS^*)^* = X\rho(A^*)^*.$$

Writing

$$X = \begin{bmatrix} X_{11} & X_{12} \\ X_{12}^* & X_{22} \end{bmatrix},$$

the above equation becomes

$$\begin{bmatrix} \pi(A)X_{11} + \delta(A)X_{12}^* & \pi(A)X_{12} + \delta(A)X_{22} \\ \pi(A)X_{12}^* & \pi(A)X_{22} \end{bmatrix}$$
$$= \begin{bmatrix} X_{11}\pi(A) + X_{12}\delta(A^*)^* & X_{12}\pi(A) \\ X_{12}^*\pi(A) + X_{22}\delta(A^*)^* & X_{22}\pi(A) \end{bmatrix}.$$

But since X is positive and invertible, X_{22} must also be positive and invertible. Equating the (1,2) entries of the above operator matrices yields

$$\begin{aligned} \delta(A) &= \delta(A)X_{22}X_{22}^{-1} = (X_{12}\pi(A) - \pi(A)X_{12})X_{22}^{-1} \\ &= X_{12}X_{22}^{-1}\pi(A) - \pi(A)X_{12}X_{22}^{-1}, \end{aligned}$$

since X_{22} commutes with A. □

Corollary 9.10 (Christensen). *Let $\mathcal{A}$ be a unital C^*-algebra, $\pi\colon \mathcal{A} \to B(\mathcal{H})$ a unital $*$-homomorphism, and $\delta\colon \mathcal{A} \to B(\mathcal{H})$ a derivation. Then δ is inner if and only if δ is completely bounded.*

Proof. By using the canonical shuffle, it is easy to see that ρ is completely bounded if and only if δ is completely bounded. But by Proposition 9.9 and Corollary 9.2, δ is inner if and only if ρ is completely bounded. □

Thus, for a given C^*-algebra $\mathcal{A}$, we see that if every bounded homomorphism of $\mathcal{A}$ is similar to a $*$-homomorphism, then every derivation of $\mathcal{A}$ is inner. Kirchberg [132] has proven the converse of this result. Namely, if a C^*-algebra has the property that every derivation into $B(\mathcal{H})$ is inner, then necessarily every bounded homomorphism is similar to a $*$-homomorphism. Thus, for C^*-algebras, Kadison's conjecture and the problem of determining if every derivation from a C^*-algebra into the bounded operators on some Hilbert space is inner are equivalent.

Turning our attention to some non-self-adjoint algebras, Theorem 9.1 can be used to give a characterization of the operators that are similar to a contraction. Let $\mathcal{P}(\mathbb{D})$ denote the algebra of polynomials with the norm it inherits as a subalgebra of $C(\mathbb{T})$.

Theorem 9.11. *Let $T \in B(\mathcal{H})$. Then T is similar to a contraction if and only if the homomorphism $\rho\colon \mathcal{P}(\mathbb{D}) \to B(\mathcal{H})$ defined by $\rho(p) = p(T)$ is completely bounded. Moreover, if this is the case, then*

$$\|\rho\|_{\mathrm{cb}} = \inf\{\|S\| \cdot \|S^{-1}\|\colon \|S^{-1}TS\| \leq 1\},$$

and the infimum is attained.

Proof. If ρ is completely bounded, then there is a similarity S, with $\|S^{-1}\| \cdot \|S\| = \|\rho\|_{\mathrm{cb}}$, such that $S^{-1}\rho(\cdot)S$ is completely contractive. Thus,

$$\|S^{-1}TS\| = \|S^{-1}\rho(z)S\| \leq \|z\| = 1,$$

where z is the coordinate function.

Conversely, if $R = S^{-1}TS$ is a contraction, then $\theta\colon \mathcal{P}(\mathbb{D}) \to B(\mathcal{H})$, given by $\theta(p) = p(R)$, is a contractive homomorphism by von Neumann's inequality, and is completely contractive by Sz.-Nagy's dilation theorem. But $\rho(\cdot) = S\theta(\cdot)S^{-1}$, and so ρ is completely bounded with

$$\|\rho\|_{\mathrm{cb}} \leq \|S\| \cdot \|S^{-1}\|.$$

The statement about the infimum follows from the corresponding statement in Theorem 9.1. □

An operator for which $\rho(p) = p(T)$ is a bounded homomorphism of $\mathcal{P}(\mathbb{D})$, that is, an operator for which the closed unit disk is a K-spectral set, is called a *polynomially bounded* operator. It is quite natural to refer to the operators for which this ρ is completely bounded as *completely polynomially bounded.* Halmos [111] conjectured that every polynomially bounded operator is similar to a contraction. The above result shows that *Halmos's conjecture* is equivalent to determining whether or not bounded homomorphisms of $\mathcal{P}(\mathbb{D})$, or equivalently $A(\mathbb{D})$, are completely bounded. Pisier [191] produced an example, in fact, a whole family of examples, of a polynomially bounded operator that is not completely polynomially bounded. Thus proving that Halmos's conjecture is not true. We shall present these examples in Chapter 10.

Holbrook [122] gave the first example of a matrix T, which is similar to a contraction such that

$$\|\rho\| < \inf\{\|S^{-1}\| \cdot \|S\|: \|S^{-1}TS\| \leq 1\},$$

where $\rho(p) = p(T)$ is the induced homomorphism of $\mathcal{P}(\mathbb{D})$. The theory of completely bounded maps at least gives a basis for explaining such phenomena, since the right-hand side of the above equation is $\|\rho\|_{cb}$ and, in general, there is no reason to expect equality of these two norms. Using the theory of completely bounded maps, there are now quite good estimates on how large the ratio of $\|\rho\|_{cb}$ to $\|\rho\|$ can be when T is an $n \times n$ matrix. See for example [193].

The above theory sheds further light on the theory of K-spectral sets. Let X be a compact set in $\mathbb{C}$, let $\mathcal{R}(X)$ be the algebra of quotients of polynomials, and let T be in $B(\mathcal{H})$ with $\sigma(T)$ contained in X. Recall that if there is a bounded homomorphism $\rho\colon \mathcal{R}(X) \to B(\mathcal{H})$ defined by $\rho(r) = r(T)$ with $\|\rho\| \leq K$, then X is called a *K-spectral set* for T. If ρ is completely bounded with $\|\rho\|_{cb} \leq K$, we shall call X a *complete K-spectral set* for T. Recall also that X is a complete spectral set for T if and only if T has a normal ∂X-dilation. Translating Theorem 9.1 into this language yields the following result.

Corollary 9.12. *A set X is a complete K-spectral set for T if and only if T is similar to an operator for which X is a complete spectral set. Moreover, there exists such a similarity satisfying $\|S^{-1}\| \cdot \|S\| \leq K$.*

In fact, we have that for ρ as above,

$$\|\rho\|_{cb} = \inf\{\|S^{-1}\| \cdot \|S\|: X \text{ is a complete spectral set for } S^{-1}TS\},$$

and the infimum is attained.

The full picture of how these concepts are related is still not complete. As discussed earlier, it is still not known, if a subset X of $\mathbb{C}$ is a spectral set for an operator, whether or not X must be a complete spectral set for the operator. However, we saw in Chapter 5 that higher-dimensional analogues of this question, when say X is a polydisk, can fail. It is known that if a set is a K-spectral set for an operator, then it need not be a complete K'-spectral set for the operator, for any K'. Halmos's conjecture was the special case of this question where $X = \mathbb{D}^-$, and Pisier's counterexample shows that the above fails even on this simple domain. However, there exist sets for which it is the case that whenever the set is a K-spectral set for an operator, then it is a complete K'-spectral set for some K'. For example, when $\mathcal{R}(X)$ is dense in $C(\partial X)$, then this is the case. But it is unknown if this is the only situation where this occurs.

Maps which are defined by some type of analytic functional calculus are often completely bounded. Recall the *Riesz functional calculus*. If G is an open set containing $\sigma(T)$ and ∂G consists of a finite number of simple, closed, rectifiable curves, then for any function in $\mathcal{R}(G)$, we have that

$$r(T) = \frac{1}{2\pi i}\int_{\partial G} r(z)(zI - T)^{-1}\,dz.$$

This functional calculus can be used to derive what is referred to as the generalized *Rota model* of an operator.

Theorem 9.13 (Herrero–Voiculescu). *Let T be in $B(\mathcal{H})$, and let G be an open set in the complex plane such that $\sigma(T)$ is contained in G and such that ∂G consists of a finite number of simple, closed, rectifiable curves. Then T is similar to an operator that has a normal ∂G-dilation. Moreover, the similarity may be chosen to satisfy*

$$\|S^{-1}\| \cdot \|S\| \le \frac{1}{2\pi}\int_{\partial G} \|(zI - T)^{-1}\|\,|dz|.$$

Proof. By Corollary 9.12, we need only prove that G^- is a complete K-spectral set for T with

$$K = \frac{1}{2\pi}\int_{\partial G} \|(zI - T)^{-1}\|\,|dz|.$$

Let $\rho\colon \mathcal{R}(G^-) \to B(\mathcal{H})$ be the homomorphism defined by $\rho(r) = r(T)$. We must prove that $\|\rho\|_{\mathrm{cb}} \le K$.

By the Riesz functional calculus,

$$\|\rho(r)\| \le \frac{1}{2\pi}\int_{\partial G} \|r(z)\cdot(zI - T)^{-1}\|\,|dz| \le K \cdot \|r\|,$$

so $\|\rho\| \le K$. Let $\hat{T}$ denote the direct sum of n copies of T and note that

$\|(z\hat{I} - \hat{T})^{-1}\| = \|(zI - T)^{-1}\|$. Thus, for $(r_{i,j})$ in $M_n(\mathcal{R}(G^-))$, we have that

$$\begin{aligned} \rho_n((r_{i,j})) &= \frac{1}{2\pi i}\int_{\partial G}(r_{i,j}(z)\cdot(zI-T)^{-1})\,dz \\ &= \frac{1}{2\pi i}\int_{\partial G}(r_{i,j}(z))\cdot(z\hat{I}-\hat{T})^{-1}\,dz, \end{aligned}$$

and so, by the same inequality as above, $\|\rho_n\| \leq K$, which proves that G^- is a complete K-spectral set for T. □

Corollary 9.14. (Rota's Theorem). *Let T be in $B(\mathcal{H})$ with $\sigma(T)$ contained in the open unit disk. Then T is similar to a contraction. Moreover, a similarity may be chosen satisfying*

$$\|S^{-1}\|\cdot\|S\| \leq \frac{1}{2\pi}\int_{\mathbb{T}}\|(zI-T)^{-1}\|\,|dz|.$$

Proof. By Corollary 9.12, T is similar to an operator with a unitary dilation, and that operator is necessarily a contraction. □

The original proof of Rota's theorem gives much more particular information on the unitary dilation and similarity than this proof does. In fact, the unitary can always be taken to be the bilateral shift, and the invertible S can be chosen to belong to the C^*-algebra generated by T (see, for example, [97] or Exercise 9.15). One difference is that the above proof can give quite different estimates on $\|S^{-1}\|\cdot\|S\|$.

Recall the spectral radius formula,

$$\sup\{|z|: z\in\sigma(T)\} = \lim_n \|T^n\|^{1/n}.$$

So if $\sigma(T)$ is contained in $\mathbb{D}$, then

$$(e^{i\theta}I - T)^{-1} = e^{-i\theta}\sum\nolimits_n (e^{-i\theta}T)^n,$$

and the latter series converges by the root test. Thus, we see that the integral estimate on $\|S^{-1}\|\cdot\|S\|$ obtained in Corollary 9.14 satisfies

$$\frac{1}{2\pi}\int_{\mathbb{T}}\|(zI-T)^{-1}\|\,|dz| \leq \sum\nolimits_n \|T^n\|.$$

The proof of Rota's theorem in Exercise 9.15 leads to an upper bound on $\|S\|\cdot\|S^{-1}\|$ of $\sum_n \|T^n\|^2$.

Notes

Kadison's study of homomorphisms of C^*-algebras [128] was motivated in part by Dixmier's study of whether or not bounded representations of groups are similar to unitary representations [72]. Dixmier's question is known to have a negative answer, that is, there exist bounded representations of groups that are not similar to unitary representations. This result was obtained by Kunze and Stein [133]. The original proof of Sz.-Nagy's result characterizing operators that are similar to unitaries (Corollary 9.8) contains the elements of Dixmier's result on amenable groups (Theorem 9.3). This proof is outlined in the exercises (Exercise 9.7) and can be found in [226].

Hadwin [109] proved that a unital homomorphism of a C^*-algebra into $B(\mathcal{H})$ is similar to a $*$-homomorphism if and only if the homomorphism is in the span of the completely positive maps from the algebra to $B(\mathcal{H})$, and conjectured that the span of the completely positive maps was the completely bounded maps. At the same time, Wittstock [242] proved his decomposition theorem, which verifies this conjecture.

Independently, Haagerup [106] proved directly that a unital homomorphism of a C^*-algebra into $B(\mathcal{H})$ is similar to a $*$-homomorphism if and only if it is completely bounded, and moreover, that there exists a similarity S such that $\|S^{-1}\| \cdot \|S\|$ is equal to the cb norm of the homomorphism. It is not clear if this sharp equality can be obtained by combining the results of Hadwin and Wittstock. Haagerup applied his result to obtain the analogous characterization of inner derivations (Corollary 9.10). The result on inner derivations had been obtained earlier by Christensen [54].

In [159], the extension theorem for completely bounded maps into $B(\mathcal{H})$ and the generalization of Stinespring's representation theorem were proven in order to extend the techniques of Hadwin and Wittstock to operator algebras. It was proven that a unital homomorphism of an operator algebra into $B(\mathcal{H})$ is similar to a completely contractive homomorphism if and only if it is completely bounded. The prime motivation was to prove that an operator is similar to a contraction if and only if it is completely polynomially bounded (Theorem 9.11), which had been conjectured by Arveson. The extension theorem had been obtained earlier by Wittstock [243].

The fact that the similarity could be chosen such that $\|S\| \cdot \|S^{-1}\|$ is equal to the cb norm of the homomorphism for a general operator algebra (Theorem 9.1) was obtained later [160]. The key new technique can be found in a paper of Holbrook [120].

The problem of characterizing the operators that are similar to contractions has been considered by Sz.-Nagy. After obtaining his characterization of the operators that are similar to unitaries, he conjectured that an operator T was similar to a contraction if and only if $\|T^n\|$ was uniformly bounded for all positive integers n. Such an operator is called *power-bounded*. Foguel [98] (see also Halmos [110]) gave an example of a power-bounded operator that is not similar to a contraction, and Halmos conjectured that the right condition was polynomially bounded.

Exercises

9.1 Verify the claim of Theorem 9.1, that $|\cdot|' = |\cdot|_n$.

9.2 Let $\mathcal{A}$ be an algebra (not necessarily unital), and let $\rho\colon \mathcal{A} \to B(\mathcal{H})$ be a homomorphism. If $\rho(\mathcal{A})\mathcal{H}$ is dense in $\mathcal{H}$, then ρ is called *nondegenerate*. Show that if $\mathcal{A}$ is unital, then ρ is nondegenerate if and only if $\rho(1) = 1$.

9.3 Let P be in $B(\mathcal{H})$ such that $P^2 = P$.

(i) Show that relative to some decomposition of $\mathcal{H}$,

$$P = \begin{bmatrix} 1 & X \\ 0 & 0 \end{bmatrix}.$$

(ii) Show that if

$$S = \begin{bmatrix} 1 & X \\ 0 & 1 \end{bmatrix},$$

then SPS^{-1} is an orthogonal projection.

(iii) Define $\rho\colon \mathbb{C} \oplus \mathbb{C} \to B(\mathcal{H})$ via $\rho(\lambda_1, \lambda_2) = \lambda_1 P + \lambda_2(1 - P)$, and show that ρ is a completely bounded homomorphism, but that in general, $\|\rho\|_{\text{cb}} < \|S^{-1}\| \cdot \|S\|$.

(iv) Compute $\|\rho\|_{\text{cb}}$.

(v) Can you construct an operator R such that $\|R\|\|R^{-1}\| = \|\rho\|_{\text{cb}}$ and RPR^{-1} is an orthogonal projection?

9.4 Let $\mathcal{B}$ be a C^*-algebra with unit, let $\mathcal{A} \subseteq \mathcal{B}$ be a subalgebra that does not contain the identity of $\mathcal{B}$, and let $\rho\colon \mathcal{A} \to B(\mathcal{H})$ be a homomorphism. Set $\mathcal{A}_1 = \{a + \lambda 1\colon a \in \mathcal{A},\ \lambda \in \mathbb{C}\}$, and define $\rho_1\colon \mathcal{A}_1 \to B(\mathcal{H})$ by $\rho_1(a + \lambda 1) = \rho(a) + \lambda \cdot 1_{\mathcal{H}}$. Prove that ρ_1 is a completely bounded homomorphism if and only if ρ is a completely bounded homomorphism.

9.5 Let $\mathcal{A}$ be a unital operator algebra, and let $\rho\colon \mathcal{A} \to B(\mathcal{H})$ be a bounded homomorphism with $\rho(1) = I$.

(i) Show that if there exists x and y in $\mathcal{H}$ such that $\rho(\mathcal{A})x = \mathcal{H}$ and $\rho(\mathcal{A})^* y = \mathcal{H}$, then ρ is completely bounded.

(ii) Show that if there exists $x_1, \ldots, x_n$ and $y_1, \ldots, y_m$ in $\mathcal{H}$ such that $\rho(\mathcal{A})x_1 + \cdots + \rho(\mathcal{A})x_n = \mathcal{H}$ and $\rho(\mathcal{A})^* y_1 + \cdots \rho(\mathcal{A})^* y_m = \mathcal{H}$, then ρ is completely bounded.

A subalgebra $\mathcal{B}$ of $B(\mathcal{H})$ with a vector x such that $\mathcal{B}x = \mathcal{H}$ is called a strictly cyclic algebra. When there exists a finite set of vectors such that $\mathcal{B}x_1 + \cdots + \mathcal{B}x_n = \mathcal{H}$, then $\mathcal{B}$ is called strictly multicyclic.

9.6 Prove that if T is a $n \times n$ matrix and X is a K-spectral set for T, then X is a complete (nK)-spectral set for T. Characterize those matrices for which $\mathbb{D}^-$ is a K-spectral set for some K in terms of their Jordan form.

9.7 (Sz.-Nagy) This exercise gives a more direct proof of Corollary 9.4. Let T be an invertible operator on $\mathcal{H}$ such that $\|T^n\| \leq M$ for all integers n, and let glim be a Banach generalized limit [34].

(i) Show that $\langle x, y\rangle_1 = \text{glim}\langle T^n x, T^n y\rangle$ defines a new inner product on $\mathcal{H}$ and that $(1/M^2)\langle x, x\rangle \leq \langle x, x\rangle_1 \leq M^2\langle x, x\rangle$.

(ii) Show that T is a unitary transformation on $(\mathcal{H}, \langle , \rangle_1)$.

(iii) Prove that there exists a similarity S on $\mathcal{H}$, with $\|S^{-1}\| \cdot \|S\| \leq M^2$, such that $S^{-1}TS$ is a unitary.

9.8 Prove that a finite direct sum of operators $T_1 \oplus \cdots \oplus T_n$ is similar to a contraction if and only if each operator in the direct sum is similar to a contraction. Moreover, prove that

$$\inf\{\|S\| \cdot \|S^{-1}\| \colon \|S^{-1}(T_1 \oplus \cdots \oplus T_n)S\| \leq 1\}$$

is achieved by an operator S that is itself a direct sum.

9.9 Let T_1, T_2, T_3 be the operators of Parrott's example, and let ρ be the contractive homomorphism of $\mathcal{P}(\mathbb{D}^3)$ defined by $\rho(z_i) = T_i$. Prove that ρ is completely bounded.

9.10 (Sz.-Nagy–Foias) An operator T in $B(\mathcal{H})$ is said to belong to class $\mathcal{C}_\rho$ if there exists a Hilbert space $\mathcal{K}$ containing $\mathcal{H}$, and a unitary U on $\mathcal{K}$, such that

$$T^n = \rho P_{\mathcal{H}} U^n\big|_{\mathcal{H}}$$

for all positive integers n. Prove that such a T is completely polynomially bounded and that there exists an invertible operator S such that $S^{-1}TS$ is a contraction with $\|S\| \cdot \|S^{-1}\| \leq 2\rho - 1$ when $\rho \geq 1$.

9.11 Let X be a compact set in the complex plane such that $\mathcal{R}(X)$ is dense in $C(\partial X)$. Prove that X is a K-spectral set for some operator T if and only if T is similar to a normal operator and $\sigma(T)$ is contained in ∂X. Show that the similarity can always be chosen such that $\|S\| \cdot \|S^{-1}\| \leq K^2$.

9.12 Let T be an operator with real spectrum. Prove that T is similar to a self-adjoint operator if and only if the Cayley transform of T, $C = (T + i)(T - i)^{-1}$, has the property that for some constant M, $\|C^n\| \leq M$ for all integers n.

9.13 Let $\mathcal{P}$ be a family of (not necessarily self-adjoint) commuting projections on a Hilbert space $\mathcal{H}$ satisfying:

(a) If $P \in \mathcal{P}$, then $(1 - P) \in \mathcal{P}$.

(b) If $P, Q \in \mathcal{P}$, then $PQ \in \mathcal{P}$.

(c) If $P, Q \in \mathcal{P}$, $PQ = 0$, then $P + Q \in \mathcal{P}$.

(d) $\|\mathcal{P}\| \doteq \sup\{\|P\| \colon P \in \mathcal{P}\}$ is finite.

(i) For $P, Q \in \mathcal{P}$, set $P \triangle Q = 1 - P - Q + 2PQ$, and show that this is an element of $\mathcal{P}$. We call $P \triangle Q$ the *symmetric difference* of P and Q.

(ii) Prove that $(\mathcal{P}, \triangle)$ is a commutative group.

(iii) Prove that $\rho(P) = 2P - 1$ defines a representation of this group and that $\|\rho\| \leq 2\|\mathcal{P}\|$.

(iv) Prove that there exists a similarity S with $\|S\| \cdot \|S^{-1}\| \leq 4\|\mathcal{P}\|^2$ such that $S^{-1}PS$ is a self-adjoint projection for all P in $\mathcal{P}$.

(v) Use (iv) to give an alternative proof of the fact that every homomorphism of a commutative, unital C^*-algebra is similar to a $*$-homomorphism. What estimate can you obtain on $\|S\| \cdot \|S^{-1}\|$?

9.14 Let $\mathcal{A}$ be a finite-dimensional C^*-algebra, and let $\rho\colon \mathcal{A} \to B(\mathcal{H})$ be a homomorphism, $\rho(1) = 1$. Show that ρ is completely bounded and that $\|\rho\|_{\text{cb}} \leq \|\rho\|^2$.

9.15 Let T be in $B(\mathcal{H})$ with $\sigma(T)$ contained in the open unit disk. Prove that $P = \sum_{k=0}^{\infty} T^{*k}T^k$ is a norm-convergent series with $\|P^{1/2}TP^{-1/2}\| \leq 1$.

Chapter 10

Polynomially Bounded and Power-Bounded Operators

Polynomially bounded and power-bounded operators have played an important role in the development of this area, and there are a number of interesting results, counterexamples, and open questions about these operators. In particular, we will present Foguel's example [98] of a power-bounded operator and Pisier's example [191] of a polynomially bounded operator that are not similar to contractions.

Recall that an operator T is *power-bounded* provided that there is a constant M such that $\|T^n\| \leq M$ for all $n \geq 0$. Clearly, if $T = S^{-1}CS$ with C a contraction, then T is power-bounded with $\|T^n\| \leq \|S^{-1}\|\|S\|$.

It is fairly easy to see (Exercise 10.1), by using the Jordan form, that a matrix $T \in M_n$ is power-bounded if and only if it is similar to a contraction. Sz.-Nagy [229] proved that the same characterization holds when T is a compact operator. This led naturally to the conjecture that an arbitrary operator is similar to a contraction if and only if it is power-bounded. Foguel provided the first example of a power-bounded operator that is not similar to a contraction.

Recall that an operator is *polynomially bounded* provided there is a constant K such that $\|p(T)\| \leq K\|p\|_\infty$ for every polynomial p, where the ∞-norm is the supremum norm over the unit disk. By von Neumann's inequality, if $T = S^{-1}CS$ with C a contraction, then T is polynomially bounded with $K = \|S^{-1}\|\|S\|$. So it was natural to conjecture that an operator is similar to a contraction if and only if it is polynomially bounded. Lebow [136] proved that Foguel's operator was not polynomially bounded, so that it could not serve as a counterexample to this stronger conjecture. This conjecture, in fact, became one of Halmos's [111] "ten problems" and remained unsolved for about 25 years until Pisier provided a counterexample.

Foguel's counterexample and Pisier's counterexample share some common features, and a number of researchers have studied operators in this general family, so we shall present some general results about operators of what are often called "Foguel type."

We begin with Sz.-Nagy's result. The reader should also recall Corollary 9.4.

Theorem 10.1 (Sz.-Nagy). *Let T be a compact operator on a Hilbert space. Then T is similar to a contraction if and only if T is power-bounded.*

Proof. Clearly every operator similar to a contraction is power bounded. If T is power bounded, say $\|T^n\| \leq M$, then by the spectral radius formula, we have that $r(T) = \lim_n \|T^n\|^{1/2} \leq \lim_n M^{1/n} = 1$. Since T is compact, there can be at most a finite number of points in the spectrum of T on the unit circle.

If one encloses these by a curve and integrates the resolvent of T about this curve, then by standard facts from the Riesz functional calculus [68], one obtains an idempotent E that commutes with T such that $\sigma(ET|_{E(\mathcal{H})}) = \sigma(T) \cap \mathbb{T}$, $\sigma((I - E)T|_{(I-E)(\mathcal{H})}) \subseteq \mathbb{D}$.

If we choose an invertible S such that $S^{-1}ES = P$ is an orthogonal projection, then P commutes with $S^{-1}TS$ and hence reduces it. Thus, $S^{-1}TS = T_1 \oplus T_2$, where $\sigma(T_1) = \sigma(T) \cap \mathbb{T}$, $\sigma(T_2) \subseteq \mathbb{D}$, and T_1 acts on a finite-dimensional space. Since $S^{-1}TS$ is power-bounded, T_1 is power-bounded and hence by Exercise 10.1 there exists S_1 such that $S_1^{-1}T_1S_1$ is a contraction. By Rota's theorem (Corollary 9.14) there exists an invertible S_2 such that $S_2^{-1}T_2S_2$ is a contraction.

Setting $R = S(S_1 \oplus S_2)$ we have that $R^{-1}TR$ is a contraction. □

We now focus on the types of operators arising in Foguel's and Pisier's counterexamples. Let $S_{\mathcal{H}}: \ell^2(\mathcal{H}) \to \ell^2(\mathcal{H})$ be the forward shift, so that $S_{\mathcal{H}}((h_0, h_1, \dots)) = (0, h_0, h_1, \dots)$, with adjoint $S_{\mathcal{H}}^*((h_0, h_1, \dots)) = (h_1, h_2, \dots)$.

Let $X = (A_{i,j}) \in B(\ell^2(\mathcal{H}))$ where $A_{ij} \in B(\mathcal{H})$ for $i, j \geq 0$.

We shall call an operator of the form

$$F = \begin{pmatrix} S_{\mathcal{H}}^* & X \\ 0 & S_{\mathcal{H}} \end{pmatrix} \quad \text{on} \quad \ell^2(\mathcal{H}) \oplus \ell^2(\mathcal{H})$$

a *Foguel operator over $\mathcal{H}$ with symbol X*. Of special interest will be when $X = (A_{i+j})$ has the Hankel form, and we call these *Foguel–Hankel operators over $\mathcal{H}$ with symbol X*. To simplify notation we generally write S for $S_{\mathcal{H}}$.

It is easily seen that for $n \geq 2$,

$$F^n = \begin{pmatrix} S^{*n} & X_n \\ 0 & S^n \end{pmatrix}, \qquad \text{where} \quad X_n = \sum_{j=0}^{n-1} S^{*j} X S^{n-1-j}.$$

Setting $X_0 = 0$, $X_1 = X$, the formula holds for all $n \geq 0$.

Note that when X has the Hankel form, then $S^*X = XS$ and consequently $X_n = nXS^{n-1} = n(A_{i+j+n-1})$.

Since $\|S^{*n}\| = \|S^n\| = 1$ for all $n \geq 0$, we see that X is power-bounded if and only if $\sup_n \|X_n\| \leq M$ for some M.

Let $\mathcal{P}$ as usual denote the space of polynomials, and define a linear map $\delta\colon \mathcal{P} \to B(\ell^2(\mathcal{H}))$ by setting $\delta(z^n) = X_n$ and extending linearly. When X has the Hankel form, we have $\delta(p) = Xp'(S)$, where p' is the derivative of p. We now have that

$$p(F) = \begin{pmatrix} p(S^*) & \delta(p) \\ 0 & p(S^n) \end{pmatrix},$$

and since $\|p(S^*)\| = \|p(S^n)\| = \|p\|_\infty$, F is polynomially bounded if and only if δ is a bounded linear map.

Analogously, applying the canonical shuffle, we have that F is completely polynomially bounded if and only if δ is a completely bounded linear map.

One final property that is useful is that δ obeys a certain derivation property. Indeed,

$$\begin{aligned}\begin{pmatrix} (p\cdot q)(S^*) & \delta(pq) \\ 0 & (p\cdot q)(S) \end{pmatrix} &= (p\cdot q)(F) = p(F)q(F) \\ &= \begin{pmatrix} p(S^*) & \delta(p) \\ 0 & p(S) \end{pmatrix}\begin{pmatrix} q(S^*) & \delta(q) \\ 0 & q(S) \end{pmatrix} \\ &= \begin{pmatrix} p(S^*)q(S^*) & p(S^*)\delta(q) + \delta(p)q(S) \\ 0 & p(S)q(S) \end{pmatrix}\end{aligned}$$

and we have

$$\delta(pq) = p(S^*)\delta(q) + \delta(p)q(S)$$

for any polynomials p and q.

We begin with Pisier's counterexample. Our presentation follows [71]. We start with a family of Foguel–Hankel operators that are easy to analyze.

To this end, set $\mathcal{H} = \ell_2$, and let $\{E_{i,j}\}_{i,j=0}^{+\infty}$ denote the usual matrix units regarded as elements of $B(\ell_2)$. To simplify notation we set $E_i = E_{i0}$. Since $E_i^* E_j = \delta_{ij} E_0$, we see that for any vector $h = (c_0, c_1, \dots) \in \ell^2$, the operator $T(h) = \sum_{i=0}^\infty c_i E_i$ is bounded and in fact

$$\|T(h)\|^2 = \|T(h)^*T(h)\| = \left\|\sum \bar{c}_i c_j E_i^* E_j\right\| = \left\|\sum |c_i|^2 E_0\right\| = \|h\|^2.$$

Thus, $T\colon \ell^2 \to B(\ell^2)$ defines an isometric linear map that identifies a vector with the infinite column matrix.

There is one subtle distinction to be made. For while ℓ^2 is a Hilbert space, $T(\ell^2)$ is an operator space. In particular, given vectors (v_{ij}) in ℓ^2, we have a well-defined norm for $(T(v_{ij}))$. To compute its norm one simply substitutes the corresponding column matrix for each vector. We shall encounter the operator space $T(\ell^2)$ again in Chapter 14, where it is called *column Hilbert space.*

Theorem 10.2. *Fix a sequence $\{a_n\}_{n=0}^{+\infty}$, and consider the Foguel–Hankel operator F over ℓ^2 with symbol $X = (a_{i+j}E_{i+j})$. Then the following are equivalent:*

(i) F is similar to a contraction,
(ii) F is polynomially bounded,
(iii) F is power-bounded,
(iv) $\sup_n n(\sum_{k=n-1}^{\infty} |a_k|^2)$ is finite.

Proof. Clearly, (i) implies (ii), and (ii) implies (iii). Recall that F is power-bounded if and only if $\sup_n \|X_n\|$ is finite, where

$$X_n = nXS^{n-1} = n(a_{i+j+n-1}E_{i+j+n-1}).$$

We have that

$$\begin{aligned} X_n^*X_n &= n^2\left(\sum_{k=0}^{\infty} \bar{a}_{i+k+n-1}a_{j+k+n-1}E_{i+k+n-1}^*E_{j+k+n-1}\right) \\ &= n^2\left(\sum_{k=0}^{\infty} \bar{a}_{i+k+n-1}a_{j+k+n-1}\delta_{ij}E_0\right). \end{aligned}$$

Thus, $X_n^*X_n$ is a diagonal operator matrix whose diagonal entries are

$$n^2\sum_{k=0}^{\infty} |a_{i+k+n-1}|^2 = n^2 \sum_{k=i+n-1}^{\infty} |a_k|^2.$$

Thus, $\|X_n^*X_n\| = n^2\sum_{k=n-1}^{\infty} |a_k|^2$, from which the equivalence of (iii) and (iv) follows.

Finally, to prove that (iv) implies (i), we consider the operator matrix $Y = (ja_{i+j-1}E_{i+j-1})$ where we set $E_{-1} = 0$. Assume for the moment that Y is bounded. We have that

$$\begin{aligned} YS - S^*Y &= \big((j+1)a_{i+(j+1)-1}E_{i+(j+1)-1}\big) - \big(ja_{(i+1)+j-1}E_{(i+1)+j-1}\big) \\ &= (a_{i+j}E_{i+j}) = X. \end{aligned}$$

Thus, if we set $R = \left(\begin{smallmatrix} I & Y \\ 0 & I \end{smallmatrix}\right)$, then $R^{-1} = \left(\begin{smallmatrix} I & -Y \\ 0 & I \end{smallmatrix}\right)$ and we have

$$\begin{aligned} R^{-1}FR &= \begin{pmatrix} I & -Y \\ 0 & I \end{pmatrix} \begin{pmatrix} S^* & X \\ 0 & S \end{pmatrix} \begin{pmatrix} I & +Y \\ 0 & I \end{pmatrix} \\ &= \begin{pmatrix} S^* & X - YS + S^*Y \\ 0 & S \end{pmatrix} = \begin{pmatrix} S^* & 0 \\ 0 & S \end{pmatrix}. \end{aligned}$$

Hence, we have that F is similar to a contraction.

It remains to prove that Y is bounded. Computing

$$\begin{aligned} Y^*Y &= (i\bar{a}_{i+j-1}E^*_{i+j-1})(ja_{i+j-1}E_{i+j-1}) \\ &= \left(\sum_{k=0}^{\infty} ij\bar{a}_{i+k-1}a_{k+j-1}E^*_{i+k-1}E_{k+j-1} \right), \end{aligned}$$

we see that Y^*Y is a diagonal operator matrix whose diagonal entries are $\sum_{k=0}^{\infty} i^2|a_{i+k-1}|^2 E_0$. Thus, $\|Y^*Y\| = \sup_n n^2 \sum_{k=0}^{\infty} |a_{k+n-1}|^2$, and so (iv) implies (i). □

Theorem 10.2 shows that the above family of operators could not possibly provide an example of a polynomially bounded operator that is not similar to a contraction. However, to obtain Pisier's counterexample we will swap the above sequence of operators $\{E_i\}$ for a different sequence $\{W_i\}$. The following propositions will allow us to prove that statements (ii), (iii), and (iv) of Theorem 10.2 are still equivalent for the Foguel–Hankel operators obtained from this new sequence of operators, but their equivalence to (i) will no longer hold.

Proposition 10.3. *Let $\mathcal{H}$ and $\mathcal{K}$ be Hilbert spaces, and let $\Phi\colon B(\mathcal{H}) \to B(\mathcal{K})$ be a bounded linear map. If $(A_{i+j}) = A$ is a bounded Hankel operator on $\ell^2(\mathcal{H})$, then $A_\Phi = (\Phi(A_{i+j}))$ is a bounded Hankel operator on $\ell^2(\mathcal{K})$. In fact, $\|A_\Phi\| \leq \|\Phi\|\|A\|$.*

Proof. By the Nehari–Page theorem (Theorem 5.10) there exists a sequence $\{A_n\}_{n=-\infty}^{-1}$ such that

$$\|A\| = \sup_{r<1} \left\| \sum_{n=-\infty}^{+\infty} r^{|n|}e^{in\theta}A_n \right\|_\infty .$$

Hence, $\sup_{r<1} \| \sum_{n=-\infty}^{+\infty} r^{|n|} e^{in\theta} \Phi(A_n)\|_\infty \leq \|\Phi\| \|A\|$. Applying the Nehari–Page theorem again, we have that

$$\|A_\Phi\| = \inf_B \sup_{r<1} \left\| \sum_{n=-\infty}^{-1} r^{|n|} e^{in\theta} B_n + \sum_{n=0}^{\infty} r^{|n|} e^{in\theta} \Phi(A_n) \right\|_\infty ,$$

where the infimum is over all sequences $B = \{B_n\}_{n=-\infty}^{-1}$ in $B(\mathcal{K})$. Hence, $\|A_\Phi\| \leq \|\Phi\| \|A\|$. □

The above result shows that even though Φ may not be a completely bounded map, for Hankel matrices it acts completely bounded.

Proposition 10.4. *Let $\mathcal{H}$ and $\mathcal{K}$ be Hilbert spaces, and let Φ: $B(\mathcal{H}) \to B(\mathcal{K})$ be a bounded linear map. Let F be a Foguel–Hankel operator over $\mathcal{H}$ with symbol $X = (A_{i+j})$, and let F_Φ be the corresponding Foguel–Hankel operator over $\mathcal{K}$ with symbol $X_\Phi = (\Phi(A_{i+j}))$. Then:*

(i) F power-bounded implies F_Φ power-bounded,
(ii) F polynomially bounded implies F_Φ polynomially bounded.

Proof. We only prove (ii); the proof of (i) is similar. Recall that F is polynomially bounded if and only if the map δ: $\mathcal{P} \to B(\ell^2(\mathcal{H}))\delta(p) = Xp'(S)$ is bounded. For the operator F_Φ, we need to consider the map $\delta_\Phi(p) = X_\Phi p'(S)$. But $Xp'(S) = (B_{i+j})$ is a Hankel operator matrix, and $X_\Phi p'(S) = (\Phi(B_{i+j}))$. Hence,

$$\|\delta_\Phi(p)\| \leq \|\Phi\| \|(B_{i+j})\| \leq \|\Phi\| \|\delta\| \|p\|_\infty.$$

Thus, δ_Φ is a bounded map with $\|\delta_\Phi\| \leq \|\Phi\| \|\delta\|$, and hence F_Φ is polynomially bounded. □

The analogous conclusion for completely polynomially bounded operators is false, and that will be the source of our counterexamples.

Our construction uses a particular choice of a sequence of operators related to the *CAR operators*. The reader already familiar with sequences of operators satisfying the CAR and their properties might wish to skip directly to Theorem 10.5 and substitute any CAR sequence for the operators $\{W_i\}$ occurring in Theorem 10.5.

A sequence $\{C_n\}_{n=0}^{\infty}$ of operators on a Hilbert space $\mathcal{H}$ is said to satisfy the *canonical anticommutation relations* (CAR) provided

(CAR a) $$C_i C_j + C_j C_i = 0$$

and

(CAR b) $$C_iC_j^* + C_j^*C_i = \delta_{ij}I$$

for all i and j, where δ_{ij} denotes the Kronecker delta.

Given such a sequence and $h = (\alpha_0, \alpha_1, \dots)$ in ℓ^2, set $\Lambda(h) = \sum_{i=0}^{\infty} \alpha_i C_i$. This defines a bounded operator with $\|\Lambda(h)\|^2 \leq \|h\|^2$, since

$$\Lambda(h)\Lambda(h)^* + \Lambda(h)^*\Lambda(h) = \sum_{i,j=0}^{\infty} \alpha_i\bar{\alpha}_j(C_jC_j^* + C_j^*C_i) = \sum_{i=0}^{\infty} |\alpha_i|^2 \cdot I,$$

by (CAR b). However, if we let $P = \Lambda(h)^*\Lambda(h)$, $Q = \Lambda(h)\Lambda(h)^*$, then $\|P\| = \|Q\|$ and $PQ = \Lambda(h)^*\Lambda(h)^2\Lambda(h)^* = 0$, since $\Lambda(h)^2 = 0$ by (CAR a). Hence $\|P\| = \|Q\| = \|P + Q\| = \sum_{i=0}^{\infty} |\alpha_i|^2$, and it follows that $\|\Lambda(h)\| = \|h\|$. Thus, Λ: $\ell^2 \to B(\mathcal{H})$ is an isometry, and $\Lambda(\ell^2)$ is another operator space, isometrically isomorphic to Hilbert space.

To construct a sequence of operators satisfying the CAR, we first need a finite sequence of operators on a finite-dimensional space that satisfy them.

To construct these operators we begin with three 2×2 matrices,

$$V = \begin{bmatrix} 1 & 0 \\ 0 & -1 \end{bmatrix}, \quad C = \begin{bmatrix} 0 & 0 \\ 1 & 0 \end{bmatrix}, \quad \text{and} \quad I_2 = \begin{bmatrix} 1 & 0 \\ 0 & 1 \end{bmatrix}.$$

We have that $V^2 = I$, $C^2 = 0$, $C^*C = E_{11}$, $CC^* = E_{22}$, $VC = -C$, and $CV = C$. For $0 \leq i \leq n-1$, we define matrices in $M_2 \otimes \cdots \otimes M_2 \cong M_{2^n}$ by setting

$$C_i = V^{\otimes(i)} \otimes C \otimes I_2^{\otimes(n-i-1)},$$

where $X^{\otimes(k)}$ denotes the tensor product of X with itself k times. Note that

(1) $C_i^2 = (V^2)^{\otimes(i)} \otimes (C^2) \otimes I_2^{\otimes(n-i-1)} = 0,$
(2) $C_i^*C_i = (V^*V)^{\otimes(i)} \otimes (C^*C) \otimes I_2^{\otimes(n-i-1)} = I_2^{\otimes(i)} \otimes E_{11} \otimes I_i^{\otimes(n-i-1)}$, and
(3) $C_iC_i^* = I_2^{\otimes(i)} \otimes E_{22} \otimes I_2^{\otimes(n-i-1)},$

while for $0 \leq i < j \leq n-1$, we have

(4) $C_iC_j = I_2^{\otimes(i)} \otimes C \otimes V^{\otimes(j-i-1)} \otimes C \otimes I_2^{\otimes(n-j-1)} = -C_jC_i,$
(5) $C_iC_j^* = I_2^{\otimes(i)} \otimes C \otimes V^{\otimes(j-i-1)} \otimes C^* \otimes I_2^{\otimes(n-j-1)} = -C_j^*C_i.$

Now (1) and (4) are seen to imply (CAR a), while (2), (3), and (5) yield (CAR b).

Given $h = (\alpha_0, \dots, \alpha_{n-1})$, set $\Lambda(h) = \sum_{i=0}^{n-1} \alpha_i C_i$. From the above relations we have that Λ is an isometry from n-dimensional Hilbert space into M_{2^n}, and $\{C_i\}$ span a Hilbert space.

Next we show that $\{C_i \otimes C_i\}$ nearly span an ℓ^1-space, that is, given scalars $a_0, \dots, a_{n-1}$ we will show that

(6) $\frac{1}{2}\sum_{i=0}^{n-1}|a_i| \le \|\sum_{i=0}^{n-1} a_i C_i \otimes C_i\| \le \sum_{i=0}^{n-1}|a_i|$.

Now up to a unitary equivalence, we have that $C_i \otimes C_i \cong (V \otimes V)^{\otimes(i)} \otimes (C \otimes C) \otimes I_4^{\otimes(n-i-1)}$ in $M_4 \otimes \cdots \otimes M_4 = M_{4^n}$. Let e_1, e_2 denote the basis for $\mathbb{C}^2$, and choose w_i, $1 \le i \le n$, constants of modulus one such that $a_i\bar{w}_i = |a_i|$. Then $x_i = (e_1 \otimes e_1 + w_i e_2 \otimes e_2)/\sqrt{2}$ will be a unit vector in $\mathbb{C}^4$, and $x = x_1 \otimes \cdots \otimes x_n$ will be a unit vector in $\mathbb{C}^{4^n}$. Since $V \otimes V(e_i \otimes e_i) = e_i \otimes e_i$, $C \otimes C(e_1 \otimes e_1) = e_2 \otimes e_2$, and $C \otimes C(e_2 \otimes e_2) = 0$, we have that

$$\left\langle \sum_{i=0}^{n-1} a_i C_i \otimes C_i x, x \right\rangle = \frac{1}{2}\sum_{i=0}^{n-1} a_i \bar{w}_i = \frac{1}{2}\sum_{i=0}^{n-1}|a_i|.$$

Thus, $\|\sum_{i=0}^{n-1} a_i C_i \otimes C_i\| \ge \frac{1}{2}\sum_{i=0}^{n-1}|a_i|$, and since $\|C_i \otimes C_i\| = 1$, the other inequality follows.

So far we have only constructed, for each n, a finite sequence $C_0, \dots, C_{n-1}$ of finite matrices satisfying the CAR. To obtain an infinite sequence one generally uses an infinite tensor product of Hilbert spaces.

To avoid discussions of infinite tensor products we use a slightly different approach. Relabel the $2^n \times 2^n$ matrices constructed above as $C_{0,n}, \dots, C_{n-1,n}$, and for $i \ge n$ set $C_{i,n} = 0$. Define bounded operators on $\sum_{n=1}^{\infty} \oplus \mathbb{C}^{2^n}$ by setting $W_i = \sum_{n=1}^{\infty} \oplus C_{i,n}$. It is readily verified that

(i) $\|\sum_{i=0}^{\infty} \alpha_i W_i\|^2 = \sum_{i=0}^{\infty}|\alpha_i|^2$ for any $(\alpha_0, \alpha_1, \dots)$ in ℓ^2,
(ii) $\frac{1}{2}\sum_{i=0}^{n-1}|a_i| \le \|\sum_{i=0}^{n-1} a_i C_{i,n} \otimes W_i\| \le \sum_{i=0}^{n-1}|a_i|$ for any $(a_0, \dots, a_{n-1})$,
(iii) $W_i W_j + W_j W_i = 0$,
(iv) $W_i W_j^* + W_j^* W_i = \delta_{ij}(I - P_i)$,

where P_i is the finite-rank projection onto $\sum_{n=1}^{i} \oplus \mathbb{C}^{2^n}$. Thus, although $\{W_i\}$ do not quite satisfy the CAR, their images in the C^*-algebra obtained by quotienting out the ideal of compact operators do satisfy them.

Theorem 10.5 (Pisier). *Fix a sequence $\{a_n\}_{n=0}^{\infty}$, and consider the Foguel–Hankel operator F over $\mathcal{H}$ with symbol $X = (a_{i+j} W_{i+j})$, where $\{W_i\}_{i=0}^{\infty}$ are the operators constructed above. Then the following are equivalent:*

(i) F is polynomially bounded,
(ii) F is power-bounded,
(iii) $\sup_n n\,(\sum_{k=n-1}^{\infty}|a_k|^2)$ is finite.

However, if $\sum_{k=0}^{\infty}(k+1)^2|a_k|^2$ is infinite, then F is not similar to a contraction.

Proof. Clearly (i) implies (ii). To see that (ii) implies (iii), recall that F is power-bounded if and only if $\sup_n \|X_n\|$ is finite, where $X_n = nXS^{n-1} = n(a_{i+j+n-1}W_{i+j+n-1})$. We have that

$$X_n^*X_n + X_nX_n^*$$
$$= n^2 \sum_{k=0}^{\infty} \bar{a}_{i+k+n-1}a_{j+k+n-1}[W_{i+k+n-1}^*W_{j+k+n-1} + W_{i+k+n-1}W_{j+k+n-1}^*].$$

By (iii), the off-diagonal terms are 0, and hence the norm of this operator matrix is just the largest norm of a diagonal entry. Since each diagonal entry is a sum of positive operators, the supremum occurs when $i = 0$. Thus,

$$\|X_n^*X_n + X_nX_n^*\| = n^2 \left\| \sum_{k=0}^{\infty} |a_{k+n-1}|^2 W_{k+n-1}^*W_{k+n-1} + W_{k+n-1}W_{k+n-1}^* \right\|$$
$$= \sup_m n^2 \left\| \sum_{k=0}^{\infty} |a_{k+n-1}|^2 (C_{k+n-1,m}^*C_{k+n-1,m} + C_{k+n-1,m}C_{k+n-1,m}^*) \right\|$$
$$= n^2 \sum_{k=0}^{\infty} |a_{k+n-1}|^2,$$

using the fact that $C_{k,m}^*C_{k,m} + C_{k,m}C_{k,m}^* = I$ for $k < m$. Hence, $\sup_n \|X_n\|$ finite implies that $\sup_n \|X_n^*X_n + X_nX_n^*\| = \sup_n n^2 \sum_{k=0}^{\infty} |a_{k+n-1}|^2$ is finite.

To prove that (iii) implies (i), we use Proposition 10.4. Define $\Phi\colon B(\ell^2) \to B(\mathcal{H})$ by $\Phi((a_{i,j})) = \sum_{i=0}^{\infty} a_{i,0}W_i$. Since the norm of a matrix is larger than the norm of any column, we have that $\|\Phi((a_{i,j}))\|^2 = \sum_{i=0}^{\infty} |a_{i,0}|^2 \leq \|(a_{ij})\|^2$ and hence $\|\Phi\| \leq 1$.

Assuming $\sup_n n \sum_{k=0}^{\infty} |a_{k+n-1}|^2$ is finite, by Theorem 10.2 we have that the Foguel–Hankel operator with symbol $(a_{i+j}E_{i+j})$ is polynomially bounded, and hence by Proposition 10.4, the Foguel–Hankel operator with symbol $(a_{i+j}\Phi(E_{i+j})) = (a_{i+j}W_{i+j})$ is polynomially bounded.

Finally, to see that F is not similar to a contraction when $\sum_{k=0}^{\infty}(k+1)^2|a_k|^2$ is infinite, it is sufficient to show that the map $\delta\colon \mathcal{P} \to B(\ell^2(\mathcal{H}))$ is not completely bounded for this operator. Note that the (0,0) entry of $\delta(z^{i+1})$ is $(i+1)a_iW_i$. Thus, if we can show that the map $\delta_0\colon \mathcal{P} \to B(\mathcal{H})$, $\delta_0(1) = 0$, $\delta_0(z^{i+1}) = (i+1)$ a_iW_i, is not completely bounded, then δ will fail to be completely bounded, too.

If we consider the $2^n \times 2^n$ matrix-valued polynomial $P(z) = \sum_{i=0}^{n}(i+1)$ $\bar{a}_iC_{i,n}z^{i+1}$, then $\delta_0^{(2^n)}(P) = \sum_{i=0}^{n}(i+1)^2|a_i|^2C_{i,n} \otimes W_i$. By the definition of

W_i and (6) for the operators $C_{i,n}$ we have that

$$\|\delta_0^{(2^n)}(P)\| \geq \left\| \sum_{i=0}^{n} (i+1)^2 |a_i|^2 C_{i,n} \otimes C_{i,n} \right\| \geq \frac{1}{2} \sum_{i=0}^{n} (i+1)^2 |a_i|^2.$$

Since Λ is an isometry, we have that $\|P\|_\infty = (\sum_{i=0}^{n} (i+1)^2 |a_i|^2)^{1/2}$. Hence, $\|\delta_0\|_{\mathrm{cb}} \geq \frac{1}{2} (\sum_{i=0}^{n} (i+1)^2 |a_i|^2)^{1/2}$, and letting n tend to infinity we have that δ_0 is not completely bounded. □

The above result leads us to Pisier's counterexample to Halmos's conjecture.

Corollary 10.6. (Pisier). *Let $a_{2^k-1} = 2^{-k}$ and $a_i = 0$ otherwise. Then the Foguel–Hankel operator with symbol $(a_{i+j} W_{i+j})$, where the W's are defined as above, is polynomially bounded, but not similar to a contraction.*

Proof. We have that $\sup_n (n+1) \sum_{k=n-1}^{\infty} |a_k|^2 = \sup_j 2^j \sum_{k=j}^{\infty} (2^{-k})^2 = 4/3$, while

$$\sum_{k=0}^{\infty} (k+1)^2 |a_k|^2 = \sum_{j=0}^{\infty} (2^j)^2 (2^{-j})^2 = +\infty,$$

and so we are done by Theorem 10.5. □

For another example, one can set $a_k = (k+1)^{-3/2}$. Then

$$\sup_n (n+1) \left(\sum_{k=n}^{\infty} |a_k|^2 \right)^{1/2} \leq \sqrt{2}$$

but

$$\sum_{k=1}^{\infty} (k+1)^2 |a_k|^2 = +\infty.$$

The same results as above hold when the sequence $\{W_n\}$ is replaced by an actual CAR sequence $\{C_n\}$.

Necessary and sufficient conditions on the sequence $\{a_n\}$ for the Foguel–Hankel operator with symbol $(a_{i+j} C_{i+j})$ to be similar to a contraction were only recently obtained. In [197], it is shown that $\sum_{k=0}^{\infty} (k+1)^2 |a_k|^2$ finite is a necessary and sufficient condition for similarity to a contraction. Earlier [11] it had been shown that if $\sum_{k=1}^{\infty} (k+1)^2 [\log(\log(k+1))]^{2+\epsilon} |a_k|^2$ is finite for any $\epsilon > 0$, then one has similarity to a contraction.

We now turn our attention to the problem of exhibiting a power-bounded operator that is not polynomially bounded. Examples of such operators generally rely on some nontrivial results about $A(\mathbb{D})$. The work of Peller [174] shows that there exist Foguel–Hankel operators with scalar symbol that are power-bounded but not polynomially bounded. However, partly for historical reasons and partly to minimize the amount of function theory that is needed, we shall present Foguel's operator [98], which was the first known example of an operator that is power-bounded, but not similar to a contraction. Later, Lebow [136] proved that Foguel's operator is not even polynomially bounded, and we reproduce this result in Theorem 10.9.

To understand Foguel's example, it is best if we start by looking at the Foguel operator

$$F_r = \begin{pmatrix} S^* & E_{r,r} \\ 0 & S \end{pmatrix},$$

where $E_{r,r}$ denotes the standard matrix unit. We have that

$$F_r^n = \begin{pmatrix} S^{*n} & X_{n,r} \\ 0 & S^n \end{pmatrix},$$

where

$$X_{n,r} = \sum_{k=0}^{n-1} S^{*k} E_{r,r} S^{n-1-k} = \sum_{k=0}^{n-1} E_{r-k,r-n+k+1},$$

and we have adopted the convention that $E_{i,j}$ denotes the standard matrix units provided that $i, j \geq 0$ and is 0 otherwise.

Note that $X_{n,r} = X_{n,r}^*$ is a matrix all of whose entries are 0 or 1 and that the 1's occur along part of the antidiagonal where $i + j = 2r - n + 1$. For $n > 2r + 1$, we have $X_{n,r} = 0$ and, in general, $X_{n,r}^2 = \sum_{\ell=r-n}^{2r+1-n} E_{\ell,\ell}$. Thus, we see that $\|X_{n,r}\| \leq 1$ for all n and hence F_r is power-bounded.

Given r and t, compute

$$X_{n,r} X_{n,t} = \sum_{k,\ell=0}^{n} E_{r-k,r-n+k+1} E_{t-n+\ell+1,t-\ell}.$$

We see that each term in this product is 0 unless $r + k = t + \ell \geq n - 1$, $r \geq k$, and $t \geq \ell$. These inequalities imply that $2r \geq t$ and $2t \geq r$. Thus, if $2r + 1 < t$ or $2t + 1 < r$ then $X_{n,r} X_{n,t} = 0$.

Theorem 10.7 (Foguel). *Let $\{k_\ell\}$ be a sequence of integers satisfying $2k_\ell + 1 < k_{\ell+1}$, and let $X = \sum_{\ell=1}^{\infty} E_{k_\ell,k_\ell}$. Then the operator $F = \begin{pmatrix} S^* & X \\ 0 & S \end{pmatrix}$ is power-bounded.*

Proof. We have that

$$F^n = \begin{pmatrix} S^{*n} & X_n \\ 0 & S^n \end{pmatrix}, \quad \text{where} \quad X_n = \sum_{\ell=1}^{\infty} X_{n,k_\ell}.$$

Since either $2k_\ell + 1 < k_m$ or $2k_m + 1 < k_\ell$ for $\ell \neq m$, we have that

$$X_n^* X_n = \sum_{\ell=1}^{\infty} X_{n,k_\ell}^* X_{n,k_\ell} = \sum_{\ell=1}^{\infty} \sum_{j=k_\ell-n}^{2k_\ell+1-n} E_{j,j}.$$

Because the intervals $\{[k_\ell - n, 2k_\ell + 1 - n]\}_{\ell=1}^{\infty}$ are all disjoint, we have that $X_n^* X_n \leq 1$. □

To prove that the operators of the above type are not polynomially bounded, we shall need a result from function theory.

Theorem 10.8. *Let $\{k_\ell\}$ be a sequence of integers satisfying $2k_\ell + 1 < k_{\ell+1}$, and given $f(z) = \sum_{k=0}^{\infty} a_k z^k$ in $A(\mathbb{D})$, define $\Gamma(f) = (a_{k_1}, a_{k_2}, \ldots)$. Then $\Gamma\colon A(\mathbb{D}) \to \ell^2$ is contractive and onto.*

Proof. Since $(\sum_{k=0}^{\infty} |a_k|^2)^{1/2} = \|f\|_2 \leq \|f\|_\infty$, Γ is contractive. Consider the matrix $B = \left(\begin{smallmatrix} 1 & -\bar{a} \\ a & 1 \end{smallmatrix}\right)$. Note that

$$B^* B = \begin{pmatrix} 1 + |a|^2 & 0 \\ 0 & 1 + |a|^2 \end{pmatrix}$$

and so we have that $\|B\| = \sqrt{1 + |a|^2}$.

Fix $h = (h_1, \ldots, h_n, 0, \ldots)$, a finitely supported vector in ℓ^2 with $\|h\|_2 = 1$, and let

$$B_\ell(e^{i\theta}) = \begin{pmatrix} 1 & -\bar{h}_\ell e^{-ik_\ell\theta} \\ h_\ell e^{ik_\ell\theta} & 1 \end{pmatrix} \qquad \text{for} \quad 1 \leq \ell \leq n.$$

Then by the above calculation, we will have that $\|B_\ell\|_\infty = \sup\{\|B_\ell(e^{i\theta})\|\colon 0 \leq \theta \leq 2\pi\} = \sqrt{1 + |h_\ell|^2}$. For each m, $1 \leq m \leq n$, let

$$F_m(e^{i\theta}) = B_1(e^{i\theta}) \cdots B_m(e^{i\theta});$$

then $\|F_m\|_\infty^2 \leq \prod_{\ell=1}^{m}(1 + |h_\ell|^2) \leq \exp(\sum_{\ell=1}^{n} |h_\ell|^2) = e$.

By induction (Exercise 10.3) one shows that

$$F_m(e^{i\theta}) = \begin{pmatrix} g_m(e^{i\theta}) & p_m(e^{-i\theta}) \\ f_m(e^{i\theta}) & q_m(e^{-i\theta}) \end{pmatrix}$$

where g_m, f_m, p_m, q_m are polynomials satisfying

(1) $g_m(0) = 1, \deg g_m \le k_m$,
(2) $q_m(0) = 1, \deg q_m \le k_m$,
(3) $\deg p_m \le k_m$,
(4) $\Gamma(f_m) = (h_1, \dots, h_m, 0, \dots)$.

Since $\|f_n\| \le \|F_n\|_\infty \le \sqrt{e}$, by the linearity of Γ we have that for every finitely supported $h \in \ell^2$ there exists $f \in A(\mathbb{D})$ with $\Gamma(f) = h$ and $\|f\| \le \sqrt{e}\|h\|$.

Given an arbitrary $h \in \ell^2$, one can choose a sequence of vectors $\{h_n\}$ with finite support such that $\sum_{n=1}^{\infty} \|h_n\| \le 2\|h\|$ and $\sum_{n=1}^{\infty} h_n = h$. For each h_n, choose $f_n \in A(\mathbb{D})$ with $\Gamma(f_n) = h_n$ and $\|f_n\| \le \sqrt{e}\|h_n\|$. Then $f = \sum_{n=1}^{\infty} f_n$ converges in $A(\mathbb{D})$, and $\Gamma(f) = h$. □

Theorem 10.9 (Lebow). *Let $\{k_\ell\}$ be a sequence of integers with $2k_\ell + 1 < k_{\ell+1}$, and let $X = \sum_{\ell=1}^{\infty} E_{k_\ell,k_\ell}$. Then the operator $F = \left(\begin{smallmatrix} S^* & X \\ 0 & S \end{smallmatrix}\right)$ is power-bounded, but not polynomially bounded.*

Proof. Let $h = (1, 1/2, 1/3, \dots)$, which is in ℓ^2. Since $2(2k_\ell + 1) + 1 < 2k_{\ell+1} + 1$, we see that the sequence $\{2k_\ell + 1\}$ satisfies the hypotheses of Theorem 10.8. Hence there exists $f(z) = \sum_{k=0}^{\infty} a_k z^k$ in $A(\mathbb{D})$ with $a_{2k_\ell+1} = 1/\ell$.

If F were polynomially bounded, then $Y = \sum_{m=1}^{\infty} a_m X_m$ would be a bounded operator, where $X_m = \sum_{\ell=1}^{\infty} X_{m,k_\ell}$.

Using the formula for $X_{n,r}$ we see that

$$\langle X_{n,r} e_0, e_0 \rangle = \begin{cases} 0 & \text{if } n \ne 2r + 1, \\ 1 & \text{if } n = 2r + 1. \end{cases}$$

Hence,

$$\langle X_m e_0, e_0 \rangle = \begin{cases} 0 & \text{if } m \ne 2k_\ell + 1 \text{ for some } \ell, \\ 1 & \text{if } m = 2k_\ell + 1, \text{ for some } \ell, \end{cases}$$

and thus $\langle Y e_0, e_0 \rangle = \sum_{\ell=1}^{\infty} a_{2k_\ell+1} = \sum_{\ell=1}^{\infty} 1/\ell$, which diverges.

Thus, Y is not a bounded operator, and consequently, F is not polynomially bounded. □

Notes

Theorem 10.5 with a CAR sequence of operators in place of the sequence $\{W_n\}$ follows from Pisier [191]. The proof given here follows that of Davidson and the author [71] closely.

Some earlier attempts at finding a counterexample focused on how large the ratio of the complete polynomial bound to the polynomial bound could be for an $n \times n$ matrix, together with obtaining asymptotic estimates of the growth of this ratio as n tended to infinity. See [184] for details of this work.

There is an extensive literature on Foguel–Hankel operators over $\mathbb{C}$ with many deep connections to function theory. See Peller [174] for the earliest published work on this family of operators, where the question arises of determining precisely for which sets of symbols these operators are, respectively, polynomially bounded or similar to contractions. Bourgain [33] proved that such operators are polynomially bounded if and only if the derivative of the symbol is in BMO. Later, Aleksandrov and Peller [4] proved that such operators are polynomially bounded if and only if they are similar to contractions.

This led naturally to the study of Foguel–Hankel operators over general Hilbert spaces. In a remarkable paper Pisier [191] generalized many of the function-theoretic results to obtain operator-valued versions, which allowed him to construct his counterexample to the Halmos's conjecture.

In an earlier unpublished work, Foias and Williams [101] studied general Foguel operators and proved that such an operator is similar to a contraction if and only if it is similar to $S \oplus S^*$. This result motivated the simplification of Pisier's original proof found by Davidson and the author [71].

The result of Foias and Williams has a homological interpretation as the vanishing of a certain Ext group for modules over the disk algebra. The study of this group of extensions was begun by Carlson and Clark [38, 39], and only later was the connection with this unpublished manuscript realized, resulting in [40]. Further contributions were made by Ferguson ([90], [91], and [93]). There are still many open questions about the computation of these Ext groups and their relations to higher-order Hankel forms. See for example [65]. In particular, the study of these groups over polydisk algebras is related to questions about the relationships between the families of "big" and "small" Hankel operators and generalizations of Nehari's theorem. See for example [96], [94], and [92].

Exercises

10.1 Let

$$J_\lambda = \begin{pmatrix} \lambda & 1 & 0 & \cdots & 0 \\ \vdots & \ddots & \ddots & \ddots & \vdots \\ 0 & & \ddots & \ddots & 0 \\ \vdots & \ddots & & \ddots & 1 \\ 0 & \dots & 0 & \dots & \lambda \end{pmatrix} \in M_k$$

be an elementary Jordan block.

(i) Show that the $(1, k)$ entry of J_λ^m is the kth derivative of z^m evaluated at λ. Deduce that J_λ is power-bounded if and only if $|\lambda| < 1$.

(ii) Prove that if $|\lambda| < 1$ then J_λ is similar to a contraction.

(iii) Let $T \in M_n$ be power-bounded. Prove that T is similar to a contraction if and only if T is power-bounded.

10.2 Let T be an operator on a Hilbert space such that $\sigma(T) = C_1 \cup C_2$ with C_1 and C_2 closed, $C_1 \subseteq \mathbb{D}$, and C_2 finite. Prove that T is similar to a contraction if and only if T is power-bounded.

10.3 (i) Show that g_1, f_1, p_1, q_1, satisfy (1)–(4) in the proof of Theorem 10.8.

(ii) Show that if q_m, f_m, p_m, q_m satisfy (1)–(4), then $g_{m+1}, f_{m+1}, p_{m+1}, q_{m+1}$ satisfy (1)–(4).

10.4 Let $\Gamma\colon A(\mathbb{D}) \to \ell^2$ be the map of Theorem 10.8. Prove that $\dot{\Gamma}\colon A(\mathbb{D})/\ker\Gamma \to \ell^2$ is a Banach space isomorphism with $\|\dot{\Gamma}\| \leq 1$, $\|\dot{\Gamma}^{-1}\| \leq \sqrt{e}$.

10.5 Let $C_1, \ldots, C_n$ denote the $2^n \times 2^n$ CAR matrices, let $E_{11}, \ldots, E_{n1}$ be the standard matrix units in M_n, and let $\Phi(\lambda_1 E_{11} + \cdots + \lambda_n E_{n1}) = \lambda_1 C_1 + \cdots + \lambda_n C_n$, so that Φ is an isometry. Prove that $\|\Phi\|_{\text{cb}} \geq \sqrt{n}/2$.

Chapter 11
Applications to K-Spectral Sets

In this chapter we apply the results of Chapter 9 to the study of multiply connected K-spectral sets. We show that for a "nice" region X with finitely many holes it is possible to write down a fairly simple characterization of the family of operators that, up to similarity, have normal ∂X-dilations. This constitutes a model theory for these operators. In contrast, if X has two or more holes, then it is still an open problem to determine whether or not every operator for which X is a spectral set has a normal ∂X-dilation, i.e., is a complete spectral set. A further difficulty with the theory of spectral sets is that it is quite difficult to determine if a given set is a spectral set for an operator. We will illustrate this difficulty in the case that X is an annulus and T is a 2×2 matrix.

Thus, even if it is eventually determined that the properties of being a spectral set and being a complete spectral set are equivalent, the use of the theory might be limited by the impossibility of recognizing operators to which it could be applied.

It is easier to determine when a "nice" set with no holes is a spectral set for an operator. If a compact subset X of the complex plane is simply connected with boundary a Jordan curve, then one can define an analytic homeomorphism f from X to the closed unit disk. In this case it is easily seen that X is a spectral set for an operator T if and only if $f(T)$ is a contraction. For this reason, criteria for operators to have a simply connected set as a spectral set are fairly readily available.

For many finitely connected sets, there is a simple criterion for the set to be a K-spectral set for an operator. To develop this criterion, we first need to extend the definition of spectral sets to include closed, possibly unbounded proper subsets of the complex plane.

To motivate this extended definition, suppose that S is an invertible operator and $\|S\| \leq R$, so that the closed disk of radius R is a spectral set for S. The fact that this disk is a spectral set for S only tells us that $\|S\| \leq R$ and loses the information that S is invertible. The statement that will capture both pieces of information about S, when appropriately defined, is that the complement of the open disk of radius R^{-1} is a spectral set for $T = S^{-1}$.

Let X be a closed, proper subset of $\mathbb{C}$, and let $\hat{X}$ denote the closure of X, regarded as a subset of the Riemann sphere. That is, $\hat{X} = X$ when X is compact, and otherwise $\hat{X}$ is X together with the point at infinity. We let $\mathcal{R}(X)$ denote the quotients of polynomials with poles off $\hat{X}$, that is, the bounded, rational functions on X with a limit at infinity. We regard $\mathcal{R}(X)$ as a subalgebra of $C(\partial\hat{X})$, which defines norms on $M_n(\mathcal{R}(X))$.

If X is a closed, proper subset of $\mathbb{C}$, and $T \in \mathcal{B}(\mathcal{H})$ with $\sigma(T) \subseteq X$, then we still seek a functional calculus, i.e., a homomorphism $\rho\colon \mathcal{R}(X) \to \mathcal{B}(\mathcal{H})$, given by $\rho(p/q) = p(T)q(T)^{-1}$. We say that X is a *(completely) K-spectral set* for T, provided that ρ is well defined and $\|\rho\| \leq K$ (respectively, $\|\rho\|_{\mathrm{cb}} \leq K$). We use the term *(complete) spectral* when $K = 1$.

Let ψ be a linear fractional transformation regarded as a map from the sphere to the sphere, let X be a closed, proper subset of $\mathbb{C}$, and suppose that the pole of ψ lies off X, so that $\psi(\hat{X}) = \psi(X)^{-} = Y$ is a compact set in $\mathbb{C}$. If f is in $\mathcal{R}(Y)$, then $f \circ \psi$ defines an element of $\mathcal{R}(X)$, and the map $\psi^*\colon \mathcal{R}(Y) \to \mathcal{R}(X)$, given by $\psi^*(f) = f \circ \psi$, defines a completely isometric algebra isomorphism between these algebras. The following results are immediate.

Proposition 11.1. *Let X be a closed, proper subset of $\mathbb{C}$, and let ψ be a linear fractional transformation with pole off X. Then X is a (complete) K-spectral set for some operator T if and only if $\psi(X)^{-}$ is a (complete) K-spectral set for $\psi(T)$.*

Proposition 11.2. *Let T be an operator. Then T is invertible with $\|T^{-1}\| \leq R$ if and only if $\{z\colon |z| \geq R^{-1}\}$ is a complete spectral set for T.*

It is now quite easy to illustrate one of the subtleties involved in the study of finitely connected spectral sets that is eliminated by the study of K-spectral sets. Let $X = \{z\colon R^{-1} \leq |z| \leq R\}$, $R > 1$, be a spectral set for some operator T, so that necessarily $\|T\| \leq R$ and $\|T^{-1}\| \leq R$. This last statement is equivalent to the statement that $X_1 = \{z\colon |z| \leq R\}$ and $X_2 = \{z\colon |z| \geq R^{-1}\}$ are both spectral sets for T. Since $X = X_1 \cap X_2$, it is natural to ask: If X_1 and X_2 are spectral sets for T, then is X a spectral set for T? The answer, as we shall show in a moment, is no. Thus, $\|T\| \leq R$ and $\|T^{-1}\| \leq R$ is not enough to guarantee that X is a spectral set for T.

On the other hand, if X_1 and X_2 are, respectively, (complete) K_1-spectral and (complete) K_2-spectral sets for T, then we shall show that X is always a (complete) K-spectral set for T, for some K (Theorem 11.5).

To construct an example where X is not spectral for T, let

$$T = \begin{bmatrix} 1 & t \\ 0 & 1 \end{bmatrix}, \qquad T^{-1} = \begin{bmatrix} 1 & -t \\ 0 & 1 \end{bmatrix}.$$

It is not difficult to calculate that, if $t = R - R^{-1}$, then $\|T\| = \|T^{-1}\| = R$. Thus, X_1 and X_2 are both spectral sets for T. However, $|z - z^{-1}| \leq R + R^{-1}$ on X, while $\|T - T^{-1}\| = 2t = 2(R - R^{-1}) > R + R^{-1}$ for $R > \sqrt{3}$. Thus, X is not a spectral set for T when $R > \sqrt{3}$. Computing the largest value of t such that X is a spectral set for T involves a considerable knowledge of the function theory of X.

The proof of Theorem 11.5 requires the introduction of the concept of *decomposability.* Let X be a closed, proper subset of $\mathbb{C}$. We call a collection $\{X_i\}$ of closed, proper subsets of $\mathbb{C}$ a *decomposition* of X provided that $X = \bigcap X_i$ and every $f \in \mathcal{R}(X)$ can be written as a uniformly convergent series $f = \sum_i f_i$, where each f_i is in $\mathcal{R}(X_i)$ and $\sum_i \|f_i\| \leq K\|f\|$, where K is independent of f. The least value of K satisfying the above inequality we call the *decomposition constant*, relative to the decomposition $\{X_i\}$. If there is a constant K such that the above inequality holds for all F in $M_n(\mathcal{R}(X))$ and all n, then we say that $\{X_i\}$ is a *complete decomposition* of X and call the least such K the *complete decomposition constant.*

Of course every set has a trivial decomposition, namely itself. However, we shall see that many sets have more interesting decompositions. Before proceeding, we point out the relevance of the above definitions.

Proposition 11.3. *Let X be a closed, proper set in $\mathbb{C}$ with (complete) decomposition $\{X_i\}$ and (complete) decomposition constant K. If T is an operator such that each X_i is a (complete) K_i-spectral set for T and $L = \sup_i K_i$ is finite, then X is a (complete) KL-spectral set for T.*

The algebra $\mathcal{R}(X)$ is called a Dirichlet algebra if $\mathcal{R}(X) + \overline{\mathcal{R}(X)}$ is dense in $C(\partial\hat{X})$. We shall call a set X for which $\mathcal{R}(X)$ is a Dirichlet algebra a *D-set.* This concept should not be confused with the concept of a *Dirichlet set.* A set X is called a Dirichlet set if every continuous function on ∂X has a harmonic extension to the interior of X. For example, an annulus is a Dirichlet set that is not a D-set.

Let X be a compact set in $\mathbb{C}$ whose boundary consists of $n + 1$ disjoint, rectifiable, simple, closed curves (i.e., Jordan curves). Such a set will be called a *nice n-holed set*. If X is a nice n-holed set, let $\{U_i\}_{i=0}^n$ denote the open components of $\mathbb{C}\backslash X$, with U_0 the unbounded component, and let $X_i = \mathbb{C}\backslash U_i$, so that X_i, $i = 0, \ldots, n$, is a closed set with X_0 compact. Note that $X = X_0 \cap X_1 \cap \cdots \cap X_n$. We call $\{X_i\}_{i=0}^n$ the *canonical decomposition* of X.

Proposition 11.4. *Let X be a nice n-holed set. Then the canonical decomposition of X is a complete decomposition of X.*

Proof. Let $\{X_j\}_{j=0}^n$ be the canonical decomposition of X, and let $\Gamma_j = \partial X_j$, with Γ_0 oriented counterclockwise, and Γ_j, $j = 1, \ldots, n$, oriented clockwise. If $F \in M_k(\mathcal{R}(X))$, then for $z \notin \Gamma_j$, set

$$F_j(x) = \frac{1}{2\pi i}\int_{\Gamma_j} F(w)(w-z)^{-1}\,dw.$$

Since $F(z) = F_0(z) + \cdots + F_n(z)$ for z in the interior of X, it is not difficult to see that each $F_j(z)$ extends to define a function in $M_k(\mathcal{R}(X_j))$, which we still denote by $F_j(z)$, and that with this extended definition, $F(z) = F_0(z) + \cdots + F_n(z)$ for all $z \in X$.

For $i \neq j$, let $d_{i,j}$ denote the minimum distance between Γ_i and Γ_j, and let ℓ_i denote the length of Γ_i times $(2\pi)^{-1}$. For $i \neq j$, and $z \in \Gamma_j$, we have that $\|F_i(z)\| \leq \ell_i d_{i,j}^{-1}\|F\|$. Thus, for $z \in \Gamma_i$,

$$\|F_i(z)\| = \left\| F(z) - \sum_{j\neq i} F_j(z) \right\| \leq \|F\| + \sum_{j\neq i} \ell_j\, d_{i,j}^{-1}\|F\|,$$

and so we have that $\|F_i\| \leq c_i\|F\|$, where c_i is a constant independent of k.

Hence, $\sum_i \|F_i\| \leq (c_0 + \cdots + c_n)\|F\|$, and we have that the canonical decomposition is a complete decomposition. □

Theorem 11.5. *Let X be a nice n-holed set with canonical decomposition $\{X_i\}_{i=0}^n$, and let $T \in B(\mathcal{H})$. The following are equivalent:*

(i) X is a (complete) K-spectral set for T, for some K,
(ii) each X_i is a (complete) K_i-spectral set for T.

Proof. A straightforward application of Propositions 11.3 and 11.4. □

In Chapter 10 we proved that the hypothesis that X is a K-spectral set for T does not imply that X is a complete K'-spectral set for T for some K' even when X is the unit disk.

There is another, more geometric version of Theorem 11.5.

Corollary 11.6. *Let X be a nice n-holed set, with canonical decomposition $\{X_i\}_{i=0}^n$, and let $T \in B(\mathcal{H})$. Assume that there exist analytic homeomorphisms $f_i\colon \hat{X}_i \to \mathbb{D}^-$, $i = 0, \ldots, n$. The following are equivalent:*

(i) there exists an invertible operator S such that $S^{-1}TS$ has a normal ∂X-dilation,

(ii) there exist invertible operators $\{S_i\}_{i=0}^n$ such that $\|S_i^{-1} f_i(T) S_i\| \le 1$ for $i = 0, \ldots, n$.

Proof. By Corollary 9.12, statements (i) and (ii) above are equivalent to statements (i) and (ii) of Theorem 11.5, respectively. □

The two results above reduce determining whether or not a nice n-holed set, with canonical decomposition $\{X_i\}_{i=0}^n$, is a complete K-spectral set for an operator T to a finite set of conditions.

As a consequence of these result we have:

Corollary 11.7. *Let T be an invertible operator, and assume that there are invertible operators S_1 and S_2 satisfying $\|S_1^{-1} T S_1\| \le R$, $\|S_2^{-1} T^{-1} S_2\| \le r^{-1}$, with $r \le R$. Then there is an invertible operator S such that $\|S^{-1} T S\| \le R$ and $\|S^{-1} T^{-1} S\| \le r^{-1}$.*

Proof. When $r < R$, this is a direct consequence of Corollary 11.6, by letting $X = \{z\colon r \le |z| \le R\}$, $f_0(z) = z$, and $f_1(z) = z^{-1}$.

If $r = R$, then $\|(T/R)^n\| \le \|S_1^{-1}\| \cdot \|S_1\|$, and $\|(T/R)^{-n}\| \le \|S_2^{-1}\| \cdot \|S_2\|$. Hence, by Corollary 9.4, there exist an invertible operator S such that $S^{-1}(T/R)S$ is a unitary. Thus, $\|S^{-1} T S\| \le R$ and $\|S^{-1} T^{-1} S\| \le r^{-1}$. □

Theorem 11.8. *Let X be a nice n-holed set with canonical decomposition $\{X_i\}_{i=0}^n$ and (complete) decomposition constant K. If $T \in B(\mathcal{H})$ and each X_i is a spectral set for T, then X is a (complete) K-spectral set for T.*

Proof. Since each X_i is a D-set, the hypothesis that X_i is a spectral set implies that T has a normal ∂X_i-dilation, by Theorem 4.4. That is, X_i is a complete spectral set for T. The remainder of the proof is a direct application of Propositions 11.3 and 11.4. □

Let $X = \{z\colon r \le |z| \le R\}$, $r < R$, be an annulus, so that the canonical decomposition of X is $X_0 = \{z\colon |z| \le R\}$ and $X_1 = \{z\colon r \le |z|\}$. We have that X_0 is a spectral set for T if and only if $\|T\| \le R$, and X_1 is a spectral set for T

if and only if $\|T^{-1}\| \leq r^{-1}$. We have seen in an earlier example that $\|T\| \leq R$ and $\|T^{-1}\| \leq r^{-1}$ is not enough, in general, to guarantee that the annulus is a spectral set for T. However, by the above theorem, the annulus will be a complete K-spectral set for T and so, up to conjugation by a similarity, T will have a normal ∂X-dilation.

It is not difficult to see from the proof of Proposition 11.4, that $2 + \frac{R+r}{R-r}$ gives an upper bound on the complete decomposition constant for the canonical decomposition of X. It is interesting to contrast this with [211, Proposition 23], where it is shown that the decomposition constant for this annulus is bounded by $2 + (\frac{R+r}{R-r})^{1/2}$. However, that proof does not appear to generalize to other sets.

Precise values for the decomposition constant and complete decomposition constant are not known, even for an annulus. Neither is it known whether these constants are achieved by operators. That is, for T an operator with $\|T\| \leq R$, $\|T^{-1}\| \leq r^{-1}$, let $\rho\colon \mathcal{R}(X) \to B(\mathcal{H})$ be the homomorphism given by $\rho(f) = f(T)$, where $X = \{z\colon r \leq |z| \leq R\}$, and set

$$K = \sup\{\|\rho\|\colon \|T\| \leq R,\ \|T^{-1}\| \leq R^{-1}\}$$

and

$$K_c = \sup\{\|\rho\|_{\mathrm{cb}}\colon \|T\| \leq R,\ \|T^{-1}\| \leq r^{-1}\}.$$

By Proposition 11.3, K and K_c will be less than or equal to the decomposition constant and the complete decomposition for X, respectively, but it is not known if these inequalities are strict or are in fact equalities.

There are sets besides nice n-holed sets for which these decomposition techniques are valuable. For example, the sets

$$X_0 = \left\{z\colon |z| \leq 1 \quad \text{and} \quad \left|z - \frac{1}{2}\right| \geq \frac{1}{2}\right\}$$

and

$$X_1 = \left\{z\colon \left|z + \frac{1}{2}\right| \geq \frac{1}{4}\right\}$$

are D-sets [68, VI.11.11], and the proof of Proposition 11.4 can be suitably modified to show that these sets define a complete decomposition of $X = X_0 \cap X_1$. Consequently, the conclusions of Theorems 11.5 and 11.8 apply to these sets as well.

Another approach to finitely connected regions involves the concept of a hypo-Dirichlet algebra. If X is a compact, Hausdorff space, then any subalgebra $\mathcal{A}$ of $C(X)$ that separates points on X and has the property that the closure of

$\mathcal{A} + \bar{\mathcal{A}}$ is of finite codimension n in $C(X)$ is called a *hypo-Dirichlet algebra on X of codimension n.*

Gamelin [103, Theorem IV.8.3] shows that if X is a compact subset of $\mathbb{C}$ whose complement has n bounded components, then $\mathcal{R}(X)$ is a hypo-Dirichlet algebra on ∂X of codimension m, with $m \leq n$. In fact, if points $\{z_j\}_{j=1}^n$ are chosen, one from each bounded component of the complement of X, then the span of the union of $\mathcal{R}(X)$, $\overline{\mathcal{R}(X)}$, and $\{\ln|z - z_j|\}_{j=1}^n$ is dense in $C(\partial X)$.

We shall show that unital, contractive homomorphisms of hypo-Dirichlet algebras are necessarily completely bounded.

Lemma 11.9. *Let $\mathcal{A}$ be a unital C^*-algebra, and let $\mathcal{S} \subseteq \mathcal{A}$ be an operator system of codimension n. Then for every $\varepsilon > 0$, there exists a completely positive map $\phi\colon \mathcal{A} \to \mathcal{S}$ and a positive linear functional s on $\mathcal{A}$ such that $\|\phi\| \leq n + 1 + \varepsilon$, $\|s\| \leq n + \varepsilon$, and $\phi(x) = x + s(x) \cdot 1$ for $x \in \mathcal{S}$.*

Proof. Let $\pi\colon \mathcal{A} \to \mathcal{A}/\mathcal{S}$ be the quotient map. It is not difficult to show that there exist self-adjoint linear functionals $\ell_1', \ell_2', \ldots, \ell_n'$ on $\mathcal{A}/\mathcal{S}$ and self-adjoint elements $h_1', h_2', \ldots, h_n'$ in $\mathcal{A}/\mathcal{S}$ that form a basis, such that $\|\ell_i'\| = \|h_i'\| = 1$ and $\ell_i(h_j') = \delta_{i,j}$, the Kronecker delta. Let $h_1, h_2, \ldots, h_n$ be self-adjoint elements in $\mathcal{A}$ such that $\pi(h_i) = h_i'$ and $\|h_i\| \leq 1 + \varepsilon/n$. Also, let $\ell_i = \ell_i' \circ \pi$, so that $\mathcal{S} = \{y \in \mathcal{A}\colon \ell_i(y) = 0, i = 1, 2, \ldots, n\}$ with $\|\ell_i\| = 1$ and $\ell_i(h_j) = \delta_{i,j}$.

By Exercise 7.14, we can write $\ell_i = f_i - g_i$, where $\|f_i\| \leq 1$, $\|g_i\| \leq 1$, and $\|f_i + g_i\| \leq 1$, with f_i and g_i positive linear functionals on $\mathcal{A}$. If for x in $\mathcal{A}$, we set

$$\gamma(x) = x - \sum_{i=1}^n \ell_i(x) h_i,$$

then γ is a projection of $\mathcal{A}$ onto $\mathcal{S}$.

Hence the completely positive map $\phi\colon \mathcal{A} \to \mathcal{A}$ defined by

$$\begin{aligned}\phi(x) &= x + \sum_{i=1}^n g_i(x)(\|h_i\| + h_i) + \sum_{i=1}^n f_i(x)(\|h_i\| - h_i) \\ &= \gamma(x) + \sum_{i=1}^n (g_i(x) + f_i(x))\|h_i\| \cdot 1\end{aligned}$$

has range in $\mathcal{S}$.

If we set $s(x) = \sum_{i=1}^n (f_i(x) + g_i(x))\|h_i\|$, then

$$\|s\| \leq \sum_{i=1}^n \|f_i + g_i\| \|h_i\| \leq n + \varepsilon,$$

and $\phi(x) = x + s(x) \cdot 1$ for x in $\mathcal{S}$. Finally, since ϕ is completely positive, we have $\|\phi\| = \|\phi(1)\| = 1 + s(1) \leq n + 1 + \varepsilon$. □

Theorem 11.10. *If $\mathcal{A} \subseteq C(X)$ is a hypo-Dirichlet algebra of codimension n, and $\rho \colon \mathcal{A} \to B(\mathcal{H})$ is a unital contraction, then $\|\rho\|_{\mathrm{cb}} \leq 2n + 1$.*

Proof. Let $\varepsilon > 0$, let $\mathcal{S}$ be the closure of $\mathcal{A} + \bar{\mathcal{A}}$ in $C(X)$, and let $\phi \colon C(X) \to \mathcal{S}$ and s be as in the previous lemma. If we extend ρ to $\tilde{\rho} \colon \mathcal{S} \to B(\mathcal{H})$ by $\tilde{\rho}(f + \bar{g}) = \rho(f) + \rho(g)^*$, then $\tilde{\rho}$ will be positive by Proposition 2.12. Thus, $\tilde{\rho} \circ \phi \colon C(X) \to B(\mathcal{H})$ is positive and hence completely positive. Finally, for f in $\mathcal{A}$, we have

$$\rho(f) = \tilde{\rho} \circ \phi(f) - s(f) \cdot 1_{\mathcal{H}},$$

so that

$$\|\rho\|_{\mathrm{cb}} \leq \|\tilde{\rho} \circ \phi\|_{\mathrm{cb}} + \|s\|_{\mathrm{cb}} = \|\tilde{\rho} \circ \phi(1)\| + \|s(1)\| \leq 2n + 1 + 2\varepsilon,$$

since $\tilde{\rho} \circ \phi$ and s are completely positive. □

Corollary 11.11. *If $\mathcal{A} \subseteq C(X)$ is a hypo-Dirichlet algebra of codimension n, and $\rho \colon \mathcal{A} \to B(\mathcal{H})$ is a unital contractive homomorphism, then ρ is similar to a homomorphism that dilates to $C(X)$. Furthermore, the similarity S may be chosen such that $\|S\| \cdot \|S^{-1}\| \leq 2n + 1$.*

Corollary 11.12. *Let X be a compact subset of $\mathbb{C}$ such that X is a spectral set for T in $B(\mathcal{H})$. If $\mathcal{R}(X)$ is a hypo-Dirichlet algebra of codimension n on ∂X, then there exists an invertible operator S in $B(\mathcal{H})$ with $\|S^{-1}\| \cdot \|S\| \leq 2n + 1$ such that $S^{-1}TS$ has a normal ∂X-dilation.*

When X is a nice n-holed set, it is interesting to contrast Corollary 11.12 with Theorem 11.8. If X is a spectral set for T, then both theorems allow us to deduce that X is a complete K-spectral set for T. Corollary 11.12 gives an upper bound on K of $2n + 1$, while the bound on K in Theorem 11.8 comes from the complete decomposition constant for X relative to the canonical decomposition of X. The proof of Proposition 11.4 gives a bound on the complete decomposition constant, but the bound one obtains in this manner is always larger than $2n + 1$. However, it should be recalled that the hypotheses of Theorem 11.8 are considerably weaker than those of Corollary 11.12.

Notes

Most of the results of this chapter were obtained by Douglas and the author in [77]. Exercise 11.3 is from [241].

Exercises

11.1 Let T be in $B(\mathcal{H})$ and let $X = \{z\colon \mathrm{Re}(z) \geq 0\}$. Prove that the following are equivalent:

- (i) X is a complete spectral set for T.
- (ii) X is a spectral set for T.
- (iii) $C = (T-1)(T+1)^{-1}$ is a contraction and $1 \notin \sigma(C)$.
- (iv) $\mathrm{Re}(T) \geq 0$.
- (v) $(T-1)^*(T-1) \leq (T+1)^*(T+1)$.

Let X be a closed, proper subset of $\mathbb{C}$, $T \in B(\mathcal{H})$, and suppose X is a spectral set for T. We say that T has a normal ∂X-dilation provided that there exists a Hilbert space $\mathcal{K}$ containing $\mathcal{H}$ and a bounded normal operator N in $B(\mathcal{K})$, with $\sigma(N) \subseteq \partial X$, such that $f(T) = P_{\mathcal{H}} f(N)|_{\mathcal{H}}$.

11.2 Let $X = \{z\colon \mathrm{Re}(z) \geq 0\}$ be a spectral set for T.

- (i) Prove that T has a normal ∂X-dilation if and only if the minimal unitary dilation U of the contraction $C = (T-1)(T+1)^{-1}$ satisfies $1 \notin \sigma(U)$.
- (ii) Give an example of an operator T such that X is a complete spectral set for T, but T has no normal ∂X-dilation.
- (iii) Let $\rho\colon \mathcal{R}(X) \to B(\mathcal{H})$ be given by $\rho(f) = f(T)$. Show that if X is a complete spectral set for T, then ρ has a $C(\partial\hat{X})$-dilation.

11.3 Let X be a proper, closed subset of $\mathbb{C}$, $\lambda \in \hat{X}$, and $T \in B(\mathcal{H})$ with $\sigma(T) \subseteq X$. Set $\mathcal{R}_\lambda(X) = \{f \in \mathcal{R}(X)\colon f(\lambda) = 0\}$.

- (i) (Williams) Show that if $\|f(T)\| \leq \|f\|$ for all $f \in \mathcal{R}_\lambda(X)$, then X is a spectral set for T.
- (ii) Show that if $\|f(T)\| \leq K\|f\|$ for all $f \in \mathcal{R}_\lambda(X)$, then X is a $(2K+1)$-spectral set.
- (iii) Give an example where $\|f(T)\| \leq K\|f\|$ for all $f \in \mathcal{R}$, but X is not a K-spectral set.
- (iv)* If $\|f(T)\| \leq \|f\|$ for all $f \in M_n(\mathcal{R}_\lambda(X))$ and all n, then is X a complete spectral set for T?

11.4 Let X_i, $i = 1, \ldots, n$, be disjoint, compact sets, and let X be their union. Show that if X is a K-spectral set for T, then T is similar to $T_1 \oplus \cdots \oplus T_n$ where X_i is K_i-spectral for T_i.

Chapter 12

Tensor Products and Joint Spectral Sets

In this chapter we outline some of the theory of tensor products of C^*-algebras, operator spaces, and operator systems, and apply this theory to multivariable dilation theory. An n-tuple of operators $(T_1, \ldots, T_n)$ is said to *doubly commute* provided that $T_iT_j = T_jT_i$ and $T_iT_j^* = T_j^*T_i$ for all $i \neq j$. We shall see that, for doubly commuting operators, a natural setting for generalizing the theory of spectral sets from a single-variable theory to a multivariable theory is the theory of tensor products of operator systems.

Let $\mathcal{A}$ and $\mathcal{B}$ be unital C^*-algebras. Then their tensor product can be made into a $*$-algebra by setting $(a \otimes b)^* = a^* \otimes b^*$ and extending linearly. By a *cross-norm*, we mean a norm $\|\cdot\|_\gamma$ on $\mathcal{A} \otimes \mathcal{B}$ with the property that $\|a \otimes b\|_\gamma = \|a\|\|b\|$ for $a \in \mathcal{A}$ and $b \in \mathcal{B}$. By a *C^*-cross-norm*, we mean a cross-norm on $\mathcal{A} \otimes \mathcal{B}$ that also satisfies the C^*-algebra axioms, $\|xy\|_\gamma \leq \|x\|_\gamma \|y\|_\gamma$ and $\|x^*x\|_\gamma = \|x\|_\gamma^2 = \|x^*\|_\gamma$ for $x, y \in \mathcal{A} \otimes \mathcal{B}$. The completion of $\mathcal{A} \otimes \mathcal{B}$ with respect to a C^*-cross-norm γ is a C^*-algebra, which we denote by $\mathcal{A} \otimes_\gamma \mathcal{B}$.

In general, there are many possible C^*-cross-norms on $\mathcal{A} \otimes \mathcal{B}$, but there are two that we shall be interested in, the *maximal* and *minimal* C^*-cross-norms.

In order to construct the minimal C^*-cross-norm, we recall the theory of tensor products of Hilbert spaces. Suppose that $\mathcal{H}$ and $\mathcal{K}$ are Hilbert spaces. It is well known that if we set $\langle h \otimes k, h' \otimes k' \rangle = \langle h, h' \rangle_{\mathcal{H}} \langle k, k' \rangle_{\mathcal{K}}$ and extend linearly, then we obtain an inner product on $\mathcal{H} \otimes \mathcal{K}$. The completion of $\mathcal{H} \otimes \mathcal{K}$ with respect to this inner product is a Hilbert space, which we still denote by $\mathcal{H} \otimes \mathcal{K}$. If T and S are operators on $\mathcal{H}$ and $\mathcal{K}$, respectively, then setting $(T \otimes_{\text{sp}} S)(h \otimes k) = (Th) \otimes (Sk)$ extends to define a bounded, linear operator on $\mathcal{H} \otimes \mathcal{K}$ with $\|T \otimes_{\text{sp}} S\| = \|T\|\|S\|$.

It is fairly easy to check that $(T_1 \otimes_{\text{sp}} S_1)(T_2 \otimes_{\text{sp}} S_2) = (T_1T_2) \otimes_{\text{sp}} (S_1S_2)$ and that $(T \otimes_{\text{sp}} S)^* = T^* \otimes_{\text{sp}} S^*$.

If $\mathcal{A}_i \subseteq B(\mathcal{H}_i)$, $i = 1, 2$, are two C^*-subalgebras, then we define the *spatial tensor product*, $\mathcal{A}_1 \otimes_{\text{sp}} \mathcal{A}_2$, to be the subspace of $B(\mathcal{H} \otimes \mathcal{K})$ spanned by the operators $T \otimes_{sp} S$ with $T \in \mathcal{A}_1$ and $S \in \mathcal{A}_2$. It can be shown that the map from the tensor product $\mathcal{A}_1 \otimes \mathcal{A}_2$ to $\mathcal{A}_1 \otimes_{sp} \mathcal{A}_2$ is one-to-one and hence setting $\|\sum T_i \otimes S_i\|_{sp} = \|\sum T_i \otimes_{sp} S_i\|$ defines a cross-norm. From this it follows easily that $\|\cdot\|_{sp}$ is a C^*-cross-norm. See [233] for proofs of these facts.

If $\pi_1 \colon \mathcal{A}_1 \to B(\mathcal{H}_1)$ and $\pi_2 \colon \mathcal{A}_2 \to B(\mathcal{H}_2)$ are unital $*$-homomorphisms, then we get a unital, $*$-preserving homomorphism $\pi_1 \otimes \pi_2 \colon \mathcal{A}_1 \otimes \mathcal{A}_2 \to B(\mathcal{H}_1 \otimes \mathcal{H}_2)$ by setting $\pi_1 \otimes \pi_2(a \otimes b) = \pi_1(a) \otimes_{sp} \pi_2(b)$. Thus, if for $x \in \mathcal{A}_1 \otimes \mathcal{A}_2$ we set $\|x\|_{\min} = \sup\{\|\pi_1 \otimes \pi_2(x)\| \colon \pi_i \colon \mathcal{A}_i \to B(\mathcal{H}_i)$ unital, $*$-homomorphism, $i = 1, 2\}$, then we obtain a C^*-cross-norm on $\mathcal{A}_1 \otimes \mathcal{A}_2$. This norm is called the *minimal* norm, and the completion of $\mathcal{A}_1 \otimes \mathcal{A}_2$ in this norm is denoted $\mathcal{A}_1 \otimes_{\min} \mathcal{A}_2$. Note that it has the property that if $\pi_i \colon \mathcal{A}_i \to B(\mathcal{H}_i)$, $i = 1, 2$, are any unital $*$-homomorphism, then $\pi_1 \otimes \pi_2 \colon \mathcal{A}_1 \otimes \mathcal{A}_2 \to B(\mathcal{H}_1 \otimes \mathcal{H}_2)$ can be extended, by continuity, to a unital $*$-homomorphism of $\mathcal{A}_1 \otimes_{\min} \mathcal{A}_2$, denoted $\pi_1 \otimes_{\min} \pi_2$.

The following result is explains the name of this C^*-cross-norm. For a proof, see [233, Theorem IV.4.19].

Theorem 12.1 (Takesaki). *Let $\mathcal{A}_1$ and $\mathcal{A}_2$ be unital C^*-algebras. If γ is a C^*-cross-norm on $\mathcal{A}_1 \otimes \mathcal{A}_2$, then $\|x\|_{\min} \le \|x\|_\gamma$ for all $x \in \mathcal{A}_1 \otimes \mathcal{A}_2$.*

Corollary 12.2. *Let $\mathcal{A}_1$ and $\mathcal{A}_2$ be unital C^*-algebras. If $\pi_i \colon \mathcal{A}_i \to B(\mathcal{H}_i)$ are one-to-one, unital $*$-homomorphisms, then for $x \in \mathcal{A}_1 \otimes \mathcal{A}_2$, $\|x\|_{\min} = \|\pi_1 \otimes \pi_2(x)\|$.*

Proof. By definition, $\|x\|_{\min} \ge \|\pi_1 \otimes \pi_2(x)\|$. But setting $\|x\|_\gamma = \|\pi_1 \otimes \pi_2(x)\|$ defines a C^*-cross-norm, from which the other inequality follows. □

Thus, when $\mathcal{A}_i \subseteq B(\mathcal{H}_i)$ are concrete C^*-subalgebras the minimal and spatial C^*-cross-norms are equal.

Suppose that $\mathcal{B}_i$, $i = 1, 2$, are unital C^*-algebras and that $\mathcal{A}_i \subseteq \mathcal{B}_i$, $i = 1, 2$, are unital C^*-subalgebras. For $x \in \mathcal{A}_1 \otimes \mathcal{A}_2 \subseteq \mathcal{B}_1 \otimes \mathcal{B}_2$, we have two possible definitions of $\|x\|_{\min}$, depending on whether we view it as an element of $\mathcal{A}_1 \otimes \mathcal{A}_2$ or of $\mathcal{B}_1 \otimes \mathcal{B}_2$. However, if we fix $\pi_i \colon \mathcal{B}_i \to B(\mathcal{H}_i)$, $i = 1, 2$, unital, one-to-one $*$-homomorphisms, then since their restrictions are also unital, one-to-one $*$-homomorphisms of $\mathcal{A}_i$, $i = 1, 2$, by Corollary 12.2 we have $\|x\|_{\min} = \|\pi_1 \otimes \pi_2(x)\|$, independent of whether we regard it as an element of $\mathcal{A}_1 \otimes \mathcal{A}_2$ or $\mathcal{B}_1 \otimes \mathcal{B}_2$.

This observation can perhaps best be summarized by saying that the natural inclusion of $\mathcal{A}_1 \otimes \mathcal{A}_2$ into $\mathcal{B}_1 \otimes \mathcal{B}_2$ extends to a $*$-isomorphism of $\mathcal{A}_1 \otimes_{\min} \mathcal{A}_2$ onto the norm closure of $\mathcal{A}_1 \otimes \mathcal{A}_2$ in $\mathcal{B}_1 \otimes_{\min} \mathcal{B}_2$. Because of this last fact, the minimal C^*-cross-norm is also often called the *injective* C^*-cross-norm.

We use this observation to define a min norm on tensor products of operator spaces. Suppose that $\mathcal{A}_i$, $i = 1, 2$, are unital C^*-algebras and that $\mathcal{S}_i \subseteq \mathcal{A}_i$, $i = 1, 2$, are subspaces. We then define the min norm on $\mathcal{S}_1 \otimes \mathcal{S}_2$ to be the restriction of the min norm on $\mathcal{A}_1 \otimes \mathcal{A}_2$ and let $\mathcal{S}_1 \otimes_{\min} \mathcal{S}_2$ denote this operator space. Note that if $\mathcal{S}_1$ and $\mathcal{S}_2$ are operator systems, then $\mathcal{S}_1 \otimes_{\min} \mathcal{S}_2$ is also an operator system.

It is important to note that since we are defining $\mathcal{S}_1 \otimes_{\min} \mathcal{S}_2$ as a concrete subspace of the C^*-algebra $\mathcal{A}_1 \otimes_{\min} \mathcal{A}_2$, it is endowed with a matrix norm and, in the case of operator systems, a matrix order.

Theorem 12.3. *Let $\mathcal{A}_i$ and $\mathcal{B}_i$ be unital C^*-algebras, let $\mathcal{S}_i \subseteq \mathcal{A}_i$ be subspaces, and let $L_i\colon \mathcal{S}_i \to \mathcal{B}_i$ be completely bounded, $i = 1, 2$. Then the linear map $L_1 \otimes L_2\colon \mathcal{S}_1 \otimes \mathcal{S}_2 \to \mathcal{B}_1 \otimes \mathcal{B}_2$, given by $(L_1 \otimes L_2)(a_1 \otimes a_2) = L_1(a_1) \otimes L_2(a_2)$, defines a completely bounded map $L_1 \otimes_{\min} L_2\colon \mathcal{S}_1 \otimes_{\min} \mathcal{S}_2 \to \mathcal{B}_1 \otimes_{\min} \mathcal{B}_2$, with $\|L_1 \otimes_{\min} L_2\|_{cb} = \|L_1\|_{cb}\|L_2\|_{cb}$. If $\mathcal{S}_1$ and $\mathcal{S}_2$ are operator systems and L_1 and L_2 are completely positive, then $L_1 \otimes_{\min} L_2$ is completely positive.*

Proof. Let $\mathcal{B}_i \subseteq B(\mathcal{H}_i)$, $i = 1, 2$. If we can show that $L_1 \otimes L_2\colon \mathcal{S}_1 \otimes \mathcal{S}_2 \to B(\mathcal{H}_1 \otimes \mathcal{H}_2)$ is completely bounded in the min norm with $\|L_1 \otimes L_2\|_{cb} = \|L_1\|_{cb}\|L_2\|_{cb}$, then we shall be done by the injectivity of the min norm. By the extension theorem for completely bounded maps, we may extend L_i to $\tilde{L}_i\colon \mathcal{A}_i \to B(\mathcal{H}_i)$ with $\|\tilde{L}_i\|_{cb} = \|L_i\|_{cb}$, $i = 1, 2$. Now applying the generalized Stinespring representation, we obtain unital $*$-homomorphisms $\pi_i\colon \mathcal{A}_i \to B(\mathcal{K}_i)$, and bounded operators $V_i\colon \mathcal{H}_i \to \mathcal{K}_i$, $W_i\colon \mathcal{H}_i \to \mathcal{K}_i$ with $\|V_i\|\|W_i\| = \|L_i\|_{cb}$, such that

$$\tilde{L}_i(a_i) = V_i^* \pi_i(a) W_i, \quad a_i \in \mathcal{A}_i, \qquad i = 1, 2.$$

Consider $V_1 \otimes V_2\colon \mathcal{H}_1 \otimes \mathcal{H}_2 \to \mathcal{K}_1 \otimes \mathcal{K}_2$, $W_1 \otimes W_2\colon \mathcal{H}_1 \otimes \mathcal{H}_2 \to \mathcal{K}_1 \otimes \mathcal{K}_2$, and $\pi_1 \otimes_{\min} \pi_2\colon \mathcal{A}_1 \otimes_{\min} \mathcal{A}_2 \to B(\mathcal{K}_1 \otimes \mathcal{K}_2)$. We have that $(V_1 \otimes V_2)^*(\pi_1 \otimes \pi_2)(a_1 \otimes a_2)(W_1 \otimes W_2) = \tilde{L}_1(a_1) \otimes \tilde{L}_2(a_2)$. Thus, $\|L_1 \otimes L_2\|_{cb} \leq \|V_1 \otimes V_2\|\,\|W_1 \otimes W_2\| = \|L_1\|_{cb}\|L_2\|_{cb}$ on $\mathcal{S}_1 \otimes \mathcal{S}_2$. We leave it to the reader to verify that $\|L_1\|_{cb}\|L_2\|_{cb} \leq \|L_1 \otimes L_2\|_{cb}$.

If $\mathcal{S}_1$ and $\mathcal{S}_2$ are operator systems and L_1 and L_2 are completely positive, then we argue as above, using the extension theorem for completely positive maps. In this case, we find that $V_i = W_i$, $i = 1, 2$, from which the result follows. □

Corollary 12.4. *Let $\mathcal{S}_1$ and $\mathcal{S}_2$ be operator spaces, $x \in \mathcal{S}_1 \otimes \mathcal{S}_2$. Then*

$$\|x\|_{\min} = \sup\{\|L_1 \otimes L_2(x)\|\colon L_i\colon \mathcal{S}_i \to B(\mathcal{H}_i),\ \|L_i\|_{cb} \leq 1,\ i = 1, 2\}.$$

Furthermore, if $\mathcal{S}_1$ *and* $\mathcal{S}_2$ *are operator systems, then we may take* L_i *to be completely positive,* $i = 1, 2$.

The above corollary shows that the min norm on tensor products of operator spaces is invariant under completely isometric isomorphisms of those operator spaces.

Let X be a compact, Hausdorff space and let $\mathcal{A}$ be a unital C^*-algebra. We let $C(X;\mathcal{A})$ denote the continuous functions from X into $\mathcal{A}$, equipped with the norm $\|F\| = \sup\{\|F(x)\|: x \in X\}$ for $F \in C(X;\mathcal{A})$. It is easy to check that if we define multiplication, addition, and the $*$-operation pointwise, then $C(X;\mathcal{A})$ is a C^*-algebra.

Now define a $*$-homomorphism from $C(X) \otimes \mathcal{A}$ into $C(X;\mathcal{A})$ by

$$\sum_{i=1}^{n} f_i \otimes a_i \to F(x) = \sum_{i=1}^{n} f_i(x)a_i.$$

It is straightforward that setting $\|\sum_{i=1}^{n} f_i \otimes a_i\|_\gamma = \|F\|$ defines a C^*-cross-norm on $C(X) \otimes \mathcal{A}$. Furthermore, a standard partition-of-unity argument shows that the image of $C(X) \otimes \mathcal{A}$ is dense in $C(X;\mathcal{A})$. Thus, the above mapping extends to a $*$-isomorphism of $C(X) \otimes_\gamma \mathcal{A}$ with $C(X;\mathcal{A})$.

We can now show that γ is actually the minimal C^*-cross-norm. To see this, note that for each fixed $x \in X$, the map $f \to f(x)$ extends to a contractive, linear map on $C(X)$. Thus by Exercise 12.1, the map $f \otimes a \to f(x)a$ extends to a contractive linear map from $C(X) \otimes_{\min} \mathcal{A}$ to $\mathcal{A}$, and so $\|\sum_{i=1}^{n} f_i \otimes a_i\|_{\min} \geq \|\sum_{i=1}^{n} f_i(x)a_i\|$. This shows that the min C^*-cross-norm is greater than the γ C^*-cross-norm, and consequently they must be equal.

We have shown the following:

Proposition 12.5. *Let X be a compact, Hausdorff space and let $\mathcal{A}$ be a unital C^*-algebra. Then the map $f \otimes a \to f(x)a$ extends to a $*$-isomorphism between $C(X) \otimes_{\min} \mathcal{A}$ and $C(X;\mathcal{A})$.*

We note that in the particular case where $\mathcal{A} = C(Y)$ for a compact Hausdorff space Y, the usual identification of a continuous function on $X \times Y$ with a continuous function from X into $C(Y)$ is a $*$-isomorphism of $C(X; C(Y))$ and $C(X \times Y)$. Thus, $C(X \times Y)$ is $*$-isomorphic to $C(X) \otimes_{\min} C(Y)$.

If $\mathcal{S}_i \subseteq C(X_i)$ is a subspace, $i = 1, 2$, then the min norm on $\mathcal{S}_1 \otimes \mathcal{S}_2$ is just the norm one obtains by viewing an element of $\mathcal{S}_1 \otimes \mathcal{S}_2$ as a function on $X_1 \times X_2$. Furthermore, if $\mathcal{S}_1$ and $\mathcal{S}_2$ are operator systems, then an element of $M_n(\mathcal{S}_1 \otimes \mathcal{S}_2)$ will be positive if and only if it is a positive matrix-valued function on $X_1 \times X_2$.

We now turn our attention to the max norm. Let $\mathcal{A}$ and $\mathcal{B}$ be unital C^*-algebras, and let $\pi_1: \mathcal{A} \to B(\mathcal{H})$, $\pi_2: \mathcal{B} \to B(\mathcal{H})$ be unital $*$-homomorphisms such that $\pi_1(a)\pi_2(b) = \pi_2(b)\pi_1(a)$ for all $a \in \mathcal{A}$ and $b \in \mathcal{B}$. We may then define a unital $*$-homomorphism $\pi: \mathcal{A} \otimes \mathcal{B} \to B(\mathcal{H})$ via $\pi(x) = \sum_{i=1}^n \pi_1(a_i)\pi_2(b_i)$, where $x = \sum_{i=1}^n a_i \otimes b_i$. Conversely, if we're given a unital $*$-homomorphism $\pi: \mathcal{A} \otimes \mathcal{B} \to B(\mathcal{H})$ and we define $\pi_1(a) = \pi(a \otimes 1)$, $\pi_2(b) = \pi(1 \otimes b)$, then we obtain a pair of unital $*$-homomorphisms of $\mathcal{A}$ and $\mathcal{B}$, respectively, with commuting ranges such that $\pi(a \otimes b) = \pi_1(a)\pi_2(b)$.

We define

$$\|x\|_{\max} = \sup\{\|\pi(x)\|: \pi: \mathcal{A} \otimes \mathcal{B} \to B(\mathcal{H}) \text{ unital } *\text{-homomorphisms}\}.$$

Thus, if $\pi_1: \mathcal{A} \to B(\mathcal{H})$, $\pi_2: \mathcal{B} \to B(\mathcal{H})$, are unital $*$-homomorphisms with commuting ranges, then we have a unital $*$-homomorphism $\pi_1 \otimes_{\max} \pi_2: \mathcal{A} \otimes_{\max} \mathcal{B} \to B(\mathcal{H})$ satisfying $\pi_1 \otimes_{\max} \pi_2(a \otimes b) = \pi_1(a)\pi_2(b)$.

Proposition 12.6. *Let $\mathcal{A}$ and $\mathcal{B}$ be unital C^*-algebras, let $x \in \mathcal{A} \otimes \mathcal{B}$, and let γ be a C^*-cross-norm on $\mathcal{A} \otimes \mathcal{B}$. Then $\|x\|_\gamma \leq \|x\|_{\max}$.*

Proof. By the Gelfand–Naimark–Segal theorem, there is a unital $*$-homomorphism $\pi: \mathcal{A} \otimes_\gamma \mathcal{B} \to B(\mathcal{H})$ with $\|x\|_\gamma = \|\pi(x)\|$. Since $\|\pi(x)\| \leq \|x\|_{\max}$ by definition, the proof is complete. □

If we have C^*-algebras $\mathcal{A}_i \subseteq \mathcal{B}_i$, $i = 1, 2$, then the natural inclusion $\mathcal{A}_1 \otimes \mathcal{A}_2 \subseteq \mathcal{B}_1 \otimes \mathcal{B}_2 \subseteq \mathcal{B}_1 \otimes_{\max} \mathcal{B}_2$ induces a C^*-cross-norm γ on $\mathcal{A}_1 \otimes \mathcal{A}_2$. The C^*-algebra $\mathcal{A}_1 \otimes_\gamma \mathcal{A}_2$ can be identified with the closure of $\mathcal{A}_1 \otimes \mathcal{A}_2$ in $\mathcal{B}_1 \otimes_{\max} \mathcal{B}_2$. Thus, by Proposition 12.6, we have a $*$-homomorphism $\mathcal{A}_1 \otimes_{\max} \mathcal{A}_2 \to \mathcal{B}_1 \otimes_{\max} \mathcal{B}_2$, which, in general, can be norm-decreasing.

Moreover, if $\pi_i: \mathcal{A}_i \to \mathcal{B}_i$, $i = 1, 2$, are onto $*$-homomorphisms, then $\pi_1 \otimes_{\max} \pi_2: \mathcal{A}_1 \otimes_{\max} \mathcal{A}_2 \to \mathcal{B}_1 \otimes_{\max} \mathcal{B}_2$ is onto, but not necessarily one-to-one. For these reasons, the max norm is often referred to as the *projective* C^*-cross-norm.

Unfortunately, the max norm on tensor products is not compatible with completely bounded maps. Huruya [123] has given an example of a completely bounded map $L: \mathcal{A}_1 \to \mathcal{A}_2$ and a C^*-algebra $\mathcal{B}$ such that the map $L \otimes \mathrm{id}: \mathcal{A}_1 \otimes \mathcal{B} \to \mathcal{A}_2 \otimes \mathcal{B}$, defined by $L \otimes \mathrm{id}(a \otimes b) = L(a) \otimes b$, does not even extend to a bounded map from $\mathcal{A}_1 \otimes_{\max} \mathcal{B}$ into $\mathcal{A}_2 \otimes_{\max} \mathcal{B}$. However, it is the case that the tensor product of completely positive maps yields a completely positive map on the tensor product in the max norm. This fact is a consequence of the following commutant lifting theorem of Arveson [6].

For a set $\mathcal{S} \subseteq B(\mathcal{H})$, let $\mathcal{S}' = \{T \in B(\mathcal{H}): TS = ST \text{ for all } S \in \mathcal{S}\}$ denote the *commutant* of $\mathcal{S}$. Note that $\mathcal{S}'$ is always an algebra and that when $\mathcal{S}$ is self-adjoint, $\mathcal{S}'$ is a C^*-algebra.

Theorem 12.7 (Arveson). *Let $\mathcal{H}$ and $\mathcal{K}$ be Hilbert spaces, let $\mathcal{B} \subseteq B(\mathcal{K})$ be a C^*-algebra containing $1_{\mathcal{K}}$, and let $V: \mathcal{H} \to \mathcal{K}$ be a bounded linear transformation with $\mathcal{B}V\mathcal{H}$ norm-dense in $\mathcal{K}$. Then for every $T \in (V^*\mathcal{B}V)'$, there exists a unique $T_1 \in \mathcal{B}'$ such that $VT = T_1V$. Furthermore, the map $T \to T_1$ is a $*$-homomorphism of $(V^*\mathcal{B}V)'$ onto $\mathcal{B}' \cap \{VV^*\}'$.*

Proof. Let $A_1, \ldots, A_n$ be in $\mathcal{B}$, and let $h_1, \ldots, h_n$ be in $\mathcal{H}$. Note that if a T_1 with the desired properties exists, then

$$T_1\left(\sum_{i=1}^{n} A_i V h_i\right) = \sum_{i=1}^{n} A_i V T h_i. \qquad (*)$$

Since the vectors appearing in the left-hand side of the above formula are dense in $\mathcal{K}$, this shows that such a T_1, provided that it exists, is necessarily unique.

Thus, we need to prove that the above formula yields a well-defined, bounded operator. Note that if P and Q are commuting positive operators and x is a vector, then $\langle PQx, x\rangle = \langle PQ^{1/2}x, Q^{1/2}x\rangle \le \|P\|\langle Qx, x\rangle$. We have that

$$\left\|\sum_{i=1}^{n} A_i V T h_i\right\|^2 = \sum_{i,j=1}^{n} \langle T^*V^*A_i^*A_jVTh_j, h_i\rangle$$
$$= \sum_{i,j=1}^{n} \langle T^*TV^*A_i^*A_jVh_j, h_i\rangle = \langle PQx, x\rangle,$$

where $Q = (V^*A_i^*A_jV)_{i,j=1}^n$, $x = h_1 \oplus \cdots \oplus h_n$, and P is the diagonal $n \times n$ operator, whose entries are T^*T. Thus, P and Q are positive and commute, and so

$$\left\|\sum_{i=1}^{n} A_i V T h_i\right\|^2 \le \|P\|\langle Qx, x\rangle = \|T\|^2 \cdot \left\|\sum_{i=1}^{n} A_i V h_i\right\|^2.$$

This equation shows that the formula $(*)$ yields a well-defined, bounded operator with $\|T_1\| \le \|T\|$.

From the formula $(*)$ it is clear that the map π, given by $\pi(T) = T_1$, is a homomorphism into $\mathcal{B}'$. To see that it is a $*$-homomorphism, calculate

$$\langle \pi(T)A_1Vh_1, A_2Vh_2\rangle = \langle V^*A_2^*A_1VTh_1, h_2\rangle$$
$$= \langle TV^*A_2^*A_1Vh_1, h_2\rangle = \langle A_1Vh_1, A_2VT^*h_2\rangle = \langle A_1Vh_1, \pi(T^*)A_2Vh_2\rangle.$$

Since linear combinations of vectors of the above form are dense in $\mathcal{K}$, we have that $\pi(T)^* = \pi(T^*)$.

To see that T_1 is also in the commutant of VV^*, observe

$$T_1 VV^* = VTV^* = V(VT^*)^* = V(T_1^* V)^* = VV^* T_1.$$

Finally, to see that π is onto $\mathcal{B}' \cap \{VV^*\}'$, let $X \in \mathcal{B}' \cap \{VV^*\}'$ and let $V^* = W^* P$ be the polar decomposition of V^*, so that X commutes with $P = (VV^*)^{1/2}$. Let $T = W^* X W$; then for $A \in \mathcal{B}$,

$$\begin{aligned} TV^*AV &= W^*XWW^*PAPW = W^*XPAPW = W^*PAPXW \\ &= V^*APWW^*XW = V^*AVT, \end{aligned}$$

so that T is in the commutant of $V^*\mathcal{B}V$. Also, $VT = PWW^*XW = PXW = XPW = XV$, so that $\pi(T) = X$. Thus, the $*$-homomorphism π is indeed onto $\mathcal{B}' \cap \{VV^*\}'$. □

Note that T is in the kernel of the above $*$-homomorphism π if and only if $VT = 0$. Thus, if V has trivial kernel, then π is a $*$-isomorphism. It is also worthwhile to note that even when the map π has a kernel, the map $\theta(X) = W^*XW$ defines a completely positive splitting of π, i.e., $\pi \circ \theta(X) = X$.

Theorem 12.8. *Let $\mathcal{A}_1, \mathcal{A}_2$, and $\mathcal{B}$ be unital C^*-algebras, and let $\theta_i \colon \mathcal{A}_i \to \mathcal{B}, i = 1, 2$, be completely positive maps with commuting ranges. Then there exists a completely positive map $\theta_1 \otimes_{\max} \theta_2 \colon \mathcal{A}_1 \otimes_{\max} \mathcal{A}_2 \to \mathcal{B}$ with $\theta_1 \otimes_{\max} \theta_2(a_1 \otimes a_2) = \theta_1(a_1)\theta_2(a_2)$.*

Proof. Clearly, we may assume that $\mathcal{B} = B(\mathcal{H})$. Let $(\pi_1, V_1, \mathcal{K}_1)$ be a minimal Stinespring representation of θ_1, and let $\gamma_1 \colon (V_1^*\pi_1(\mathcal{A}_1)V_1)' \to \pi_1(\mathcal{A}_1)' \cap \{V_1V_1^*\}'$ be the $*$-homomorphism of Theorem 12.7. We then have that $\pi_1 \colon \mathcal{A}_1 \to B(\mathcal{K}_1)$, $\tilde{\theta}_2 = \gamma_1 \circ \theta_2 \colon \mathcal{A}_2 \to B(\mathcal{K}_1)$ are completely positive maps with commuting ranges.

Let $V_1^* = W_1^* P_1$ be the polar decomposition of V_1^*. By the remarks following Theorem 12.7, $\theta_2(a_2) - W_1^*\tilde{\theta}_2(a_2)W_1$ is in the kernel of γ_1 and consequently, $V_1(\theta_2(a_2) - W_1^*\tilde{\theta}_2(a_2)W_1) = 0$. Thus, $V_1\theta_2(a_2) = V_1W_1^*\tilde{\theta}_2(a_2)W_1 = P_1\tilde{\theta}_2(a_2)W_1 = \tilde{\theta}_2(a_2)P_1W_1 = \tilde{\theta}_2(a_2)V_1$, since $\tilde{\theta}_2(a_2)$ commutes with $V_1V_1^* = P_1^2$ and hence with P_1. Hence, we have that

$$V_1^*\pi_1(a_1)\tilde{\theta}_2(a_2)V_1 = V_1^*\pi_1(a_1)V_1\theta_2(a_2) = \theta_1(a_1)\theta_2(a_2).$$

Repeating the above argument, we let $(\pi_2, \mathcal{K}_2, V_2)$ be a minimal Stinespring representation of $\tilde{\theta}_2$ and let $\gamma_2 \colon (V_2^*\pi_2(\mathcal{A}_2)V_2)' \to \pi_2(\mathcal{A}_2)' \cap \{V_2V_2^*\}'$. Then

$\tilde{\pi}_1 = \gamma_2 \circ \pi_1\colon \mathcal{A}_1 \to B(\mathcal{K}_2)$ is a $*$-homomorphism whose range commutes with $\pi_2(\mathcal{A}_2)$. Also $V_2^* \pi_2(a_2) \tilde{\pi}_1(a_1) V_2 = \tilde{\theta}_2(a_2)\pi(a_1)$.

Finally, by the universal property of the max norm, we have a $*$-homomorphism, $\pi\colon \mathcal{A}_1 \otimes_{\max} \mathcal{A}_2 \to B(\mathcal{K}_2)$ with $\pi(a_1 \otimes a_2) = \tilde{\pi}_1(a_1)\pi_2(a_2)$. Let $V\colon \mathcal{H} \to \mathcal{K}_2$ be defined by $V = V_2 V_1$, so that $\theta\colon \mathcal{A}_1 \otimes_{\max} \mathcal{A}_2 \to B(\mathcal{H})$, defined by $\theta(x) = V^* \pi(x) V$, is completely positive. Finally,

$$\theta(a_1 \otimes a_2) = V_1^* V_2^* \tilde{\pi}_1(a_1)\pi_2(a_2) V_2 V_1 = V_1^* \pi_1(a_1) \tilde{\theta}_2(a_2) V_1 = \theta_1(a_1)\theta_2(a_2).$$

Thus, $\theta_1 \otimes_{\max} \theta_2 = \theta$ is the desired completely positive map. □

Because of the different properties of the max and min norm, it is important to know when they coincide, that is, when there is a unique C^*-cross-norm on $\mathcal{A} \otimes \mathcal{B}$. A C^*-algebra $\mathcal{A}$ that has the property that the max and min C^*-cross-norms coincide for every unital C^*-algebra $\mathcal{B}$ is called *nuclear*. There is a deep and elegant theory characterizing these C^*-algebras. See Lance [135] for an excellent survey. For our current purposes it will be enough to know that commutative C^*-algebras are nuclear. It is also valuable to note that M_n is nuclear (Exercise 12.4), so that the norm we defined on $M_n(\mathcal{A}) = M_n \otimes \mathcal{A}$ is the unique C^*-cross-norm.

Proposition 12.9. *Let X be a compact Hausdorff space. Then $C(X)$ is nuclear.*

Proof. Let $\mathcal{B}$ be a unital C^*-algebra, and let $\pi_1\colon C(X) \to B(\mathcal{H})$, $\pi_2\colon \mathcal{B} \to B(\mathcal{H})$, be $*$-homomorphisms with commuting ranges. It will be sufficient to fix $\sum_{i=1}^n f_i \otimes b_i$ in $C(X) \otimes \mathcal{B}$ and show that its max and min norm coincide. By Proposition 12.5 we have that

$$\left\| \sum_{i=1}^n f_i \otimes b_i \right\|_{\min} = \sup \left\{ \left\| \sum_{i=1}^n f_i(x) b_i \right\| : x \in X \right\}.$$

Let E be the $B(\mathcal{H})$-valued spectral measure associated with the $*$-homomorphism π_1. If B is a Borel set in X, then the projection $E(B)$ will commute with $\pi_2(\mathcal{B})$, since $\pi_2(\mathcal{B})$ commutes with $\pi_1(C(X))$. This implies that

$$\left\| \sum_{i=1}^n \pi_1(f_i)\pi_2(b_i) \right\| = \sup \left\{ \left\| E(B) \left(\sum_{i=1}^n \pi_1(f_i)\pi_2(b_i) \right) \right\|, \right.$$
$$\left. \left\| (E(X \backslash B) \left(\sum_{i=1}^n \pi_1(f_i)\pi_2(b_i) \right) \right\| \right\}.$$

Consequently, if $\{U_\lambda\}$ is any open cover of X,

$$\left\|\sum_{i=1}^{n} \pi_1(f_i)\pi_2(b_i)\right\| = \sup_{\lambda} \left\{ \left\| E(U_\lambda)\left(\sum_{i=1}^{n} \pi_1(f_i)\pi_2(b_i)\right)\right\| \right\}.$$

Now fix $\varepsilon > 0$, and for each $x \in X$, choose an open neighborhood U_x of x such that $|f_i(x) - f_i(y)| < \varepsilon$ for $y \in U_x, i = 1, \ldots, n$. This implies that $\|E(U_x)\pi_1(f_i) - f_i(x)E(U_x)\| < \varepsilon$. Thus,

$$\left\| E(U_x)\left(\sum_{i=1}^{n}(\pi_1(f_i) - f_i(x))\pi_2(b_i)\right)\right\| < \varepsilon(\|b_i\| + \cdots + \|b_n\|).$$

Since the collection $\{U_\lambda\}$ forms an open cover, we have

$$\begin{aligned}\left\|\sum_{i=1}^{n} \pi_1(f_i)\pi_2(b_i)\right\| &\le \sup\left\{\left\|\sum_{i=1}^{n} f_i(x)\pi_2(b_i)\right\| : x \in X\right\} \\ &\quad + \varepsilon(\|b_1\| + \cdots + \|b_n\|) \\ &\le \left\|\sum_{i=1}^{n} f_i \otimes b_i\right\|_{\min} + \varepsilon(\|b_1\| + \cdots + \|b_n\|).\end{aligned}$$

Finally, using the facts that ε was arbitrary and that the max norm is the supremum over all such π_1 and π_2, we have $\|\sum_{i=1}^{n} f_i \otimes b_i\|_{\max} \le \|\sum_{i=1}^{n} f_i \otimes b_i\|_{\min}$, which completes the proof. □

We are now in a position to discuss some of the applications of the tensor theory to operator theory. Recall that a set of operators $\{T_i\}$ is said to doubly commute if $T_i^*T_j = T_jT_i^*$ and $T_iT_j = T_jT_i$ for $i \neq j$. This is equivalent to requiring that the C^*-algebras generated by each of these operators commutes with the C^*-algebra generated by any of the other operators, but does not require that each of these C^*-algebras be commutative.

Theorem 12.10 (Sz.-Nagy–Foias). *Let $\{T_i\}_{i=1}^n$ be a doubly commuting family of contractions on a Hilbert space $\mathcal{H}$. Then there exists a Hilbert space $\mathcal{K}$ containing $\mathcal{H}$ as a subspace, and a doubly commuting family of unitary operators $\{U_i\}_{i=1}^n$ on $\mathcal{K}$, such that*

$$T_1(k_1)\cdots T_n(k_n) = P_{\mathcal{H}}U_1^{k_1}\cdots U_n^{k_n}\big|_{\mathcal{H}}, \qquad \text{where} \quad T(k) = \begin{cases} T^k, & k \ge 0 \\ T^{*-k}, & k < 0. \end{cases}$$

Moreover, if $\mathcal{K}$ is the smallest reducing subspace for the family $\{U_i\}_{i=1}^n$ containing $\mathcal{H}$, then $\{U_i\}_{i=1}^n$ is unique up to unitary equivalence. That is, if $\{U_i'\}_{i=1}^n$ and

$\mathcal{K}'$ are another such set and space, then there is a unitary $W: \mathcal{K} \to \mathcal{K}'$ leaving $\mathcal{H}$ fixed such that $WU_iW^* = U_i'$, $i = 1, \ldots, n$.

Proof. First, assume that $n = 2$. Then we have completely positive maps $\theta_i: C(\mathbb{T}) \to B(\mathcal{H})$ defined by $\theta_i(p + \bar{q}) = p(T_i) + q(T_i)^*$, $i = 1, 2$. Since the range of θ_1 commutes with the range of θ_2, there is a completely positive map $\theta_1 \otimes_{\max} \theta_2: C(\mathbb{T}) \otimes_{\max} C(\mathbb{T}) \to B(\mathcal{H})$ satisfying $\theta_1 \otimes_{\max} \theta_2(f_1 \otimes f_2) = \theta_1(f_1)\theta_2(f_2)$.

However, we have that $C(\mathbb{T}) \otimes_{\max} C(\mathbb{T})$ is $*$-isomorphic to $C(\mathbb{T} \times \mathbb{T})$. Thus, we have a completely positive map $\theta: C(\mathbb{T}^2) \to B(\mathcal{H})$ with $\theta(z_1^{k_1} z_2^{k_2}) = T_1(k_1)T_2(k_2)$, where z_1 and z_2 are the coordinate functions on $\mathbb{T}^2$. The result now follows for $n = 2$ by considering the Stinespring representation of θ.

For $n > 2$, by using the associativity of the tensor product (Exercise 12.7), and arguing as above, one obtains a completely positive map $\theta: C(\mathbb{T}^n) \to B(\mathcal{H})$ with $\theta(z_1^{k_1} \cdots z_n^{k_n}) = T_1(k_1) \cdots T_n(k_n)$. □

The above result, for $n = 2$, is weaker than Ando's Dilation Theorem, since that result does not assume that the operators doubly commute.

For operator systems $\mathcal{S}_i \subseteq \mathcal{A}_i$, $i = 1, 2$, we wish to define a commutative max norm. Because of the projective properties of the max norm, it is *not* sufficient to just consider the norm induced by the inclusion $\mathcal{S}_1 \otimes \mathcal{S}_2 \subseteq \mathcal{A}_1 \otimes_{\max} \mathcal{A}_2$. Instead we take Theorem 12.8 as our defining property. If $\theta_i: \mathcal{S}_i \to B(\mathcal{H})$, $i = 1, 2$, are maps with commuting ranges, we always have a well-defined map $\theta_1 \otimes \theta_2: \mathcal{S}_1 \otimes \mathcal{S}_2 \to B(\mathcal{H})$. For $(x_{i,j}) \in M_n(\mathcal{S}_1 \otimes \mathcal{S}_2)$, we set

$$\|(x_{i,j})\|_{\max} = \sup\{\|(\theta_1 \otimes \theta_2(x_{i,j}))\|: \theta_\ell: \mathcal{S}_\ell \to B(\mathcal{H}),\ \ell = 1, 2\},$$

where θ_1 and θ_2 are unital, completely positive maps with commuting ranges and $\mathcal{H}$ is an arbitrary Hilbert space. By considering the direct sum of sufficiently many of the maps $\theta_1 \otimes \theta_2$, we can obtain a unital map $\gamma: \mathcal{S}_1 \otimes \mathcal{S}_2 \to B(\mathcal{H})$ with the property that $\|(x_{i,j})\|_{\max} = \|(\gamma(x_{i,j}))\|$ for all $(x_{i,j}) \in M_n(\mathcal{S}_1 \otimes \mathcal{S}_2)$ and all n.

We then define $\mathcal{S}_1 \otimes_{\max} \mathcal{S}_2$ to be the concrete operator system that is the image of $\gamma(\mathcal{S}_1 \otimes \mathcal{S}_2)$.

Note that if $\mathcal{S}_1$ and $\mathcal{S}_2$ were actually C^*-algebras, then by Theorem 12.8, the norms on $M_n(\mathcal{S}_1 \otimes \mathcal{S}_2)$ as above would correspond to the original definition. That is, given two C^*-algebras, their maximal tensor product as operator systems coincides with their maximal tensor product as C^*-algebras.

We summarize the properties of $\mathcal{S}_1 \otimes_{\max} \mathcal{S}_2$ as follows:

Proposition 12.11. *Let $\mathcal{S}_1$ and $\mathcal{S}_2$ be operator systems, $\mathcal{B}$ be a C^*-algebra, and $\theta_i \colon \mathcal{S}_i \to \mathcal{B}$, $i = 1, 2$, be completely positive maps with commuting ranges. Then:*

(i) there exists a completely positive map $\theta_1 \otimes_{\max} \theta_2 \colon \mathcal{S}_1 \otimes_{\max} \mathcal{S}_2 \to \mathcal{B}$ with $\theta_1 \otimes_{\max} \theta_2(a_1 \otimes a_2) = \theta_1(a_1)\theta_2(a_2)$;

(ii) an element $(x_{i,j})$ in $M_n(\mathcal{S}_1 \otimes_{\max} \mathcal{S}_2)$ is positive if and only if the matrix $(\theta_1 \otimes_{\max} \theta_2(x_{i,j}))$ is positive in $M_n(\mathcal{B})$ for every pair θ_1, θ_2 of completely positive maps with commuting ranges and every C^*-algebra $\mathcal{B}$.

Proof. Exercise 12.5. □

As with C^*-algebras, it is important to know when the min and max norms coincide on operator systems. If $\mathcal{A}$ is a nuclear C^*-algebra, then for every operator system $\mathcal{S}$, the min and max norms on $\mathcal{A} \otimes \mathcal{S}$ coincide (Exercise 12.6). Thus, we define an operator system $\mathcal{S}$ to be *nuclear* if for every operator system $\mathcal{T}$, the min and max norms coincide on $\mathcal{S} \otimes \mathcal{T}$.

We now have all the necessary concepts to discuss joint spectral sets and joint dilations. Let $X_i \subseteq \mathbb{C}$ be compact, $i = 1, \dots, n$. We set $X = X_1 \times \cdots \times X_n$ and define $\partial_d X = \partial X_1 \times \cdots \times \partial X_n$. We let $\mathcal{R}_d(X)$ denote the subalgebra of $C(\partial_d X)$ spanned by functions of the form $r_1(z_1) \cdots r_n(z_n)$, $r_i \in \mathcal{R}(X_i)$, and let $\mathcal{R}(X)$ denote the subalgebra of $C(\partial_d X)$ consisting of the rational functions, $\mathcal{R}(X) = \{p(z_1, \dots, z_n)/q(z_1, \dots, z_n) \colon p, q$ are polynomials, $q \neq 0$ on $X\}$.

The algebra $\mathcal{R}_d(X)$ is contained in $\mathcal{R}(X)$, is algebraically isomorphic to the tensor product $\mathcal{R}(X_1) \otimes \cdots \otimes \mathcal{R}(X_n)$, and, in general, is not dense in $\mathcal{R}(X)$. Suppose that $T_i \in B(\mathcal{H})$, X_i is a K_i-spectral set for T_i, $i = 1, \dots, n$, the set $\{T_i\}_{i=1}^n$ commutes, and we have a well-defined homomorphism, $\rho \colon \mathcal{R}_d(X) \to B(\mathcal{H})$, given by $\rho(r_1 \cdots r_n) = r_1(T_1) \cdots r_n(T_n)$. In this case, we call X a *joint K-spectral set* for $\{T_i\}_{i=1}^n$ provided that $\|\rho\| \leq K$, and a *complete joint K-spectral set* provided that $\|\rho\|_{\mathrm{cb}} \leq K$.

If there exists a family of commuting normals $\{N_i\}_{i=1}^n$ on a Hilbert space $\mathcal{K}$, containing $\mathcal{H}$ with $\sigma(N_i) \subseteq \partial X_i$, then we call $\{N_i\}_{i=1}^n$ a *joint normal $\partial_d X$-dilation* of $\{T_i\}_{i=1}^n$ provided that $r_1(T_1) \cdots r_n(T_n) = P_{\mathcal{H}} r_1(N_1) \cdots r_n(N_n)|_{\mathcal{H}}$ for all $r_i \in \mathcal{R}(X_i)$. The following is immediate.

Proposition 12.12. *Let $\{T_i\}_{i=1}^n$ be a family of commuting operators. Then $\{T_i\}_{i=1}^n$ has a joint normal $\partial_d X$-dilation if and only if X is a complete joint spectral set for $\{T_i\}_{i=1}^n$. There exists an invertible S with $\|S^{-1}\| \cdot \|S\| \leq K$ such* *that $\{S^{-1}T_iS\}_{i=1}^n$ has a joint normal $\partial_d X$-dilation if and only if X is a complete joint K-spectral set for $\{T_i\}_{i=1}^n$.*

Theorem 12.13. *Let $\{T_i\}_{i=1}^n$ be doubly commuting operators, and let X_i be a complete spectral set for T_i. If $\mathcal{R}(X_i)+\overline{\mathcal{R}(X_i)}$ is dense in $C(\partial X_i)$ for $i = 1, \ldots, n-1$, then $\{T_i\}_{i=1}^n$ has a joint normal $\partial_d X$-dilation.*

Proof. Assume $n=2$, and let $\mathcal{S}_i$ denote the closure of $\mathcal{R}(X_i)+\overline{\mathcal{R}(X_i)}$ in $C(\partial X_i)$, so that $\mathcal{S}_1 = C(\partial X_1)$. By hypothesis, we have completely positive maps $\theta_i\colon \mathcal{S}_i \to B(\mathcal{H})$ satisfying $\theta_i(f+\bar{g}) = f(T_i)+g(T_i)^*$, with commuting ranges. Thus, we have a completely positive map $\theta_1 \otimes_{\max} \theta_2\colon \mathcal{S}_1 \otimes_{\max} \mathcal{S}_2 \to B(\mathcal{H})$. But $\mathcal{S}_1 \otimes_{\max} \mathcal{S}_2 = \mathcal{S}_1 \otimes_{\min} \mathcal{S}_2$ and $\mathcal{S}_1 \otimes_{\min} \mathcal{S}_2$ is completely isometrically embedded in $C(\partial X_1) \otimes_{\min} C(\partial X_2) = C(\partial_d X)$.

Thus, for $(f_{i,j}) \in M_n(\mathcal{R}_d(X))$, we have $\|(\rho(f_{i,j}))\| = \|(\theta_1 \otimes_{\max} \theta_2(f_{i,j}))\| \le \|(f_{i,j})\|_{\max} = \|(f_{i,j})\|_{\min}$, where $\rho\colon \mathcal{R}_d(X) \to B(\mathcal{H})$ is the homomorphism defined by $\rho(r_1(z_1)r_2(z_2)) = r_1(T_1)r_2(T_2)$. But since $\|(f_{i,j})\|_{\min}$ is just the norm of $(f_{i,j})$ in $M_n(C(\partial_d X))$, we have that X is a complete joint spectral set for $\{T_1, T_2\}$.

The proof for $n > 2$ is analogous. □

Lemma 12.14. *Let X be a compact Hausdorff space, $\mathcal{S} \subseteq C(X)$ an operator system of codimension n, and $\mathcal{T}$ another operator system. Then for $(a_{i,j}) \in M_k(\mathcal{S} \otimes \mathcal{T})$, one has $\|(a_{i,j})\|_{\max} \le (2n+1)\|(a_{i,j})\|_{\min}$.*

Proof. By Lemma 11.9, for any $\varepsilon > 0$ there is a completely positive map $\phi\colon C(X) \to \mathcal{S}$ with the property that for $f \in \mathcal{S}$, $\phi(f) = f + s(f)\cdot 1$, where s is a positive linear functional with $\|s\| \le n+\varepsilon$, $\|\phi\| \le n+1+\varepsilon$. Let $a = \sum_{i=1}^k f_i \otimes b_i \in \mathcal{S} \otimes \mathcal{T}$, and note that we have a completely positive map $\phi \otimes \mathrm{id}\colon C(X) \otimes_{\max} \mathcal{T} \to \mathcal{S} \otimes_{\max} \mathcal{T}$ with $\|\phi \otimes \mathrm{id}\| = \|\phi \otimes \mathrm{id}(1 \otimes 1)\| \le n+1+\varepsilon$. Note that a can also be regarded as an element of $C(X) \otimes \mathcal{T}$. But since $C(X)$ is nuclear and since the min norm is injective, the max and min norms of a in $C(X) \otimes \mathcal{T}$ and the min norm of a in $\mathcal{S} \otimes \mathcal{T}$ all coincide. Thus, we have that $\|\phi \otimes \mathrm{id}(a)\|_{\max} \le (n+1+\varepsilon)\|a\|_{\min}$. But $\phi \otimes \mathrm{id}(a) = a + \sum_{i=1}^k s(f_i)1 \otimes b_i$, and so

$$\|a\|_{\max} \le \|\phi \otimes \mathrm{id}(a)\|_{\max} + \left\| \sum s(f_i)b_i \right\| \le (n+1+\varepsilon)\|a\|_{\min} + (n+\varepsilon)\|a\|_{\min}.$$

The last inequality follows by noting that $s \otimes_{\min} \mathrm{id}\colon \mathcal{S} \otimes_{\min} \mathcal{T} \to \mathcal{T}$ satisfies $\|s \otimes_{\min} \mathrm{id}\| = \|s\|$. Thus, $\|a\|_{\max} \le (2n+1)\|a\|_{\min}$, since ε was arbitrary.

The proof for matrices is similar and uses the complete positivity of all the maps. □

Theorem 12.15. *Let $\{T_i\}_{i=1}^{n+1}$ be doubly commuting operators with X_i a complete spectral set for T_i. If $\mathcal{R}(X_i)$ is a hypo-Dirichlet algebra of codimension k_i on $\partial X_i, i = 1, \dots, n$, then there exists a similarity S with $\|S\| \cdot \|S^{-1}\| \leq (2k_1 + 1) \cdots (2k_n + 1)$ such that $\{S^{-1}T_iS\}_{i=1}^{n+1}$ has a joint, normal $\partial_d X$-dilation.*

Proof. First assume that $n = 1$. Let $\mathcal{S}_i = \mathcal{R}(X_i) + \overline{\mathcal{R}(X_i)}$, and let $\phi_i \colon \mathcal{S}_i \to B(\mathcal{H})$ be defined by $\phi_i(f + \bar{g}) = f(T_i) + g(T_i)^*$. Then ϕ_1 and ϕ_2 have commuting ranges, and so there is a completely positive map $\phi_1 \otimes_{\max} \phi_2 \colon \mathcal{S}_1 \otimes_{\max} \mathcal{S}_2 \to B(\mathcal{H})$ and $\|\phi_1 \otimes_{\max} \phi_2\|_{cb} = 1$. But by the above lemma, the map $\theta_1 \otimes \theta_2 \colon \mathcal{S}_1 \otimes \mathcal{S}_2 \to B(\mathcal{H})$ will extend to be completely bounded in the min norm, with $\|\theta_1 \otimes_{\min} \theta_2\| \leq 2k_1 + 1$. Thus, X is a complete, joint $(2k_1 + 1)$-spectral set for $\{T_1, T_2\}$, from which the result follows.

To argue for an arbitrary n, note that by repeated applications of Proposition 12.11 and the above lemma, the identity map on $\mathcal{S}_1 \otimes \cdots \otimes \mathcal{S}_n$ extends to a completely bounded map from $\mathcal{S}_1 \otimes_{\min} \cdots \otimes_{\min} \mathcal{S}_n$ to $\mathcal{S}_1 \otimes_{\max} \cdots \otimes_{\max} \mathcal{S}_n$, with $\|\mathrm{id} \otimes \cdots \otimes \mathrm{id}\|_{cb} \leq (2k_1 + 1) \cdots (2k_n + 1)$. The proof is now completed as in the $n = 1$ case. □

Theorem 12.16. *Let $\{T_i\}_{i=1}^{n}$ be doubly commuting operators with X_i a spectral set for T_i. If $\mathcal{R}(X_i)$ is a hypo-Dirichlet algebra of codimension k_i on ∂X_i, then there exists a similarity S with $\|S\| \cdot \|S^{-1}\| \leq (2k_1 + 1) \cdots (2k_n + 1)$ such that $\{S^{-1}T_iS\}_{i=1}^{n}$ has a joint, normal $\partial_d X$-dilation.*

Proof. Fix $\varepsilon > 0$, and let $\phi_i \colon C(\partial X_i) \to \mathcal{S}_i$, where $\mathcal{S}_i$ is the closure of $\mathcal{R}(X_i) + \overline{\mathcal{R}(X_i)}$, and s_i be as in Lemma 11.9. Since X_i is spectral for T_i, we have positive maps $\theta_i \colon \mathcal{S}_i \to B(\mathcal{H})$ with commuting ranges. Hence $\alpha_i = \theta_i \circ \phi_i \colon C(\partial X_i) \to B(\mathcal{H})$ are completely positive. Set $\beta_i = \theta_i \circ s_i$. Again using the nuclearity of $C(\partial X_i)$, we may form the min tensor of any combination of the maps α_i and β_i and obtain a completely positive map on $C(\partial_d X)$. But for f_i in $\mathcal{S}_i$, $\theta_i(f_i) = \alpha_i(f_i) - \beta_i(f_i)$. Hence on $\mathcal{S}_1 \otimes \cdots \otimes \mathcal{S}_n$, $\theta_1 \otimes \cdots \otimes \theta_n = (\alpha_1 - \beta_1) \otimes \cdots \otimes (\alpha_n - \beta_n)$, which, when the right hand side is expanded, expresses $\theta_1 \otimes \cdots \otimes \theta_n$ as a difference of sums of completely positive maps on $C(\partial_d X)$. Computing the sum of the norms of each of these maps and letting ε tend to 0 yields the result. □

We close this chapter with one final result. Although the hypotheses of the theorem look quite restrictive, many operators that arise in operator theory have the property that the C^*-algebra that they generate is nuclear. For example, Toeplitz operators, subnormal operators, and essentially normal operators usually generate nuclear C^*-algebras.

Theorem 12.17. *Let $\{T_i\}_{i=1}^{n+1}$ be doubly commuting operators with X_i completely K_i-spectral for T_i. If the C^*-algebra generated by each T_i is nuclear, $i = 1, \ldots, n$, then there exists a similarity S with $\|S\| \cdot \|S^{-1}\| \leq K_1 \cdots K_{n+1}$ such that $\{S^{-1}T_iS\}_{i=1}^{n+1}$ has a joint normal $\partial_d X$-dilation.*

Proof. By hypothesis, the map $\rho_i \colon \mathcal{R}(X_i) \to B(\mathcal{H})$ given by $\rho_i(f) = f(T_i)$ is completely bounded. Extend ρ_i to $\theta_i \colon C(\partial X_i) \to B(\mathcal{H})$ with $\|\theta_i\|_{\text{cb}} = \|\rho_i\|_{\text{cb}}$. We may then form

$$\theta_1 \otimes_{\min} \cdots \otimes_{\min} \theta_n = \theta \colon C(\partial_d X) \to B(\mathcal{H}) \otimes_{\min} \cdots \otimes_{\min} B(\mathcal{H}),$$

and we shall have that $\|\theta\|_{\text{cb}} \leq \|\rho_1\|_{\text{cb}} \cdots \|\rho_{n+1}\|_{\text{cb}}$. However, for $f \in \mathcal{R}_d(X)$, $\theta(f) = \rho_1 \otimes \cdots \otimes \rho_{n+1}(f)$ is in $C^*(T_1) \otimes \cdots \otimes C^*(T_{n+1})$, and the min norm on this latter algebra is the restriction of the min norm on $B(\mathcal{H}) \otimes_{\min} \cdots \otimes_{\min} B(\mathcal{H})$.

Hence we have that

$$\rho = \rho_1 \otimes \cdots \otimes \rho_{n+1} \colon \mathcal{R}_d(X) \to C^*(T_1) \otimes_{\min} \cdots \otimes_{\min} C^*(T_{n+1})$$

is completely bounded with $\|\rho\|_{\text{cb}} \leq \|\theta\|_{\text{cb}}$. Now the min and max norms will agree on this latter tensor product, and the tensor product in the max norm maps completely contractively to the C^*-algebra generated by $T_1, \ldots, T_{n+1}$. Thus, we have that the map $\tilde{\rho} \colon \mathcal{R}_d(X) \to B(\mathcal{H})$ defined by $\tilde{\rho}(f_1(z_1) \cdots f_{n+1}(z_{n+1})) = f_1(T_1) \cdots f_{n+1}(T_{n+1})$ is completely bounded, from which the result follows. □

There is a parallel theory for the maximal tensor product of operator algebras studied in [169]. Given unital operator algebras $\mathcal{A}_i$, $i = 1, 2$, and unital completely contractive homomorphisms $\pi_i \colon \mathcal{A}_i \to B(\mathcal{H})$, $i = 1, 2$, with commuting ranges, one defines a homomorphism $\pi \colon \mathcal{A}_1 \otimes \mathcal{A}_2 \to B(\mathcal{H})$ by setting $\pi(a_1 \otimes a_2) = \pi_1(a_1)\pi_2(a_2)$. Conversely, given any unital homomorphism $\pi \colon \mathcal{A}_1 \otimes \mathcal{A}_2 \to B(\mathcal{H})$, if we set $\pi_1(a_1) = \pi(a_1 \otimes 1)$ and $\pi_2(a_2) = \pi(1 \otimes a_2)$, then π_1 and π_2 are unital homomorphisms of $\mathcal{A}_1$ and $\mathcal{A}_2$ with commuting ranges.

Consequently we call a unital homomorphism $\pi \colon \mathcal{A}_1 \otimes \mathcal{A}_2 \to B(\mathcal{H})$ *admissible* provided that π_1 and π_2 are completely contractive.

For $(x_{ij}) \in M_n(\mathcal{A}_1 \otimes \mathcal{A}_2)$, we define

$$\|(x_{ij})\|_{\max} = \sup\{\|(\pi(x_{ij}))\| \colon \pi \,\text{admissible}\}$$

and let $\mathcal{A}_1 \otimes_{\max} \mathcal{A}_2$ denote $\mathcal{A}_1 \otimes \mathcal{A}_2$ endowed with this family of matrix norms. By considering the direct sum of a sufficiently large set of admissible homomorphisms, one obtains a completely isometric unital homomorphism $\pi: \mathcal{A}_1 \otimes_{\max} \mathcal{A}_2 \to B(\mathcal{H})$ for some Hilbert space $\mathcal{H}$. Thus $\mathcal{A}_1 \otimes_{\max} \mathcal{A}_2$ can be identified, completely isometrically isomorphically, with the concrete operator algebra $\pi(\mathcal{A}_1 \otimes_{\max} \mathcal{A}_2)$.

The universal operator algebra for n-tuples of commuting contractions $(\mathcal{P}_n, \|\cdot\|_u)$, discussed in Chapter 5, can be identified with the maximal tensor product $\mathcal{P}(\mathbb{D}) \otimes_{\max} \cdots \otimes_{\max} \mathcal{P}(\mathbb{D})$ (n copies), where $\mathcal{P}(\mathbb{D})$ is the operator algebra of polynomials in one variable equipped with the supremum norm over the unit disk.

In Chapter 16 we will give an abstract characterization of operator algebras, in analogy with the Gelfand–Naimark–Segal characterization of C^*-algebras. This characterization will allow us to discuss abstract operator algebras like $\mathcal{A}_1 \otimes_{\max} \mathcal{A}_2$, even in cases where we may not be able to explicitly exhibit a completely isometric representation as an algebra of operators on a Hilbert space.

Notes

For a more thorough treatment of tensor products of operator systems, see Choi and Effros [49]. In particular, they obtain an abstract characterization of operator systems, which we shall present in Chapter 13.

Lance's survey article [135] and Takesaki's text [233] are two excellent sources for a further introduction to tensor products and nuclearity.

See Dash [70] and Pott [196] for further results on joint spectral sets.

Power and the author [169] prove that for any unital operator algebra $\mathcal{A}$, we have that $\mathcal{A} \otimes_{\min} \mathcal{P}(\mathbb{D}) = \mathcal{A} \otimes_{\max} \mathcal{P}(\mathbb{D})$ if and only if $\mathcal{A} \otimes_{\min} \mathcal{T}_n = \mathcal{A} \otimes_{\max} \mathcal{T}_n$ for all n, where $\mathcal{T}_n$ denotes the algebra of $n \times n$ upper triangular matrices.

Exercises

12.1 Let $\mathcal{A}_i, i = 1, 2$, be unital C^*-algebras, and let $f\colon \mathcal{A}_1 \to \mathbb{C}$ be a bounded linear functional. Prove that there exists a completely bounded map $F\colon \mathcal{A}_1 \otimes_{\min} \mathcal{A}_2 \to \mathcal{A}_2$ with $\|F\|_{\text{cb}} = \|f\|$ such that $F(a_1 \otimes a_2) = f(a_1)a_2$. If f is positive, prove that F is completely positive.

12.2 Let $\mathcal{A}$ and $\mathcal{B}$ be unital C^*-algebras. Prove that if there exists a constant c such that $\|x\|_{\max} \le c\|x\|_{\min}$ for all $x \in \mathcal{A} \otimes \mathcal{B}$, then $\|x\|_{\max} = \|x\|_{\min}$.

12.3 Let $\mathcal{A}$ and $\mathcal{B}$ be unital C^*-algebras. Verify the following containments:

$$\left\{ \sum_{i,j} a_{i,j} \otimes b_{i,j} \colon (a_{i,j}) \in M_n(\mathcal{A})^+, (b_{i,j}) \in M_n(\mathcal{B})^+ \right\}$$

$$\subseteq (\mathcal{A} \otimes \mathcal{B}) \cap (\mathcal{A} \otimes_{\max} \mathcal{B})^+ \subseteq (\mathcal{A} \otimes \mathcal{B}) \cap (\mathcal{A} \otimes_{\min} \mathcal{B})^+.$$

12.4 Prove that every finite-dimensional C^*-algebra is nuclear.

12.5 Let $\mathcal{S}_i$, $i = 1, 2$, be operator systems, and let $\mathcal{B}$ be a unital C^*-algebra.

(i) Prove that if $\theta_i \colon \mathcal{S}_i \to \mathcal{B}$, $i = 1, 2$, are unital, completely positive maps with commuting ranges, then there exists a unital, completely contractive map $\theta_1 \otimes_{\max} \theta_2 \colon \mathcal{S}_1 \otimes_{\max} \mathcal{S}_2 \to \mathcal{B}$ with $\theta_1 \otimes_{\max} \theta_2(a \otimes b) = \theta_1(a)\theta_2(b)$.

(ii) Deduce that $\theta_1 \otimes_{\max} \theta_2$ is also completely positive.

(iii) Let $\theta_i \colon \mathcal{S}_i \to B(\mathcal{H})$, $i = 1, 2$, be completely positive maps with commuting ranges, and let $\theta_i(1) = P_i$. Prove that there exist unital, completely positive maps $\tilde{\theta}_i \colon \mathcal{S}_i \to B(\mathcal{H})$, $i = 1, 2$, with commuting ranges such that $\theta_i = P_i^{1/2} \tilde{\theta}_i P_i^{1/2}$, $i = 1, 2$.

(iv) Prove Proposition 12.11.

12.6 Let $\mathcal{A}$ be a C^*-algebra. Prove that if $\mathcal{A} \otimes_{\max} \mathcal{B} = \mathcal{A} \otimes_{\min} \mathcal{B}$ for every C^*-algebra $\mathcal{B}$, then $\mathcal{A} \otimes_{\max} \mathcal{S} = \mathcal{A} \otimes_{\min} \mathcal{S}$ for every operator system $\mathcal{S}$.

12.7 Let $\mathcal{S}_i$, $i = 1, 2, 3$, be operator systems, and let γ denote either the min or max norm.

(i) Prove that $\mathcal{S}_1 \otimes_\gamma (\mathcal{S}_2 \otimes_\gamma \mathcal{S}_3)$ and $(\mathcal{S}_1 \otimes_\gamma \mathcal{S}_2) \otimes_\gamma \mathcal{S}_3$ are completely isometrically isomorphic.

(ii) Prove that if $\mathcal{S}_1$ and $\mathcal{S}_2$ are nuclear, then $\mathcal{S}_1 \otimes_{\max} \mathcal{S}_2 = \mathcal{S}_1 \otimes_{\min} \mathcal{S}_2$ is nuclear.

12.8 (Holbrook–Sz.-Nagy) Let S and T be operators that doubly commute.

(i) Prove that if $S \in \mathcal{C}_\rho$ and $T \in \mathcal{C}_\sigma$, then $ST \in \mathcal{C}_{\rho\sigma}$.

(ii) Prove that $w(ST) \le \|S\| w(T)$.

(iii) Prove that $w(ST) \le 2w(S)w(T)$, and give an example to show that this inequality is sharp.

12.9 Let $\mathcal{S}_1$ and $\mathcal{S}_2$ be operator systems, let $(p_{ij}) \in M_n(\mathcal{S}_1)^+$, and let $(q_{k\ell}) \in M_m(\mathcal{S}_2)^+$. Prove that $(p_{ij} \otimes q_{k\ell}) \in M_{nm}(\mathcal{S}_1 \otimes_{\max} \mathcal{S}_2)^+$.

12.10 Let $\mathcal{S}_1$ and $\mathcal{S}_2$ be operator systems, and let $\theta_i \colon \mathcal{S}_i \to B(\mathcal{H})$, $i = 1, 2$, be unital complete isometries with commuting ranges. Define $\theta(S_1 \otimes S_2) = \theta_1(S_1)\theta_2(S_2)$. Prove that for all n,

$$M_n(\mathcal{S}_1 \otimes_{\max} \mathcal{S}_2)^+ \subseteq \{(x_{ij}) \in M_n(\mathcal{S}_1 \otimes \mathcal{S}_2) \colon (\theta(x_{ij})) \ge 0\}$$
$$\subseteq M_n(\mathcal{S}_1 \otimes_{\min} \mathcal{S}_2)^+.$$

Chapter 13

Abstract Characterizations of Operator Systems and Operator Spaces

The Gelfand–Naimark–Segal theorem gives an abstract characterization of the Banach $*$-algebras that can be represented $*$-isomorphically as C^*-subalgebras of $B(\mathcal{H})$ for some Hilbert space $\mathcal{H}$. Thus, the GNS theorem frees us from always having to regard C^*-algebras as concrete subalgebras of some Hilbert space. At the same time we may continue to regard them as concrete C^*-subalgebras when that might aid us in a proof. For example, proving that the quotient of a concrete C^*-subalgebra of some $B(\mathcal{H})$ by a two-sided ideal can again be regarded as a C^*-subalgebra of some $B(\mathcal{K})$ would be quite difficult without the GNS theorem. On the other hand defining the norm on $M_n(\mathcal{A})$ and many other constructions are made considerably easier by regarding $\mathcal{A}$ as a concrete C^*-subalgebra of some $B(\mathcal{H})$.

In this chapter we shall develop the Choi–Effros [49] abstract characterization of operator systems and Ruan's [203] abstract characterization of operator spaces. In analogy with the GNS theory, these characterizations will free us from being forced to regard operator spaces and systems as concrete subspaces of operators.

We begin with the theory of abstract operator systems. We wish to characterize operator systems up to complete order isomorphism. To this end, let $\mathcal{S}$ be a complex vector space, and assume that there exists a conjugate linear map $s \to s^*$ on $\mathcal{S}$ with $(s^*)^* = s$ for all s in $\mathcal{S}$. We call such a space a $*$-*vector space*. We let $\mathcal{S}_h = \{s \in \mathcal{S}: s = s^*\}$ and note that every element x in $\mathcal{S}$ can be written $x = h + ik$ with $h = (x + x^*)/2$ and $k = (x - x^*)/2i$ both in $\mathcal{S}_h$. For (x_{ij}) in $M_n(\mathcal{S})$ we set $(x_{ij})^* = (x_{ji}^*)$, so that $M_n(\mathcal{S})$ is also a $*$-vector space.

Clearly, if we wish $\mathcal{S}$ to be an operator system, then for each n we shall need a distinguished cone $\mathcal{C}_n$ in $M_n(\mathcal{S})_h$ that plays the role of the "positive" operators. Moreover, as n varies, these sets should have certain relationships. For example, if p_1 is "positive" in $\mathcal{S}$, then $\begin{pmatrix} p_1 & 0 \\ 0 & 0 \end{pmatrix}$ should be "positive" in $M_2(\mathcal{S})$.

Note that, since $\mathcal{S}$ is a vector space, there is a natural action of scalar matrices, of the appropriate sizes, on matrices over $\mathcal{S}$. Namely, for $A = (a_{ij})$ an $m \times n$ matrix and $X = (x_{ij})$ an $n \times k$ matrix with entries in $\mathcal{S}$, we define an $m \times k$ matrix over $\mathcal{S}$ by $A \cdot X = (\sum_{\ell=1}^{n} a_{i\ell} x_{\ell j})$. We define multiplication on the right similarly. Thus, for example,

$$\text{if} \quad A = (1, 0), \quad \text{then} \quad A p_1 A^* = \begin{pmatrix} p_1 & 0 \\ 0 & 0 \end{pmatrix}.$$

These considerations motivate the following definition. Given a $*$-vector space $\mathcal{S}$, we say that $\mathcal{S}$ is *matrix ordered* provided that:

(i) for each n we are given a cone $\mathcal{C}_n$ in $M_n(\mathcal{S})_h$,
(ii) $\mathcal{C}_n \cap (-\mathcal{C}_n) = \{0\}$ for all n,
(iii) for every n and m and A an $n \times m$ matrix, we have that $A^* \mathcal{C}_n A \subseteq \mathcal{C}_m$.

We call the collection $\{\mathcal{C}_n\}$ a *matrix order* on $\mathcal{S}$.

We adopt the same terminology for maps between matrix ordered *-vector spaces as for operator spaces. Thus, given two matrix-ordered *-vector spaces $\mathcal{S}$ and $\mathcal{S}'$ with cones $\mathcal{C}_n$ and $\mathcal{C}'_n$, we call a linear map $\phi\colon \mathcal{S} \to \mathcal{S}'$ *completely positive* provided that $(x_{ij}) \in \mathcal{C}_n$ implies that $(\phi(x_{ij})) \in \mathcal{C}'_n$. Similarly, we call ϕ a *complete order isomorphism* provided that ϕ is invertible with both ϕ and ϕ^{-1} completely positive.

Finally, we need to axiomatize the role that 1 plays in an operator system. Let $\mathcal{S}$ be a matrix-ordered $*$-vector space. We call $e \in \mathcal{S}_h$ an *order unit* for $\mathcal{S}$ provided that for every $x \in \mathcal{S}_h$ there exists a positive real r such that $re + x \in \mathcal{C}_1$. We call an order unit e *Archimedean* if $re + x \in \mathcal{C}_1$ for all $r > 0$ implies that $x \in \mathcal{C}_1$. We call e a *matrix order unit* provided that

$$I_n = \begin{pmatrix} e & 0 & \cdots & 0 \\ 0 & \ddots & \ddots & \vdots \\ \vdots & \ddots & \ddots & 0 \\ 0 & \cdots & 0 & e \end{pmatrix}$$

is an order unit for $M_n(\mathcal{S})$ for all n, and an *Archimedean matrix order unit* provided each I_n is Archimedean.

We begin with a few trivial observations. Since there exists $r > 0$ such that $re + e \in \mathcal{C}_1$, it follows that $e = (r+1)^{-1}(re + e)$ is in $\mathcal{C}_1$, because $\mathcal{C}_1$ is a cone. Similarly, $I_n \in \mathcal{C}_n$. If $re + x \in \mathcal{C}_1$, then for any $s \geq r$ we have $se + x = (s - r)e + (re + x) \in \mathcal{C}_1$. Finally, if $x \in \mathcal{S}_h$, then there exists $r > 0$ such that $re \pm x \in \mathcal{C}_1$ and hence $x = (re + x)/2 - (re - x)/2$. Thus, $\mathcal{S}_h = \mathcal{C}_1 - \mathcal{C}_1$, which is the condition needed for $\mathcal{C}_1$ to be a *full cone* in $\mathcal{S}$. Similarly, $\mathcal{C}_n$ is a full cone in $M_n(\mathcal{S})$ for all n.

Our goal is to prove the following:

Theorem 13.1 (Choi–Effros). *If $\mathcal{S}$ is a matrix-ordered $*$-vector space with an Archimedean matrix order unit e, then there exists a Hilbert space $\mathcal{H}$, an operator system $\mathcal{S}_1 \subseteq B(\mathcal{H})$, and a complete order isomorphism $\varphi\colon \mathcal{S} \to \mathcal{S}_1$ with $\varphi(e) = I_{\mathcal{H}}$. Conversely, every concrete operator system $\mathcal{S}$ is a matrix-ordered $*$-vector space with an Archimedean matrix order unit $e = I_H$.*

Before proving this theorem we need two preliminary results.

Recall the correspondence between linear functionals $s\colon M_n(\mathcal{S}) \to \mathbb{C}$ and linear maps $\phi\colon \mathcal{S} \to M_n$ of Chapter 6. We omit the factor of n used in Chapter 6. Thus, given ϕ, we define $s_\phi\colon M_n(\mathcal{S}) \to \mathbb{C}$ via

$$s_\phi((x_{ij})) = \sum_{i,j=1}^{n} \langle \phi(x_{i,j})e_j, e_i \rangle,$$

where $\{e_1, \ldots, e_n\}$ is the canonical basis for $\mathbb{C}^n$. Conversely, given $s\colon M_n(\mathcal{S}) \to \mathbb{C}$, we define $\phi_s\colon \mathcal{S} \to M_n$ via $\phi_s(x) = (s_{ij}(x))$, where $s_{ij}(x) = s(x \otimes E_{i,j})$ and $E_{i,j}$ are the canonical matrix units for M_n. These two operations are mutual inverses, i.e., $\phi_{(s_\phi)} = \phi$ and $s_{(\phi_s)} = s$.

Proposition 13.2. *Let $\mathcal{S}$ be a matrix ordered $*$-vector space, let $s\colon M_n(\mathcal{S}) \to \mathbb{C}$, and let $\phi\colon \mathcal{S} \to M_n$ with $\phi = \phi_s$. Then the following are equivalent:*

(i) $s(\mathcal{C}_n) \geq 0$,
(ii) $\phi\colon \mathcal{S} \to M_n$ *is n-positive,*
(iii) $\phi\colon \mathcal{S} \to M_n$ *is completely positive.*

Proof. The proof is similar to the proof of Theorem 6.1. It is routine to verify that (iii) implies (ii) and that (ii) implies (i).

So we prove that (i) implies (iii). First note that, if $h = (\beta_1, \ldots, \beta_n)$ and $k = (\alpha_1, \ldots, \alpha_n)$, then $\langle \phi(x)h^t, k^t \rangle = s((\bar{\alpha}_i x \beta_j)) = s(k^* x h)$. Thus, if $X = (x_{ij}) \in \mathcal{C}_m$ and $v_1^t, \ldots, v_m^t \in \mathbb{C}^n$, then

$$\sum_{i,j=1}^{m} \langle \phi(x_{ij})v_j^t, v_i^t \rangle = \sum_{i,j=1}^{m} s(v_i^* x_{ij} v_j) = s(A^* X A),$$

where

$$A = \begin{pmatrix} v_1 \\ \vdots \\ v_m \end{pmatrix}$$

is $m \times n$. Hence, $(\phi(x_{ij}))$ is positive, and so ϕ is completely positive. □

Proposition 13.3. *Let $\mathcal{S}$ be a matrix-ordered $*$-vector space with Archimedean matrix order unit e and for $X \in M_n(\mathcal{S})$ set*

$$\|X\|_n = \inf\left\{r : \begin{pmatrix} rI_n & X \\ X^* & rI_n \end{pmatrix} \in \mathcal{C}_{2n}\right\}.$$

Then $\|\cdot\|_n$ is a norm on $M_n(\mathcal{S})$, and $\mathcal{C}_n$ is a closed subset of $M_n(\mathcal{S})$ in the topology induced by this norm.

Proof. We only do the case $n = 1$.

First we prove that $\|x\|_1 \geq 0$. If

$$\begin{pmatrix} re & x \\ x^* & re \end{pmatrix} \in \mathcal{C}_2,$$

set $A = \begin{pmatrix} 1 & 0 \\ 0 & -1 \end{pmatrix}$; then

$$\begin{pmatrix} re & -x \\ -x^* & re \end{pmatrix} = A^* \begin{pmatrix} re & x \\ x^* & re \end{pmatrix} A \in \mathcal{C}_2.$$

Adding these two matrices yields $2rI_2 \in \mathcal{C}_2$. Now using $I_2 \in \mathcal{C}_2$, $\mathcal{C}_2$ a cone, and $\mathcal{C}_2 \cap (-\mathcal{C}_2) = \{0\}$ yields $r \geq 0$.

Next we show that $\|x\|_1 = 0$ implies $x = 0$. If $\|x\|_1 = 0$, then

$$\begin{pmatrix} re & x \\ x^* & re \end{pmatrix} \in \mathcal{C}_2 \quad \text{for all} \quad r > 0.$$

Hence

$$(1, \bar{\lambda}) \begin{pmatrix} re & x \\ x^* & re \end{pmatrix} \begin{pmatrix} 1 \\ \lambda \end{pmatrix} = r(1 + |\lambda|^2)e + \lambda x + (\lambda x)^* \in \mathcal{C}_1$$

for every complex number λ and any $r > 0$. By the Archimedean property, $\lambda x + (\lambda x)^* \in \mathcal{C}_1$. Setting $\lambda = \pm 1$ yields $x + x^* = 0$, and setting $\lambda = \pm i$ yields $ix + (ix)^* = 0$. Hence, $x = 0$.

Similar tricks show that $\|\lambda x\|_1 = |\lambda| \|x\|_1$, $\|x + y\|_1 \leq \|x\|_1 + \|y\|_1$, and $\|x^*\|_1 = \|x\|_1$, so we leave them to the reader (Exercise 13.5).

Finally, we show that $\mathcal{C}_1$ is closed. Let $x_n \in \mathcal{C}_1$, $x \in \mathcal{S}$ with $\|x - x_n\|_1 \to 0$. Since $x_n = x_n^*$, we have $x = x^*$. Given any $r > 0$, choose n so that $\|x - x_n\|_1 < r$. Then

$$\begin{pmatrix} re & x - x_n \\ x - x_n & re \end{pmatrix} \in \mathcal{C}_2,$$

and picking $A = \begin{pmatrix} 1 \\ 1 \end{pmatrix}$ yields $2re + 2x - 2x_n \in \mathcal{C}_1$. Hence, $re + x \in \mathcal{C}_1$, since $x_n \in \mathcal{C}_1$ and $\mathcal{C}_1$ is a cone. Thus, $x \in \mathcal{C}_1$ by the Archimedean property. □

We shall refer to the norm given in Proposition 13.3 as the *matrix norm induced by the matrix order*. Because of the link between norm and order for operators on a Hilbert space, viz., $\|T\| \leq 1$ if and only if $\left(\begin{smallmatrix} I & T \\ T^* & I \end{smallmatrix}\right) \geq 0$, we see that any complete order isomorphism ϕ onto a concrete operator system with $\phi(e) = I$ must be a complete isometry of this induced matrix norm.

Proof of Theorem 13.1. Assume $\mathcal{S}$ is a matrix-ordered $*$-vector space with an Archimedean matrix order unit e. Let $\mathcal{P}_n = \{\phi\colon \mathcal{S} \to M_n \mid \phi$ completely positive, $\phi(e) = I\}$, and define $J = \sum_{n=1}^{\infty} \sum_{\phi \in \mathcal{P}_n} \oplus\, \phi\colon \mathcal{S} \to \sum_{n=1}^{\infty} \sum_{\phi \in \mathcal{P}_n} \oplus\, M_n^{\phi}$, where the latter direct sum is in the ℓ^{∞} sense.

Since the latter direct sum is a C^*-algebra, to prove that $\mathcal{S}$ is completely order-isomorphic to an operator system, it will be enough to prove that J is a complete order isomorphism between $\mathcal{S}$ and $J(\mathcal{S})$. To prove this we must show that for $(x_{ij}) \in M_n(\mathcal{S})$, we have $(x_{ij}) \in \mathcal{C}_n$ if and only if $(J(x_{ij})) \geq 0$ in the latter C^*-algebra.

Clearly, by the choice of ϕ's, (x_{ij}) in $\mathcal{C}_n$ implies $(J(x_{ij})) \geq 0$. To complete the proof it will be enough to show that if (x_{ij}) is *not* in $\mathcal{C}_n$ then there exists k and $\phi \in \mathcal{P}_k$ with $(\phi(x_{ij})) \not\geq 0$.

Since $\mathcal{C}_n$ is closed in the norm topology of $M_n(\mathcal{S})$, by the Krein–Milman theorem (for cones) there exists a linear functional $s\colon M_n(\mathcal{S}) \to \mathbb{C}$ with $s(\mathcal{C}_n) \geq 0$ but $s((x_{ij})) < 0$. Consider $\phi_s\colon \mathcal{S} \to M_n$; then

$$\sum_{i,j=1}^{n} \langle \phi_s(x_{ij}) e_j, e_i \rangle = s((x_{ij})) < 0.$$

Hence, $(\phi_s(x_{ij}))$ is not positive.

All that remains is to replace ϕ_s with a unital completely positive map.

Let $\phi_s(e) = P \in M_n^+$. If P is invertible, we may choose A so that $A^* P A = I$. Setting $\psi(x) = A^* \phi_s(x) A$, we have that $\psi(e) = I$,

$$\sum_{i,j=1}^{n} \langle (\psi(x_{ij})) A^{-1} e_j, A^{-1} e_i \rangle < 0,$$

and we are done.

If P is not invertible, let $\|x\| \leq 1$. Then $\left(\begin{smallmatrix} e & x \\ x^* & e \end{smallmatrix}\right) \in \mathcal{C}_2$ and hence

$$\begin{pmatrix} P & \phi_s(x) \\ \phi_s(x)^* & P \end{pmatrix} \geq 0.$$

From this it follows that if $Ph = 0$ then $\phi_s(x)h = 0$.

Let Q be the projection onto $\ker(P)^{\perp}$, so that $Q\phi_s(x)Q = \phi_s(x)$. Let $\operatorname{rank}(Q) = k$, and choose an $n \times k$ matrix A and a $k \times n$ matrix B so that

$A^*PA = I_k$, $AB = Q$. Setting $\psi(x) = A^*\phi_s(x)A$, we have that $\psi\colon \mathcal{S} \to M_k$ is completely positive with $\psi(e) = I_k$ and

$$\sum_{i,j=1}^{n} \langle \psi(x_{ij})Be_j, Be_i\rangle = \sum_{i,j=1}^{m} \langle \phi_s(x_{ij})Qe_j, Qe_i\rangle = \sum_{i,j=1}^{n} \langle \phi_s(x_{ij})e_j, e_i\rangle < 0.$$

Thus, $\psi \in \mathcal{P}_k$ with $(\psi(x_{ij})) \not\geq 0$.

We leave the proof of the converse statement to the reader. □

For an illustration of the uses of the abstract characterization of operator systems, we return to the question examined in Chapter 6 of determining when an operator system $\mathcal{S}$ has the property that every positive map with domain $\mathcal{S}$ is automatically completely positive.

Recall that if $\mathcal{T}$ is any other operator system and $\theta\colon \mathcal{S} \to \mathcal{T}$ is a positive map, then θ is necessarily positive on the subset $\mathcal{S}^+ \otimes M_n^+$ of $M_n(\mathcal{S})^+$ and hence on its closure. Let $\mathcal{C}_n$ denote the closure of $\mathcal{S}^+ \otimes M_n^+$, with $\mathcal{C}_1 = \mathcal{S}^+$.

Clearly, for every n and m, $\mathcal{C}_n$ is a cone, $\mathcal{C}_n \cap (-\mathcal{C}_n) = (0)$, and given any $m \times n$ matrix B, we have that $B^*\mathcal{C}_m B \subset \mathcal{C}_n$. Thus, the collection $\{\mathcal{C}_n\}_{n\geq 1}$ is a (possibly new) matrix order on $\mathcal{S}$.

We let $\mathcal{S}_M$ denote this new matrix-ordered space.

We leave it as an exercise (Exercise 13.4) to show that the Archimedean matrix order unit e is also an Archimedean matrix order unit for $\mathcal{S}_M$. Hence, $\mathcal{S}_M$ is an abstract operator system.

Clearly, the identity map from $\mathcal{S}$ to $\mathcal{S}_M$ is a positive map. If the identity map from $\mathcal{S}$ to $\mathcal{S}_M$ is completely positive, then $\mathcal{C}_n = M_n(\mathcal{S})^+$ for all n, and consequently every positive map with domain $\mathcal{S}$ is completely positive. Conversely, if every positive map with domain $\mathcal{S}$ is completely positive, then we must have that the identity map from $\mathcal{S}$ to $\mathcal{S}_M$ is completely positive and hence $\mathcal{C}_n = M_n(\mathcal{S})^+$.

Thus, we have obtained a new proof of the equivalence of (i) and (iii) in Corollary 6.7. Note that this new proof avoids the need for a separating-functional argument. This essentially happens because the separating-functional argument is built into the proof of the characterization of operator systems.

We now turn our attention to the abstract characterization of operator spaces. It will be convenient to consider rectangular, as well as square, matrices. Given a vector space V, we let $M_{m,n}(V)$ denote the vector space of $m \times n$ matrices over V.

We call V a *matrix normed space* provided that we are given norms $\|\cdot\|_{m,n}$ on $M_{m,n}(V)$ such that whenever $A = (a_{ij}) \in M_{p,m}$, $X = (x_{ij}) \in M_{m,n}(V)$, and

$B = (b_{i,j}) \in M_{n,q}$, then $\|A \cdot X \cdot B\|_{p,q} \le \|A\| \|X\|_{m,n} \|B\|$, where

$$A \cdot X \cdot B = \left(\sum_{k=1}^{m} \sum_{\ell=1}^{n} a_{ik} x_{k\ell} b_{\ell j} \right) \in M_{p,q}(V)$$

and $\|A\|$, $\|B\|$ are the operator norms on $M_{p,m} = B(\mathbb{C}^m, \mathbb{C}^p)$ and $M_{n,q} = B(\mathbb{C}^q, \mathbb{C}^n)$, respectively.

The above axioms guarantee, among other things, that if a matrix is enlarged by adding rows and columns of 0's, then its norm will remain unchanged (Exercise 13.1). We abbreviate $M_{n,n}(V) = M_n(V)$ and $\|\cdot\|_{n,n} = \|\cdot\|_n$.

Let V be a matrix-normed space, and for $X \in M_{m,n}(V)$, $Y \in M_{p,q}(V)$ define $X \oplus Y = \left(\begin{smallmatrix} X & 0 \\ 0 & Y \end{smallmatrix}\right) \in M_{m+p,n+q}(V)$, where the 0's indicate matrices of 0's of appropriate sizes. We call the matrix norm on V an *L^∞-matrix norm*, and call V an *L^∞-matrix-normed space* provided that $\|X \oplus Y\|_{m+p,n+q} = \max\{\|X\|_{m,n}, \|Y\|_{p,q}\}$ for all m, n, p, q, X, and Y.

Note that for any matrix-normed space, $\max\{\|X\|_{m,n}, \|Y\|_{p,q}\} \le \|X \oplus Y\|_{m+p,n+q}$, so it suffices to prove the other inequality.

Theorem 13.4 (Ruan's theorem). *Let V be a matrix-normed space. Then there exists a Hilbert space $\mathcal{H}$ and a complete isometry $\varphi \colon V \to B(\mathcal{H})$ if and only if V is an L^∞-matrix-normed space.*

Proof. It is easy to see that every concrete subspace of some $B(\mathcal{H})$ is an L^∞-matrix-normed space. Thus, if V is completely isometric to such a subspace, it must be an L^∞-matrix-normed space.

So assume that V is an L^∞-matrix-normed space. We shall create an abstract operator system $\mathcal{S}$, i.e., a matrix-ordered $*$-vector space with an Archimedean matrix order unit e, that contains V as a vector subspace and such that the matrix norm induced by the order structure restricts to the matrix norm on V.

By Theorem 13.1, $\mathcal{S}$ will be completely order-isomorphic to an operator system. But this complete order isomorphism will necessarily be a complete isometry of this induced matrix norm. Thus, if we can create such an "abstract" operator system $\mathcal{S}$, the representation of $\mathcal{S}$ as a "concrete" operator system will give us the desired complete isometry of V. We now construct $\mathcal{S}$. The idea is to "abstractly" recreate the operator system $\mathcal{S}_{\mathcal{M}}$ of Lemma 8.1 that comes from a concrete operator space $\mathcal{M}$.

To this end, let V^* denote a complex conjugate copy of the vector space V, i..e, $V^* = \{v^* \colon v \in V\}$ with $v_1^* + v_2^* = (v_1 + v_2)^*$ and $\lambda v^* = (\bar{\lambda} v)^*$. As a vector space, $\mathcal{S}$ will be $\mathbb{C} \oplus \mathbb{C} \oplus V \oplus V^*$, but it is easiest to understand the matrix order if we write

$$\mathcal{S} = \left\{ \begin{pmatrix} \lambda & v \\ w^* & \mu \end{pmatrix} : \lambda, \mu \in \mathbb{C},\ v \in V,\ w^* \in V^* \right\}.$$

We make $\mathcal{S}$ a $*$-vector space by setting

$$\begin{pmatrix} \lambda & v \\ w^* & \mu \end{pmatrix}^* = \begin{pmatrix} \bar{\lambda} & w \\ v^* & \bar{\mu} \end{pmatrix}$$

so that

$$\mathcal{S}_h = \left\{ \begin{pmatrix} r_1 & v \\ v^* & r_2 \end{pmatrix} : r_1,\ r_2 \in \mathbb{R},\ v \in V \right\}.$$

We set

$$\mathcal{C}_1 = \left\{ \begin{pmatrix} r_1 & v \\ v^* & r_2 \end{pmatrix} : r_1,\ r_2 \geq 0,\ \|v\|^2 \leq r_1 r_2 \right\}.$$

This definition is motivated by the following fact for $\mathcal{S}_{\mathcal{M}}$, when $\mathcal{M}$ is a concrete operator space: For $r_1, r_2 > 0$, $\left(\begin{smallmatrix} r_1 & m \\ m^* & r_2 \end{smallmatrix}\right) \geq 0$ if and only if

$$\begin{pmatrix} r_1^{-1/2} & 0 \\ 0 & r_2^{-1/2} \end{pmatrix} \begin{pmatrix} r_1 & m \\ m^* & r_2 \end{pmatrix} \begin{pmatrix} r_1^{-1/2} & 0 \\ 0 & r_2^{-1/2} \end{pmatrix}$$
$$= \begin{pmatrix} 1 & r_1^{-1/2} r_2^{-1/2} m \\ r_1^{-1/2} r_2^{-1/2} m^* & 1 \end{pmatrix} \geq 0,$$

and this latter occurs if and only if $\|r_1^{-1/2} r_2^{-1/2} m\| \leq 1$.

To define $\mathcal{C}_n$, note that after a canonical shuffle we may write

$$M_n(\mathcal{S}) = \left\{ \begin{pmatrix} A & X \\ Y^* & B \end{pmatrix} : A, B \in M_n,\ X \in M_n(V),\ Y^* \in M_n(V^*) \right\}$$

and that $M_n(\mathcal{S})_h$ consists of the above matrices with $A = A^*$, $B = B^*$, and $Y^* = X^* = (x_{ji}^*)$. Thus, we define $\mathcal{C}_n$ as the set

$$\left\{ \begin{pmatrix} P & X \\ X^* & Q \end{pmatrix} : P, Q \geq 0 \text{ and } \left\|(P + \varepsilon I)^{-1/2} X (Q + \varepsilon I)^{-1/2}\right\| \leq 1 \right.$$
$$\left. \text{for all real } \varepsilon > 0 \right\}.$$

We now show that $\mathcal{S}$ is matrix-ordered and that $e = \left(\begin{smallmatrix} 1 & 0 \\ 0 & 1 \end{smallmatrix}\right)$ is an Archimedean matrix order unit.

To this end, suppose that $S = (S_{ij})$ is in $M_n(\mathcal{S})$, that $A = (a_{ij}) \in M_{m,n}$, and that

$$S_{ij} = \begin{pmatrix} \lambda_{ij} & x_{ij} \\ y_{ij}^* & \mu_{ij} \end{pmatrix},$$

with $S \cong \left(\begin{smallmatrix} L & X \\ Y^* & M \end{smallmatrix}\right)$ after the canonical shuffle. Then

$$A \cdot S = \left(\sum a_{i\ell} S_{\ell j}\right) \cong \begin{pmatrix} AL & AX \\ AY^* & AM \end{pmatrix}.$$

Thus, if $S \cong \left(\begin{smallmatrix} P & X \\ X^* & Q \end{smallmatrix}\right)$ is in $\mathcal{C}_n$, we must show that

$$ASA^* \cong \begin{pmatrix} APA^* & AXA^* \\ AX^*A^* & AQA^* \end{pmatrix}$$

is in $\mathcal{C}_m$. We have $AX^*A^* = (AXA^*)^*$, and it remains to show that for $\varepsilon > 0$,

$$\left\|(APA^* + \varepsilon I)^{-1/2}(AXA^*)(AQA^* + \varepsilon I)^{-1/2}\right\| \le 1.$$

Since

$$\begin{aligned}
&(APA^* + \varepsilon I)^{-1/2}(AXA^*)(AQA^* + \varepsilon I)^{-1/2} \\
&\quad = \left[(APA^* + \varepsilon I)^{-1/2} A (P + \delta I)^{+1/2}\right] \\
&\quad\quad \times \left[(P + \delta I)^{-1/2} X (Q + \delta I)^{-1/2}\right] \\
&\quad\quad \times \left[(Q + \delta I)^{1/2} \cdot A^*(AQA^* + \varepsilon I)^{-1/2}\right],
\end{aligned}$$

it will be enough to show that for $\delta > 0$ suitably chosen, each factor in brackets has norm less than 1.

By hypothesis $\|(P + \delta I)^{-1/2} X (Q + \delta I)^{-1/2}\| \le 1$ for any $\delta > 0$.

Now $\|(APA^* + \varepsilon I)^{-1/2} A (P + \delta I)^{1/2}\| \le 1$ if and only if $A(P + \delta I)A^* \le APA^* + \varepsilon I$, which holds for any $\delta \le \varepsilon / \|A^*A\|$. Similarly, it can be seen that $\|(Q + \delta I)^{1/2} A^*(AQA^* + \varepsilon I)^{-1/2}\| \le 1$ for any $\delta \le \varepsilon/\|A^*A\|$. Thus, for any $A \in M_{m,n}$ we have that $A\mathcal{C}_n A^* \subseteq \mathcal{C}_m$.

We use this fact to see that $\mathcal{C}_n$ is a cone. Clearly, $t\mathcal{C}_n \subseteq \mathcal{C}_n$ for $t \ge 0$. If $S_i, i = 1, 2$, are in $\mathcal{C}_n$, then the L^∞ condition implies that $S_1 \oplus S_2$ is in $\mathcal{C}_{2n}$. Taking $A = (I_n, I_n)$ we have that $A(S_1 \oplus S_2)A^* = S_1 + S_2 \in \mathcal{C}_n$, and so $\mathcal{C}_n$ is a cone.

We claim that $e = \left(\begin{smallmatrix} 1 & 0 \\ 0 & 1 \end{smallmatrix}\right)$ is a matrix order unit. We must show that if $\left(\begin{smallmatrix} H & X \\ X^* & K \end{smallmatrix}\right)$ is in $M_n(\mathcal{S})_h$, then there exists $r \ge 0$ so that

$$r \begin{pmatrix} I_n & 0 \\ 0 & I_n \end{pmatrix} + \begin{pmatrix} H & X \\ X^* & K \end{pmatrix}$$

is in $\mathcal{C}_n$. Let $r_1 = \max\{\|H\|, \|K\|\}$, $r_2 = \|X\|$; then

$$\begin{pmatrix} r_1 I_n + H & 0 \\ 0 & r_1 I_n + K \end{pmatrix} \in \mathcal{C}_n \quad \text{and} \quad \begin{pmatrix} r_2 I_n & X \\ X^* & r_2 I_n \end{pmatrix} \in \mathcal{C}_n,$$

and so $r = r_1 + r_2$ suffices.

The fact that e is Archimedean is trivial from the use of $\varepsilon > 0$ in the definition of $\mathcal{C}_n$.

Finally, if we embed V into $\mathcal{S}$ via $v \to \left(\begin{smallmatrix} 0 & v \\ 0 & 0 \end{smallmatrix}\right)$, then this mapping is easily seen to be a complete isometry from V into $\mathcal{S}$ equipped with the matrix norm induced by the order.

Thus, we have produced the desired "abstract" operator system $\mathcal{S}$, and the proof is complete. □

In Chapter 14 we shall take a closer look at abstract operator spaces, focusing on a few important examples and classes of operator spaces that serve as an introduction to the field.

Notes

Since the time that Ruan's theorem first appeared, it has been customary to identify L^∞-matrix-normed spaces and operator spaces as one and the same thing. In fact, many books start by defining the term "operator space" to mean an L^∞-matrix-normed space and then prove that such objects can be represented completely isometrically as spaces of operators on some Hilbert space. From this viewpoint, a theory of operator spaces can be developed that parallels much of the classical theory of Banach spaces. Two excellent such texts are Pisier's [193] *An Introduction to the Theory of Operator Spaces* (Cambridge University Press) and Effros and Ruan's [88] *Operator Spaces* (Oxford University Press). In keeping with standard usage, from this point on we shall generally refer to L^∞-matrix-normed spaces simply as operator spaces. Occasionally, we shall use the terms *abstract operator space* to emphasize that we are dealing with an L^∞-matrix-normed space, and *concrete operator space* for an actual subspace of the space of operators on a Hilbert space.

Similarly, we shall identify matrix-ordered $*$-vector spaces with Archimedean matrix order units and operator systems, using the terms *abstract operator system* and *concrete operator system* when we wish to emphasize the difference.

Exercises

13.1 Let V be a matrix-normed space, let $X \in M_{m,n}(V)$, and let $Y \in M_{p,q}(V)$ be a matrix obtained from X by introducing finitely many rows and columns of 0's. Prove that $\|X\|_{m,n} = \|Y\|_{p,q}$.

13.2 Let V be a vector space, and assume that we are given a sequence of norms $\|\cdot\|_n$ on $M_n(V)$ satisfying:

(i) $\|AXB\|_n \leq \|A\|\|X\|_n\|B\|$ for $X \in M_n(V)$, $A \in M_n$, and $B \in M_n$;

(ii) for $X \in M_n(V)$, $\|X \oplus 0\|_{m+n} = \|X\|_n$, where 0 denotes an $m \times m$ matrix of 0's.

For $X \in M_{m,n}(V)$ set $\|X\|_{m,n} = \|\hat{X}\|_\ell$, where $\ell = \max\{m, n\}$ and $\hat{X}$ is the matrix obtained by adding sufficiently many rows or columns to X to make it square. Prove that $(V, \|\cdot\|_{m,n})$ is a matrix-normed space. These alternate axioms are often given as the axioms for a matrix-normed space, and consequently no mention is made of the norms of rectangular matrices.

13.3 Let V be an operator space, let W be a closed subspace, and let $q \colon V \to V/W$ denote the quotient map $q(v) = v + W$. Prove that if we define norms on $M_{n,m}(V/W)$ by setting

$$\|(q(v_{ij}))\|_{n,m} = \inf\{\|(v_{i,j} + w_{i,j})\|_{n,m} \colon w_{i,j} \in W\},$$

then V/W is an operator space.

13.4 Prove that $\mathcal{S}_M$ is an operator system.

13.5 Verify the claims of Proposition 13.3.

Chapter 14
An Operator Space Bestiary

In the last chapter we obtained an abstract characterization of operator spaces that allows us to define these spaces without a concrete representation. This result has had a tremendous impact and has led to the development of a general theory of operator spaces that parallels in some ways the development of the theory of Banach spaces.

In this chapter we give the reader a brief introduction to some of the basics of this theory, focusing on some of the more important operator spaces that we will encounter in later chapters.

We have already encountered one example of the power of this axiomatic characterization. In Exercise 13.3, it was shown that if V is an operator space and $W \subseteq V$ a closed subspace, then V/W is an operator space, where the matrix norm structure on V/W comes from the identification $M_{m,n}(V/W) = M_{m,n}(V)/M_{m,n}(W)$. Yet in most concrete situations it is difficult to actually exhibit a concrete completely isometric representation of V/W as operators on a Hilbert space.

The first natural question in the area is as follows: If V is, initially, just a normed space, then is it always possible to assign norms $\|\cdot\|_{m,n}$ to $M_{m,n}(V)$ for all m and n in such a fashion that V becomes an operator space? The answer to this question is yes. Consider the dual space V^* of V, and let V_1^* denote its unit ball, equipped with the weak* topology. The continuous functions on this compact, Hausdorff space $C(V_1^*)$ constitute an abelian C^*-algebra, and the map $j\colon V \to C(V_1^*)$ defined by $j(v)(f) = f(v)$ for $f \in V_1^*$ is a linear isometry. Since subspaces of C^*-algebras are operator spaces, identifying V with $j(V)$ induces a particular family of norms on $M_{m,n}(V)$ that makes V an operator space. Thus, for (v_{ij}) in $M_{m,n}(V)$ we have that

$$\|(v_{ij})\|_{m,n} = \|(j(v_{ij}))\|_{m,n} = \sup\{\|(f(v_{ij}))\|_{M_{m,n}}\colon f \in V_1^*\},$$

where $\|(f(v_{ij}))\|_{M_{m,n}}$ indicates the norm of the scalar $m \times n$ matrix $(f(v_{ij}))$ viewed as a linear transformation from n-dimensional Hilbert space to m-dimensional Hilbert space.

Clearly, we need a way to distinguish between the ordinary normed space V and the operator space $j(V)$. It is standard to let MIN(V) denote the operator space $j(V) \subseteq C(V_1^*)$. This operator space is often called the *minimal operator space of* V, for reasons made apparent below. For an arbitrary $m \times n$ matrix over V, we simply write $\|(v_{ij})\|_{\mathrm{MIN}(V)}$ to denote its norm in $M_{m,n}(\mathrm{MIN}(V))$.

Theorem 14.1. *Let V be a normed space, $\mathcal{H}$ be a Hilbert space, and let $\varphi: V \to B(\mathcal{H})$ be an isometric map. Then for (v_{ij}) in $M_{m,n}(V)$ we have*

(i) $\|(v_{ij})\|_{\mathrm{MIN}(V)} = \sup\{\|\sum_{i=1}^m \sum_{j=1}^n \alpha_i v_{ij} \beta_j\|_V\colon \sum_{i=1}^m |\alpha_i|^2 \leq 1, \sum_{j=1}^n |\beta_j|^2 \leq 1\}$, *where the supremum is over all α_i, β_j in $\mathbb{C}$ satisfying the inequalities;*

(ii) $\|(\varphi(v_{ij}))\|_{B(\mathcal{H}^{(n)},\mathcal{H}^{(m)})} \geq \|(v_{ij})\|_{\mathrm{MIN}(V)}$.

Proof. We have that

$$\begin{aligned}\|(v_{ij})\|_{\mathrm{MIN}(V)} &= \sup\{\|(f(v_{ij}))\|_{M_{m,n}}\colon f \in V_1^*\}\\ &= \sup\left\{\left|f\left(\sum_{i=1}^m\sum_{j=1}^n \alpha_i v_{ij}\beta_j\right)\right| : f \in V_1^*, \sum_{i=1}^m |\alpha_i|^2 \leq 1, \sum_{j=1}^n |\beta_j|^2 \leq 1\right\}\\ &= \sup\left\{\left\|\sum_{i=1}^m\sum_{j=1}^n \alpha_i\beta_j v_{ij}\right\|_V : \sum_{i=1}^m |\alpha_i|^2 \leq 1, \sum_{j=1}^n |\beta_j|^2 \leq 1\right\},\end{aligned}$$

which proves (i).

To see (ii), note that

$$\begin{aligned}&\|(\varphi(v_{ij}))\|_{B(\mathcal{H}^{(n)},\mathcal{H}^{(m)})}\\ &\quad\geq \sup\left\{\left|\left\langle(\varphi(v_{ij}))\begin{pmatrix}\alpha_1 h\\ \vdots\\ \alpha_m h\end{pmatrix}, \begin{pmatrix}\beta_1 k\\ \vdots\\ \beta_n k\end{pmatrix}\right\rangle\right| :\right.\\ &\qquad\left. h, k \in \mathcal{H}, \|h\| = 1, \|k\| = 1, \sum_{i=1}^m |\alpha_i|^2 \leq 1, \sum_{j=1}^n |\beta_j|^2 \leq 1\right\}\end{aligned}$$

$$= \sup\left\{\left|\left\langle\varphi\left(\sum_{i=1}^{m}\sum_{j=1}^{n}\alpha_i\beta_j v_{ij}\right)h, k\right\rangle\right| : \right.$$

$$\left. h, k \in \mathcal{H}, \|h\| = 1, \|k\| = 1, \sum_{i=1}^{m}|\alpha_i|^2 \le 1, \sum_{j=1}^{n}|\beta_j|^2 \le 1\right\}$$

$$= \sup\left\{\left\|\varphi\left(\sum_{i=1}^{m}\sum_{j=1}^{n}\alpha_i\beta_j v_{ij}\right)\right\|_{B(\mathcal{H})} : \sum_{i=1}^{m}|\alpha_i|^2 \le 1, \sum_{j=1}^{n}|\beta_j|^2 \le 1\right\}$$

$$= \|(v_{ij})\|_{\mathrm{MIN}(V)},$$

by (i) and the fact that φ is an isometry. □

The beauty of (i) is that it gives a formula "internal" to V to compute MIN(V).

Now that we have a "minimal" way to represent a normed space as an operator space, it is not difficult to create a "maximal" representation. We shall denote this operator space by MAX(V).

To define MAX(V), given (v_{ij}) in $M_{m,n}(V)$, we set

$$\|(v_{ij})\|_{\mathrm{MAX}(V)} = \sup\{\|(\varphi(v_{ij}))\|_{B(\mathcal{H}^{(n)},\mathcal{H}^{(m)})} : \varphi : V \to B(\mathcal{H}) \text{ isometric}\},$$

where the supremum is taken over all Hilbert spaces $\mathcal{H}$ and all linear isometries $\varphi: V \to B(\mathcal{H})$. This operator space is called the *maximal operator space of* V.

The construction of MIN(V) guarantees that the collection of such isometries is nonempty and hence $\|v\|_V = \|v\|_{\mathrm{MAX}(V)}$. Since every "abstract" operator space, can be represented as a "concrete" operator space, any L^∞-matrix-norm structure that we could endow V with must be smaller than MAX(V). The following gives a more concrete realization of these matrix norms.

Theorem 14.2. *Let V be a normed space and let (v_{ij}) be in $M_{m,n}(V)$. Then $\|(v_{ij})\|_{\mathrm{MAX}(V)} = \inf\{\|A\|\|B\|: A \in M_{m,k}, B \in M_{k,n}, y_i \in V, \|y_i\| \le 1, 1 \le i \le k$, and $(v_{ij}) = A\,\mathrm{Diag}(y_1, \ldots, y_k)B\}$, where $\mathrm{Diag}(y_1, \ldots, y_k)$ represents the $k \times k$ diagonal matrix with entries $y_1, \ldots, y_k$, and the infimum is taken over all ways to represent (v_{ij}) as such a product.*

Proof. Let $\|(v_{ij})\|_{m,n}$ denote the infimum appearing on the right hand side of the above equation. If $\varphi: V \to B(H)$ is any linear map and $(v_{ij}) = A\,\mathrm{Diag}(y_1, \ldots, y_k)B$ is any such factorization, then

$$(\varphi(v_{ij})) = A\,\mathrm{Diag}(\varphi(y_1), \ldots, \varphi(y_k))B,$$

and consequently,

$$\|(\varphi(v_{ij}))\|_{B(\mathcal{H}^{(n)},\mathcal{H}^{(m)})} \le \|A\|\|B\| \cdot \max\{\|\varphi(y_1)\|, \dots, \|\varphi(y_k)\|\}.$$

Hence, $\|(v_{ij})\|_{\text{MAX}(V)} \le \|(v_{ij})\|_{m,n}$.

To prove the other inequality, since MAX(V) is the largest of all possible L^∞-matrix norms, it will suffice to prove that V equipped with $\{\|\cdot\|_{m,n}\}$ is an L^∞-matrix-normed space.

First, to verify that $\|\cdot\|_{m,n}$ is a norm on $M_{m,n}(V)$, let (v_{ij}) and (w_{ij}) in $M_{m,n}(V)$ be given, and fix $\epsilon > 0$. Then there exist integers k_1, k_2, elements $\|y_i\| \le 1$, $1 \le i \le k_1$, $\|x_j\| \le 1$, $1 \le j \le k_2$, of V, and matrices $A_1 \in M_{m,k_1}$, $B_1 \in M_{k_1,n}$, $A_2 \in M_{m,k_2}$, $B_2 \in M_{k_2,n}$ such that $(v_{ij}) = A_1 \operatorname{Diag}(y_1, \dots, y_{k_1})B_1$ and $(w_{ij}) = A_2 \operatorname{Diag}(x_1, \dots, x_{k_2})B_2$ with $\|A_1\|\|B_1\| \le \|(v_{ij})\|_{m,n} + \epsilon$ and $\|A_2\|\|B_2\| \le \|(w_{ij})\|_{m,n} + \epsilon$. Replacing A_1, B_1 by rA_1, $r^{-1}B_1$ with $r = \sqrt{\|B_1\|/\|A_1\|}$, we may assume that $\|A_1\| = \|B_1\|$ and similarly that $\|A_2\| = \|B_2\|$. We have that

$$(v_{ij}) + (w_{ij}) = (A_1, A_2) \operatorname{Diag}(y_1, \dots, y_{k_1}, x_1, \dots, x_{k_2}) \begin{pmatrix} B_1 \\ B_2 \end{pmatrix}$$

and hence

$$\begin{aligned}
\|(v_{ij}) + (w_{ij})\|_{m,n} &\le \|(A_1, A_2)\| \left\| \begin{pmatrix} B_1 \\ B_2 \end{pmatrix} \right\| \\
&\le \|A_1A_1^* + A_2A_2^*\|^{1/2}\|B_1^*B_1 + B_2^*B_2\|^{1/2} \\
&\le (\|A_1\|^2 + \|A_2\|^2)^{1/2}(\|B_1\|^2 + \|B_2\|^2)^{1/2} \\
&\le \|(v_{ij})\|_{m,n} + \|(w_{ij})\|_{m,n} + 2\epsilon.
\end{aligned}$$

Since ϵ was arbitrary, the triangle inequality follows. It is clear that $\|\lambda(v_{ij})\|_{m,n} = |\lambda|\|(v_{ij})\|_{m,n}$ and hence $\|\cdot\|_{m,n}$ is a norm.

If $(v_{ij}) = A \operatorname{Diag}(y_1, \dots, y_k)B$, and C and D are scalar matrices of appropriate sizes, then $C(v_{ij})D = (CA) \operatorname{Diag}(y_1, \dots, y_k)(BD)$. Thus,

$$\|C(v_{ij})D\| \le (\|C\|\|D\|)(\|A\|\|B\|),$$

and taking the infimum over all such representations of (v_{ij}) yields $\|C(v_{ij})D\| \le \|C\|\|D\|\|(v_{ij})\|_{m,n}$. Hence $(V, \{\|\cdot\|_{m,n}\})$ is a matrix norm.

Finally, to see the L^∞ condition, let $(v_{ij}) \in M_{m,n}(V)$, $(w_{ij}) \in M_{p,q}(V)$, and factor $(v_{ij}) = A_1 \operatorname{Diag}(y_1, \dots, y_{k_1})B_1$, $(w_{ij}) = A_2 \operatorname{Diag}(x_1, \dots, x_{k_2})B_2$ with $\|A_1\| = \|B_1\| \le \sqrt{\|(v_{ij})\|_{m,n} + \epsilon}$ and $\|A_2\| = \|B_2\| \le \sqrt{\|(w_{ij})\|_{p,q} + \epsilon}$. Then

$(v_{ij}) \oplus (w_{ij}) = (A_1 \oplus A_2)\,\mathrm{Diag}(y_1, \ldots, y_{k_1}, x_1, \ldots, x_{k_2})(B_1 \oplus B_2)$ and hence

$$\|(v_{ij}) \oplus (w_{ij})\|_{m+p,n+q} \le \max\{\|(v_{ij})\|_{m,n} + \epsilon,\ \|(w_{ij})\|_{p,q} + \epsilon\},$$

from which the L^∞ condition follows.

This completes the proof. □

Note that while the original definition of MAX(V) was "extrinsic" in the sense that it required looking at all representations of V as operators on a Hilbert space, Theorem 14.2 gives an "intrinsic" characterization of this norm. Also, since the original formula for MAX(V) involved a supremum while Theorem 14.2 involves an infimum, we have methods for obtaining upper and lower bounds on norms in MAX(V).

The main idea of the above proof was to "guess" a factorization formula for MAX(V) and then to verify it by showing that it satisfied Ruan's axioms. Such factorization formulas will appear frequently in later chapters and are one of the main tools of this field.

Since every operator space structure on a normed space V lies between MIN(V) and MAX(V), it is natural to wonder to what extent MIN(V) and MAX(V) differ. One way to measure this difference is to regard the identity map on V as a map i: MIN(V) $\to$ MAX(V) and attempt to compute its cb norm. This number is denoted $\alpha(V)$, that is,

$$\alpha(V) = \|i\|_{\mathrm{cb}} = \sup\left\{\frac{\|(v_{ij})\|_{\mathrm{MAX}(V)}}{\|(v_{ij})\|_{\mathrm{MIN}(V)}} : (v_{ij}) \in M_{m,n}(V),\ \ m, n \text{ arbitrary}\right\}.$$

Thus, $\alpha(V) = 1$ if and only if the identity map is a complete isometry from MIN(V) to MAX(V), which is equivalent to there existing a unique operator space structure on V. Currently, there are only two normed spaces known with $\alpha(V) = 1$, namely, ℓ_2^∞ and ℓ_2^1, which denote $\mathbb{C}^2$ equipped with the norms $\|(x_1, x_2)\|_\infty = \max\{|x_1|, |x_2|\}$ and $\|(x_1, x_2)\|_1 = |x_1| + |x_2|$, respectively. It is still unknown if there exist any other normed spaces V for which $\alpha(V) = 1$, but if they exist, they must be of dimension 2, since it is known that $\alpha(V) > 1$ whenever $\dim(V) \ge 3$. See [163] and [193] for these results.

Exact values of $\alpha(V)$ are still unknown for most important finite-dimensional normed spaces. While an extensive development of this constant is beyond our current interests, we do record some of the important facts taken from [163], [164], and [193]. We adopt the convention that ℓ_n^p denotes $\mathbb{C}^n$ equipped with the p-norm.

Theorem 14.3. *Let V be a normed space and V^* denote its dual space. Then*

(i) $\alpha(V) = \alpha(V^*)$,

(ii) if V is n-dimensional, then $\sqrt{n}/2 \leq \alpha(V) \leq n$,
(iii) if V is infinite-dimensional, then $\alpha(V) = +\infty$,
(iv) for all n, $\alpha(\ell_n^2) \leq n/\sqrt{2}$, *while* $(n+1)/2 \leq \alpha(\ell_n^2)$ *for n odd and* $\sqrt{n^2+2n}/2 \leq \alpha(\ell_n^2)$ *for n even,*
(v) for all $n \geq 2$, $\sqrt{n/2} \leq \alpha(\ell_n^1) \leq \sqrt{n-1}$,
(vi) if $\dim(V) = \dim(W)$, *then* $\alpha(V) \leq d(V, W)\alpha(W)$, *where* $d(V, W) = \inf\{\|T\|\|T^{-1}\|: T: V \to W$ *is invertible*$\}$.

Theorem 14.3(iii) implies that if $\dim(V) = +\infty$, then there always exists a Hilbert space $\mathcal{H}$ and a bounded linear map φ: MIN$(V) \to B(\mathcal{H})$ that is not completely bounded. When $\dim(V) < +\infty$ and φ: MIN$(V) \to B(\mathcal{H})$, then (Exercise 14.2) $\|\varphi\|_{\text{cb}} \leq \alpha(V)\|\varphi\|$, so that $\alpha(V)$ gives the best estimate of the ratio of the cb norm to the norm.

Combining the lower bounds on $\alpha(\ell_n^2)$ with the fact that $d(\ell_n^2, V) \leq \sqrt{n}$ for every normed space V of dimension n and applying Theorem 14.3(vi) yields the fact, cited earlier, that $\alpha(V) > 1$ for $\dim(V) \geq 3$.

One difficulty in computing $\alpha(V)$ is that concrete representations of MAX(V) are known for few normed spaces V. For example, no concrete representation is known for MAX(ℓ_n^2), which is the main obstruction to computing an exact value for $\alpha(\ell_n^2)$.

One of the rare exceptions is ℓ_n^1. If $\mathbb{F}_n$ denotes the free group on n generators $\{u_1, \ldots, u_n\}$ and $C_u^*(\mathbb{F}_n)$ is the universal C^*-algebra of this group, then (Exercise 14.3) it can be shown that the map φ: MAX$(\ell_n^1) \to C_u^*(\mathbb{F}_n)$ given by $\varphi((\lambda_1, \ldots, \lambda_n)) = \lambda_1 u_1 + \cdots + \lambda_n u_n$ is a complete isometry.

By comparison, if $\mathbb{T}^n = \mathbb{R}^n/\mathbb{Z}^n$ denotes the standard n-torus, so that $C(\mathbb{T}^n)$ is the abelian C^*-algebra generated by $\{z_1, \ldots, z_n\}$ – where $z_k = e^{it_k}$, $1 \leq k \leq n$, is unitary – then the map ψ: MIN$(\ell_n^1) \to C(\mathbb{T}^n)$ given by $\psi((\lambda_1, \ldots, \lambda_n)) = \lambda_1 z_1 + \cdots + \lambda_n z_n$ is a complete isometry (Exercise 14.4).

Thus, MIN(ℓ_n^1) is represented by the "universal" n-tuple for commuting unitaries, while MAX(ℓ_n^1) is represented by the "universal" n-tuple for noncommuting unitaries.

We shall now take a closer look at MAX(V) and MIN(V) for a general finite-dimensional normed space V.

To this end, let $B \subseteq \mathbb{C}^n$ be the closed unit ball of some norm on $\mathbb{C}^n$; equivalently, let B be a bounded, closed, absorbing, absolutely convex set. If we set

$$\|x\| = \inf\{t: t^{-1}x \in B,\ t > 0\},$$

then $\|\cdot\|$ is the norm on $\mathbb{C}^n$ and B is the (closed) unit ball of this norm. We shall write $(\mathbb{C}^n, B)$ when we want to indicate this normed space. Thus, for example

$$(\mathbb{C}^n, \mathbb{D}^{n^-}) = \ell_n^\infty \quad \text{and} \quad (\mathbb{C}^n, \mathbb{B}_n^-) = \ell_n^2,$$

where $\mathbb{B}_n = \{(\lambda_1, \dots, \lambda_n) \colon |\lambda_1|^2 + \cdots + |\lambda_n|^2 < 1\}$ is the (complex) Euclidean ball. The *polar* of B is the set

$$B^* = \{y \in \mathbb{C}^n \colon |y \cdot x| \leq 1 \text{ for all } x \in B\},$$

where $y \cdot x$ is the usual dot product. The polar of B is easily seen to be the closed unit ball of the dual of $(\mathbb{C}^n, B)$, so that $(\mathbb{C}^n, B)^* = (\mathbb{C}^n, B^*)$.

Let $\{e_1, \dots, e_n\}$ denote the standard basis for $\mathbb{C}^n$. Given $(v_{ij}) \in M_{p,q}(\mathbb{C}^n)$, let $A_\ell = (v_{ij} \cdot e_\ell)$, $1 \leq \ell \leq n$, denote the scalar $p \times q$ matrices obtained by taking the coefficients of the vectors (v_{ij}) with respect to the ℓth basis vector. With respect to the identification $M_{p,q}(\mathbb{C}^n) = M_{p,q} \otimes \mathbb{C}^n$ we have that $(v_{ij}) = A_1 \otimes e_1 + \cdots + A_n \otimes e_n$. In this manner we also identify $M_{p,q}(\mathbb{C}^n)$ with n-tuples of $p \times q$ scalar matrices, and also write $(v_{ij}) = (A_1, \dots, A_n)$.

Now to specify an operator space structure on $(\mathbb{C}^n, B)$ it is enough to define the unit ball of $M_{p,q}(\mathbb{C}^n)$ for all p and q and then verify that the resulting families of norms satisfy Ruan's axioms. Thus for a clearer understanding of $\mathrm{MIN}((\mathbb{C}^n, B))$ and $\mathrm{MAX}((\mathbb{C}^n, B))$ we wish to identify the n-tuples of matrices $(A_1, \dots, A_n)$ that belong to their unit balls.

Theorem 14.4. *Let $B \subseteq \mathbb{C}^n$ be the unit ball of some norm on $\mathbb{C}^n$, let $V = (\mathbb{C}^n, B)$ and let $A_1, \dots, A_n$ be $p \times q$ matrices. We have:*

(i) $\|A_1 \otimes e_1 + \cdots + A_n \otimes e_n\|_{\mathrm{MIN}(V)} \leq 1$ *if and only if* $\|\mu_1 A_1 + \cdots + \mu_n A_n\|_{M_{p,q}} \leq 1$ *for all* $(\mu_1, \dots, \mu_n) \in B^*$,

(ii) $\|A_1 \otimes e_1 + \cdots + A_n \otimes e_n\|_{\mathrm{MAX}(V)} \leq 1$ *if and only if* $\|A_1 \otimes C_1 + \cdots + A_n \otimes C_n\|_{M_{p,q}(B(\mathcal{H}))} \leq 1$ *for all Hilbert spaces $\mathcal{H}$ and n-tuples $(C_1, \dots, C_n)$ of operators on $\mathcal{H}$ satisfying $\|\lambda_1 C_1 + \cdots + \lambda_n C_n\| \leq 1$ for all $(\lambda_1, \dots, \lambda_n) \in B$.*

Proof. We first prove (ii). Note that $\varphi \colon V \to B(\mathcal{H})$ is contractive if and only if the n-tuple $C_i = \varphi(e_i)$ satisfies $\|\lambda_1 C_1 + \cdots + \lambda_n C_n\| \leq 1$ for all $(\lambda_1, \dots, \lambda_n) \in B$. Now if $(v_{ij}) = A_1 \otimes e_1 + \cdots + A_n \otimes e_n$, then

$$(\varphi(v_{ij})) = A_1 \otimes \varphi(e_1) + \cdots + A_n \otimes \varphi(e_n) = A_1 \otimes C_1 + \cdots + A_n \otimes C_n.$$

Since $\|A_1 \otimes e_1 + \cdots + A_n \otimes e_n\|_{\mathrm{MAX}(V)} \leq 1$ if and only if $\|A_1 \otimes \varphi(e_1) + \cdots + A_n \otimes \varphi(e_n)\| \leq 1$ for all Hilbert spaces $\mathcal{H}$ and $\varphi \colon V \to B(\mathcal{H})$ contractive, (ii) follows.

To prove (i), recall that $\mathrm{MIN}(V)$ is the operator space structure obtained by the embedding $j \colon V \to C(V_1^*)$. But the unit ball of V^* is just the polar B^* and

$j(e_i)((\mu_1, \ldots, \mu_n)) = \mu_i$. Hence,

$$\begin{aligned}\|A_1 \otimes e_1 + \cdots + A_n e_n\|_{\mathrm{MIN}(V)} \\ &= \|A_1 \otimes j(e_1) + \cdots + A_n \otimes j(e_n)\|_{M_{p,q}(C(B^*))} \\ &= \sup\{\|\mu_1 A_1 + \cdots + \mu_n A_n\|_{M_{p,q}} \colon (\mu_1, \ldots, \mu_n) \in B^*\},\end{aligned}$$

and (i) follows. □

As an application of these concepts we consider a generalization of the classical Schwarz inequality from complex analysis. Recall that the Schwarz inequality says that if $f\colon \mathbb{D} \to \mathbb{D}$ is analytic with $f(0) = 0$, then $|f'(0)| \leq 1$. Moreover, by considering functions of the form $f(z) = \alpha z$ we see that the set of possible values of $f'(0)$ is exactly $\mathbb{D}^-$. Suppose now that $G \subseteq \mathbb{C}^n$ is the open unit ball of some norm on $\mathbb{C}^n$. Given $F = (f_{ij})\colon G \to M_{p,q}$ analytic, we let $DF(0)$ denote the n-tuple of $p \times q$ matrices

$$DF(0) = \left(\left(\frac{\partial f_{ij}(0)}{\partial z_1}\right), \ldots, \left(\frac{\partial f_{ij}(0)}{\partial z_n}\right)\right).$$

We are interested in determining the set of all possible n-tuples $DF(0)$ for $F\colon G \to \mathrm{ball}(M_{p,q})$ with $F(0) = 0$. A description of these n-tuples can be found in most books on several complex variables; see [205] for example. But those descriptions should be compared with the following for clarity and the ease with which it can be recalled.

Proposition 14.5 (Generalized Schwarz lemma). *Let $G \subseteq \mathbb{C}^n$ be the open unit ball of some norm; let $V = (\mathbb{C}^n, G^-)$. There exists $F\colon G \to \mathrm{ball}(M_{p,q})$ analytic, $F(0) = 0$, such that $DF(0) = (A_1, \ldots, A_n)$ if and only if $\|A_1 \otimes e_1 + \cdots + A_n \otimes e_n\|_{\mathrm{MIN}(V^*)} \leq 1$.*

Proof. Assume $\|A_1 \otimes e_1 + \cdots + A_n \otimes e_n\|_{\mathrm{MIN}(V^*)} \leq 1$. Since $V^* = (\mathbb{C}^n, G^*)$ and $(G^*)^* = G^-$, by Theorem 14.4 (i) we have $\|A_1\lambda_1 + \cdots + A_n\lambda_n\| \leq 1$ for all $(\lambda_1, \ldots, \lambda_n) \in G^-$. Define $F\colon G \to M_{p,q}$ by $F((z_1, \ldots, z_n)) = A_1 z_1 + \cdots + A_n z_n$. Then $F(G) \subseteq \mathrm{ball}(M_{p,q})$, $F(0) = 0$, and $DF(0) = (A_1, \ldots, A_n)$.

Conversely, assume $F\colon G \to \mathrm{ball}(M_{p,q})$ is analytic, $F(0) = 0$, and $DF(0) = (A_1, \ldots, A_n)$. Fix unit vectors $h \in \mathbb{C}^q$, $k \in \mathbb{C}^p$ and a point $(\lambda_1, \ldots, \lambda_n) \in G$, and define $f\colon \mathbb{D} \to \mathbb{C}$ via $f(z) = \langle F(\lambda_1 z, \ldots, \lambda_n z)h, k\rangle$. Since $f(0) = 0$ and $f(\mathbb{D}) \subseteq \mathbb{D}$, by Schwarz's lemma we have that $|f'(0)| = |\langle (A_1\lambda_1 + \cdots + A_n\lambda_n)h, k\rangle| \leq 1$. Since h and k are arbitrary unit vectors, we have that $\|A_1\lambda_1 + \cdots + A_n\lambda_n\| \leq 1$ for all $(\lambda_1, \ldots, \lambda_n) \in G$. Thus, $\|A_1 \otimes e_1 + \cdots + A_n \otimes e_n\|_{\mathrm{MIN}(V^*)} \leq 1$ by another application of Theorem 14.4 (i). □

The following is yet another way, in operator space language, to interpret the generalized Schwarz lemma. Let $H^\infty(G)$ denote the space of bounded analytic functions on G. If for $(f_{ij}) \in M_{p,q}(H^\infty(G))$ we set

$$\|(f_{ij})\|_\infty = \sup\{\|(f_{ij}(z))\|_{M_{p,q}}: z \in G\},$$

then $H^\infty(G)$ is an operator space. Let $H_0^\infty(G) = \{f \in H^\infty(G): f(0) = 0\}$ and $H_{00}^\infty(G) = \{f \in H^\infty(G): f(0) = 0, Df(0) = 0\}$ denote the corresponding operator subspaces. The map $D: H_0^\infty(G) \to \mathbb{C}^n$, $f \to Df(0)$ has kernel $H_{00}^\infty(G)$ and so induces a quotient map $\dot{D}: H_0^\infty(G)/H_{00}^\infty(G) \to \mathbb{C}^n$. It is now easy to see that Proposition 14.5 is equivalent to the statement that $\dot{D}$ is a complete isometry between the quotient operator space $H_0^\infty(G)/H_{00}^\infty(G)$ and $\mathrm{MIN}(V^*)$.

The space $H_0^\infty(G)/H_{00}^\infty(G)$ is often called the *cotangent space of G at 0.* Thus, the generalized Schwarz lemma is simply one way to describe what the natural operator space structure is on the cotangent space.

The generalized Schwarz lemma and Theorem 14.4 also make it possible to generalize the phenomenon in Parrott's example.

Theorem 14.6. *Let $G \subseteq \mathbb{C}^n$ be the open unit ball of a norm on $\mathbb{C}^n$, and let $V = (\mathbb{C}^n, G^-)$. Then there exists a Hilbert space $\mathcal{H}$ and a unital contractive homomorphism $\rho: H^\infty(G) \to B(\mathcal{H})$ with $\|\rho\|_{\mathrm{cb}} \geq \alpha(V)$.*

Proof. Given a Hilbert space $\mathcal{H}$ and operators $A_1, \ldots, A_n \in B(\mathcal{H})$, we define $\rho: H^\infty(G) \to B(\mathcal{H} \oplus \mathcal{H})$ via

$$\rho(f) = \begin{pmatrix} f(0)I & \frac{\partial f(0)}{\partial z_1} A_1 + \cdots + \frac{\partial f(0)}{\partial z_n} A_n \\ 0 & f(0)I \end{pmatrix}.$$

Using the product rule, it is easily checked that ρ is a homomorphism.

We claim that if $\|A_1\mu_1 + \cdots + A_n\mu_n\| \leq 1$ for all $(\mu_1, \ldots, \mu_n) \in G^*$, then $\|\rho\| \leq 1$.

To prove this claim, first consider the case where $f(0) = 0$, $\|f\|_\infty < 1$. Then $Df(0) = (\mu_1, \ldots, \mu_n) \in G^*$ by Proposition 14.5, and so $\|\rho(f)\| = \|A_1\mu_1 + \cdots + A_n\mu_n\| \leq 1$. For arbitrary $f \in H^\infty(G)$, with $\|f\|_\infty < 1$, let $\alpha = f(0)$ and let $\varphi_\alpha(z) = (z - \alpha)(1 - \bar{\alpha}z)^{-1}$. It is easily seen that $\sigma(\rho(f)) = f(0) = \alpha$ and that

$$\varphi_\alpha(\rho(f)) = (\rho(f) - \alpha I)(I - \bar{\alpha}\rho(f))^{-1} = \rho((f - \alpha 1)(1 - \bar{\alpha} f)) = \rho(\varphi_\alpha \circ f).$$

Since $\varphi_\alpha \circ f(0) = 0$ and $\|\varphi_\alpha \circ f\|_\infty \leq 1$, we have that $\|\varphi_\alpha(\rho(f))\| \leq 1$. Clearly, $\rho(f) = \varphi_{-\alpha}(\varphi_\alpha(\rho(f)))$, and hence we may apply von Neumann's inequality to deduce that $\|\rho(f)\| = \|\varphi_{-\alpha}(\varphi_\alpha(\rho(f)))\| \leq 1$.

Applying Theorem 14.4 (i), we see that $\|A_1 \otimes e_1 + \cdots + A_n \otimes e_n\|_{\mathrm{MIN}(V)} \leq 1$ implies $\|\rho\| \leq 1$.

Given any $p \times q$ matrices $B_1, \ldots, B_n$ such that $\|B_1\lambda_1 + \cdots + B_n\lambda_n\| \leq 1$, we have that $F((z_1, \ldots, z_n)) = B_1 z_1 + \cdots + B_n z_n \in M_{p,q}(H^\infty(G))$ and $\|F\| \leq 1$. Since $DF(0) = (B_1, \ldots, B_n)$, setting $F = (f_{ij})$, we have

$$(\rho(f_{ij})) = \begin{pmatrix} 0 & A_1 \otimes B_1 + \cdots + A_n \otimes B_n \\ 0 & 0 \end{pmatrix}.$$

Thus, $\|\rho\|_{\mathrm{cb}} \leq \|A_1 \otimes B_1 + \cdots + A_n \otimes B_n\|$. Taking the supremum over all such $(B_1, \ldots, B_n)$ and applying Theorem 14.4 (ii), we obtain $\|\rho\|_{\mathrm{cb}} \geq \|A_1 \otimes e_1 + \cdots + A_n \otimes e_n\|_{\mathrm{MAX}(V)}$. Since this holds for any

$$\|A_1 \otimes e_1 + \cdots + A_n \otimes e_n\|_{\mathrm{MIN}(V)} \leq 1,$$

we may take a sequence of such A's, with $\|A_1^{(k)} \otimes e_1 + \cdots + A_n^{(k)} \otimes e_n\|_{\mathrm{MIN}(V)} \leq 1$ and $\lim_k \|A_1^{(k)} \otimes e_1 + \cdots + A_n^{(k)} \otimes e_n\|_{\mathrm{MAX}(V)} = \alpha(V)$. Finally, setting $A_\ell = \sum_{k=1}^{\infty} \oplus A_\ell^{(k)}$ and using these operators to define a homomorphism ρ, we have that $\|\rho\| = 1$ while $\|\rho\|_{\mathrm{cb}} \geq \alpha(V)$. □

Applying this construction to the polydisk $G = \mathbb{D}^n$, we obtain an n-tuple of commuting contractions that induce a contractive homomorphism ρ of $H^\infty(\mathbb{D}^n)$ with $\|\rho\|_{\mathrm{cb}} \geq \alpha(\ell_n^\infty) \geq \sqrt{n/2}$. When $n = 3$, we find that this construction essentially reduces to the example of Parrott presented in Chapter 7.

As mentioned earlier, $\alpha(V) > 1$ whenever $\dim(V) \geq 3$. Thus, for every $G \subseteq \mathbb{C}^n$, $n \geq 3$ as above, there exists a unital contractive homomorphism $\rho\colon H^\infty(G) \to B(\mathcal{H})$ with $\|\rho\|_{\mathrm{cb}} > 1$.

If we let $\mathcal{P}(G) \subset H^\infty(G)$ denote the subalgebra generated by the polynomials in the coordinate functions, then it is not hard to see (Exercise 14.10) that the analogue of Theorem 14.6 holds with $\mathcal{P}(G)$ in place of $H^\infty(G)$. Thus, for such a set G there always exists an n-tuple of commuting operators such that G^- is a joint spectral set for the n-tuple, but not a complete spectral set.

These considerations show how special Ando's dilation theorem is, since the bidisk is one of the very few possible domains for which contractive homomorphisms can be completely contractive.

An important feature of normed spaces is that the space $B(V, W)$ of bounded linear maps between two normed spaces V and W is again a normed space and that, in particular, the dual $V^* = B(V, \mathbb{C})$ of a normed space V is again a normed space. This feature is shared by operator spaces and completely bounded maps. That is, if E and F are operator spaces, then there is a natural way to make $\mathrm{CB}(E, F)$, and hence E^*, into operator spaces.

To see how this is done suppose that $(\varphi_{ij}) \in M_{m,n}(\mathrm{CB}(E, F))$. We identify (φ_{ij}) with a map $\Phi\colon E \to M_{m,n}(F)$ by setting $\Phi(e) = (\varphi_{ij}(e))$, and define a norm on $M_{m,n}(\mathrm{CB}(E, F))$ by setting

$$\|(\varphi_{ij})\|_{M_{m,n}(\mathrm{CB}(E,F))} = \|\Phi\|_{\mathrm{CB}(E,M_{m,n}(F))}.$$

In short, we identify $M_{m,n}(\mathrm{CB}(E, F)) = \mathrm{CB}(E, M_{m,n}(F))$. It is not difficult to see that

$$\begin{aligned}&\|(\varphi_{ij})\|_{M_{m,n}(CB(E,F))}\\&\quad = \sup\{\|(\varphi_{ij}(e_{k\ell}))\|_{M_{mp,nq}(F)}\colon (e_{k\ell}) \in M_{p,q}(E),\ \|(e_{k\ell})\| \le 1\},\end{aligned}$$

where the matrix $(\varphi_{ij}(e_{k\ell}))$ can be interpreted as the $p \times q$ matrix of $m \times n$ matrices

$$\begin{pmatrix} \Phi(e_{11}) & \dots & \Phi(e_{1q}) \\ \vdots & & \vdots \\ \Phi(e_{p1}) & \dots & \Phi(e_{pq}) \end{pmatrix},$$

or as the $m \times n$ matrix of $p \times q$ matrices

$$\begin{pmatrix} (\varphi_{11}(e_{k\ell})) & \dots & (\varphi_{1n}(e_{k\ell})) \\ \vdots & & \vdots \\ (\varphi_{m1}(e_{k\ell})) & \dots & (\varphi_{mn}(e_{k\ell})) \end{pmatrix},$$

since these two matrices differ by a canonical shuffle.

Proposition 14.7 (Ruan). *Let E and F be operator spaces, and equip matrices over* $\mathrm{CB}(E, F)$ *with the norms obtained by the identifications*

$$M_{m,n}(\mathrm{CB}(E, F)) = \mathrm{CB}(E, M_{m,n}(F)).$$

Then $\mathrm{CB}(E, F)$ *is an operator space.*

Proof. We leave it to the reader (Exercise 14.5) to verify that $\mathrm{CB}(E, F)$ satisfies the axioms of an L^∞-matrix-normed space. □

One special case of the above result is when $F = \mathbb{C}$. Recall that for every linear $\varphi\colon E \to \mathbb{C}$ we have $\|\varphi\| = \|\varphi\|_{\mathrm{cb}}$. Thus, $E^* = B(E, \mathbb{C}) = \mathrm{CB}(E, \mathbb{C})$, isometrically. Hence, by the above result there is a natural way to endow the Banach space dual of an operator space with a matrix-normed structure such that the dual of an operator space is again an operator space. In short, we identify $M_{m,n}(E^*) = \mathrm{CB}(E, M_{m,n})$. Thus, if $(\varphi_{ij}) \in M_{m,n}(E^*)$, then we identify it with the map $\Phi\colon E \to M_{m,n}$ given by $\Phi(e) = (\varphi_{ij}(e))$ and set $\|(\varphi_{ij})\| = \|\Phi\|_{\mathrm{cb}}$.

In the future we shall refer to the Banach space dual, endowed with this matrix-normed structure, as *the dual operator space*. We shall write E^* for the dual operator space of E as well as for the Banach space dual of E, and hope it will be clear from the context which dual we are referring to.

Proposition 14.8. *Let E be an operator space, and let $j\colon E \to E^{**}$ be the canonical embedding $j(e)(\varphi) = \varphi(e)$ for $\varphi \in E^*$. Then j is a complete isometry.*

Proof. If $X = (x_{ij}) \in M_{m,n}(E)$, then $(j(x_{ij})) \in M_{m,n}(E^{**})$ is identified with the map $\hat{X}\colon E^* \to M_{m,n}$ defined by $\hat{X}(\varphi) = (\varphi(x_{ij}))$. Thus, we must prove $\|\hat{X}\|_{\text{cb}} = \|(x_{ij})\|$. Now if $(\varphi_{k,\ell}) \in M_{pq}(E^*)$, then

$$\|(\hat{X}(\varphi_{k,\ell}))\| = \|(\varphi_{k\ell}(x_{ij}))\| \leq \|\Phi\|_{\text{cb}}\|(x_{ij})\|$$

where $\Phi\colon E \to M_{pq}$ is given by $\Phi(x) = (\varphi_{k\ell}(x))$. Since $\|(\varphi_{k\ell})\| = \|\Phi\|_{\text{cb}}$, we have that $\|\hat{X}\|_{\text{cb}} \leq \|(x_{ij})\|$.

Now to prove the other inequality we shall need Ruan's theorem. So let $\gamma\colon E \to B(\mathcal{H})$ be a complete isometry. Pick subspaces $\mathcal{H}_q \subseteq \mathcal{H}_{q+1}$ with $\dim(\mathcal{H}_q) = q$ such that $\|(x_{ij})\| = \|(\gamma(x_{ij}))\| = \sup_q \|(P_q\gamma(x_{ij})P_q)\|$, where $P_q\colon \mathcal{H} \to \mathcal{H}_q$ denotes the projection. Choosing an orthonormal set $\{e_k\}_{k=1}^{\infty}$ such that $\text{span}\{e_1, \ldots, e_q\} = \mathcal{H}_q$, we obtain linear functionals $\gamma_{k\ell}(x) = \langle\gamma(x)e_\ell, e_k\rangle$ such that $P_q\gamma(x)P_q \cong (\gamma_{k,\ell}(x))_{k,\ell=1}^q$. Hence $(\gamma_{k,\ell})_{k,\ell=1}^q \in M_q(E^*)$, $\|(\gamma_{k,\ell})\| \leq \|\gamma\|_{\text{cb}} = 1$, and

$$\|\hat{X}\|_{\text{cb}} \geq \sup_q \|(\hat{X}(\gamma_{k,\ell}))\| = \sup_q \|(P_q\gamma(x_{ij})P_q)\| = \|(x_{ij})\|. \qquad \square$$

Blecher [19] proves the deeper fact that the image of the unit ball of $M_{m,n}(E)$ is weak*-dense in the unit ball of $M_{m,n}(E^{**})$ for all m and n.

Not surprisingly, some of the most important operator spaces are those that as normed spaces are Hilbert spaces. We have already met two ways to make n-dimensional Hilbert space ℓ_n^2 into an operator space: $\text{MIN}(\ell_n^2)$ and $\text{MAX}(\ell_n^2)$. By Exercise 14.5, $\text{MIN}(\ell_n^2)^* = \text{MAX}(\ell_n^2)$ and $\text{MAX}(\ell_n^2)^* = \text{MIN}(\ell_n^2)$, so these are a dual pair. Another important dual pair are *row* and *column Hilbert* space.

Consider the identification of ℓ_n^2 with the operator space $M_{n,1}$ of the $n \times 1$ matrices, i.e., column vectors of length n. A $p \times q$ matrix of vectors (v_{ij}) is normed by writing it as a $pn \times q$ scalar matrix under the identification $M_{p,q}(M_{n,1}) = M_{pn,q}$. When the Hilbert space ℓ_n^2 is equipped with this operator space structure, we call it *column Hilbert space*, and it is generally denoted C_n instead of $M_{n,1}$.

Similarly, if we identify ℓ_n^2 with the operator space $M_{1,n}$, i.e., row vectors of length n, then a $p \times q$ matrix of vectors (v_{ij}) is normed by writing it as a $p \times qn$

scalar matrix under the identification $M_{p,q}(M_{1,n}) = M_{p,qn}$. The Hilbert space ℓ_n^2 equipped with this operator space structure is called *row Hilbert space*, and it is denoted R_n instead of $M_{1,n}$.

For a quick example of how these two structures differ, let $\{e_1, \ldots, e_n\}$ denote the canonical basis vectors for ℓ_n^2, and consider the $1 \times n$ matrix of vectors $(e_1, \ldots, e_n)$. We have $M_{1,n}(C_n) = M_{n,n}$, and this is identified with the identity matrix, while $M_{1,n}(R_n) = M_{1,n^2}$ and hence is identified with a row vector of length n^2.

Hence we have $\|(e_1, \ldots, e_n)\|_{M_{1,n}(C_n)} = 1$, while $\|(e_1, \ldots, e_n)\|_{M_{1,n}(R_n)} = \sqrt{n}$, and thus the natural isometric identification of C_n with R_n is not a complete isometry. In fact, if an $n \times n$ matrix A is regarded as a linear map from C_n to R_n, then $\|A\|_{\text{cb}}$ is equal to the Hilbert–Schmidt norm of A. For a proof of this fact, see [140].

We now wish to examine these two operator space structures from another viewpoint. Formally, C_n can be thought of as the image of ℓ_n^2 under the isometric map $\varphi\colon \ell_n^2 \to C_n$ given by $\varphi(e_i) = E_{i,1}$. Given a matrix $(v_{k\ell}) \in M_{p,q}(\ell_n^2)$, let $A_i = (\langle v_{k\ell}, e_i\rangle)$, $1 \le i \le n$, denote the scalar $p \times q$ matrices obtained by taking the coefficients of $(v_{k\ell})$ with respect to the ith basis vector. The identification $M_{p,q}(\ell_n^2) = M_{p,q} \otimes \ell_n^2$ identifies $(v_{k\ell}) = \sum_{i=1}^n A_i \otimes e_i$, and

$$(\varphi(v_{k\ell})) = \mathrm{id}_{M_{p,q}} \otimes \varphi\left(\sum_{i=1}^n A_i \otimes e_i\right) = \sum_{i=1}^n A_i \otimes \varphi(e_i)$$

$$= \sum_{i=1}^n A_i \otimes E_{i1} \quad \text{in} \quad M_{p,q} \otimes M_{n,1} = M_{pn,q}.$$

Thus $(\varphi(v_{k\ell}))$ is identified with the $pn \times q$ matrix

$$\begin{pmatrix} A_1 \\ \vdots \\ A_n \end{pmatrix},$$

and we have

$$\|(v_{k\ell})\|_{M_{p,q}(C_n)}^2 = \left\|\left(\sum_{i=1}^n A_i \otimes E_{i1}\right)^* \left(\sum_{j=1}^n A_j \otimes E_{ji}\right)\right\|$$

$$= \left\|\sum_{i=1}^n A_i^* A_i \otimes E_{11}\right\| = \left\|\sum_{i=1}^n A_i^* A_i\right\|.$$

A similar analysis for R_n shows that if $(v_{k\ell}) = \sum_{i=1}^n A_i \otimes e_i$ is in $M_{p,q} \otimes \ell_n^2$, then its image in R_n is $\sum_{i=1}^n A_i \otimes E_{1i}$, which can be identified with the $p \times qn$

matrix $(A_1, \ldots, A_n)$, and

$$\|(v_{k\ell})\|^2_{M_{p,q}(R_n)} = \left\| \left(\sum_{i=1}^n A_i \otimes E_{1i}\right)\left(\sum_{j=1}^n A_j \otimes E_{1j}\right)^* \right\| = \left\|\sum_{i=1}^n A_i A_i^*\right\|.$$

Given $v \in R_n$, define $\gamma(v)\colon C_n \to \mathbb{C}$, via $\gamma(v)(w) = v \cdot w$, where $v \cdot w$ denotes the matrix product of the row vector v with the column vector w. This gives a map $\gamma\colon R_n \to C_n^*$, which is clearly an isometry.

Proposition 14.9. *The map $\gamma\colon R_n \to C_n^*$ is a complete isometry.*

Proof. Let $(v_{ij}) \in M_{p,q}(R_n) = M_{p,qn}$. Then $\Phi = (\gamma(v_{ij}))\colon C_n \to M_{p,q}$ is the map given by $\Phi(w) = (v_{ij} \cdot w)$, and we must prove that $\|\phi\|_{\text{cb}} = \|(v_{ij})\|_{M_{p,qn}}$.

First note that if we form the $qn \times q$ matrix

$$\hat{w} = \begin{pmatrix} w & 0 & \ldots & 0 \\ 0 & w & \ddots & \vdots \\ \vdots & \ddots & \ddots & 0 \\ 0 & \ldots & 0 & w \end{pmatrix},$$

then $\|\hat{w}\| = \|w\|$ and $\Phi(w) = (v_{ij}) \cdot \hat{w}$ is the actual matrix product. Thus, $\|\Phi(w)\|_{M_{p,q}} \leq \|(v_{ij})\|_{M_{p,q}(R_n)} \cdot \|w\|$.

Now if $(w_{k,\ell}) \in M_{r,s}(C_n)$, then

$$(\Phi(w_{k,\ell})) = ((v_{ij}) \cdot \hat{w}_{k,\ell}) = \begin{pmatrix} (v_{ij}) & 0 & \\ & \ddots & \\ 0 & & (v_{ij}) \end{pmatrix} \cdot \begin{pmatrix} \hat{w}_{11} & \ldots & \hat{w}_{1s} \\ \vdots & & \vdots \\ \hat{w}_{r1} & \ldots & \hat{w}_{rs} \end{pmatrix},$$

and hence

$$\|(\Phi(w_{k\ell}))\| \leq \|(v_{ij})\|_{M_{p,q}(R_n)} \|(\hat{w}_{k\ell})\|_{M_{r,s}(M_{qn,q})}.$$

A moment's reflection and a canonical shuffle show that $\|(\hat{w}_{k\ell})\|_{M_{r,s}(M_{qn,q})} = \|(w_{k\ell})\|_{M_{r,s}(C_n)}$.

Hence, $\|\Phi\|_{\text{cb}} \leq \|(v_{ij})\|_{M_{p,q}(R_n)}$.

To see that $\|\Phi\|_{\text{cb}} \geq \|(v_{ij})\|_{M_{p,q}(R_n)}$, consider $(E_{11}, \ldots, E_{n1}) \in M_{1,n}(C_n)$. Then $\|(E_{11}, \ldots, E_{n1})\| = 1$, and so

$$\begin{aligned} \|\Phi\|_{\text{cb}} &\geq \|(\Phi(E_{11}), \ldots, \Phi(E_{n1}))\|_{M_{p,qn}} = \|((v_{ij} \cdot E_{11}), \ldots, (v_{ij} \cdot E_{n1}))\|_{M_{p,qn}} \\ &= \|(A_1, \ldots, A_n)\|_{M_{p,qn}}. \end{aligned}$$

But the scalar matrices $A_1, \ldots, A_n$ are easily seen to be the coefficients of $(v_{ij}) = \sum_{i=1}^n A_i \otimes e_i$, and so this last quantity is $\|(v_{ij})\|_{M_{p,q}(R_n)}$ by the above calculation. □

In a similar fashion the matrix product pairing defines a complete isometry of C_n onto R_n^*.

It is valuable to look at the above pairing from the tensor viewpoint. Keeping the notation as above, if $\Phi(w) = (v_{ij} \cdot w)$ with $(v_{ij}) = \sum_{i=1}^n A_i \otimes e_i$ and $(w_{k\ell}) = \sum_{i=1}^n B_i \otimes e_i$, then $(\Phi(w_{k\ell})) = \sum_{i=1}^n A_i \otimes B_i$ where $A_i \otimes B_i$ is the Kronecker tensor product of matrices discussed in Chapter 3.

One of the most important properties of a Hilbert space is that it is conjugate linearly isometrically isomorphic to its own dual. Thus, given a Hilbert space $\mathcal{H}$, one would like an operator space structure on $\mathcal{H}$ such that the map from $\mathcal{H}$ to $\mathcal{H}^*$ is a conjugate linear complete isometry. Pisier [176] proved that such an operator space structure exists and is unique. The resulting operator space is called *the operator Hilbert space* and is denoted $\mathcal{OH}$.

If $\{e_1, \ldots, e_n\}$ denotes the standard basis for ℓ_n^2 and $\sum_{i=1}^n A_i \otimes e_i$ is in $M_{p,q}(\ell_n^2)$, then

$$\left\| \sum_{i=1}^n A_i \otimes e_i \right\|_{\mathcal{O}\ell_n^2} = \left\| \sum_{i=1}^n A_i \otimes \bar{A}_i \right\|_{M_{p^2,q^2}},$$

where $\bar{A}$ denotes the complex conjugate of the matrix A, and $A \otimes \bar{A}$ is the $p^2 \times q^2$ Kronecker tensor encountered earlier. That is, $A \otimes \bar{A}$ is the matrix whose $((i_1, i_2), (j_1, j_2))$ entry is $a_{i_1,j_1}\bar{a}_{i_2,j_2}$ for $1 \le i_1, i_2 \le p$ and $1 \le j_1, j_2 \le q$.

Pisier's proof rests on a pretty Cauchy–Schwarz-type inequality for Kronecker tensors due to Haagerup [108] namely,

$$\left\| \sum_{i=1}^n A_i \otimes B_i \right\| \le \left\| \sum_{i=1}^n A_i \otimes \bar{A}_i \right\|^{1/2} \left\| \sum_{i=1}^n B_i \otimes \bar{B}_i \right\|^{1/2}.$$

In addition to the five operator space structures that we have already discussed on Hilbert space, there are many more. Among these the most natural to study are the *homogeneous operator space structures.* An operator space X is called *homogeneous* provided every bounded linear map $T\colon X \to X$ is completely bounded and $\|T\|_{\text{cb}} = \|T\|$.

For example, MIN(X) and MAX(X) are homogeneous operator spaces for every normed space (Exercise 14.1). In addition, row, column, and Pisier's operator Hilbert space are homogeneous operator spaces.

There exist many "exotic" homogeneous operator space structures on Hilbert space. Zhang [248] exhibits, for each integer n, two homogeneous operator space structures on ℓ^2_n, say $\mathcal{H}_1$ and $\mathcal{H}_2$, such that if $T\colon \mathcal{H}_1 \to \mathcal{H}_2$ is linear with $\operatorname{rank}(T) \leq n-1$, then $\|T\|_{\text{cb}} = \|T\|$, but $\|I\|_{\text{cb}} \neq \|I\|$, where I denotes the identity. In addition, every pair of $(n-1)$-dimensional subspaces of $\mathcal{H}_1$ and $\mathcal{H}_2$ are completely isometrically isomorphic, yet $\mathcal{H}_1$ and $\mathcal{H}_2$ are *not* completely isometrically isomorphic.

For each *symmetrically normed ideal* $\mathcal{J}$ in the sense of Gohberg and Krein [105], Mathes and the author [142] exhibit two homogeneous operator space structures on ℓ^2 say $\mathcal{H}_1$ and $\mathcal{H}_2$, such that $T\colon \mathcal{H}_1 \to \mathcal{H}_2$ is completely bounded if and only if $T \in \mathcal{J}$ and the completely bounded norm of T is equivalent to its norm in $\mathcal{J}$.

Thus, in particular, for $1 < p < +\infty$ there exist homogeneous operator space structures on ℓ^2 such that $T\colon \mathcal{H}_1 \to \mathcal{H}_2$ is completely bounded if and only if it belongs to the *Schatten p-ideal* $\mathcal{C}_p$ with $\|T\|_{\text{cb}}$ equivalent to $(\sum_{n=1}^{\infty} s_n^p)^{1/p}$, where $\{s_n\}$ is the sequence of singular values of T.

A characterization of the set of symmetric norms that one can achieve exactly as the cb norm between two homogeneous operator spaces is not known.

It is also not very well understood how nonhomogeneous an operator space norm can be made. That is, given an operator space X, how small a subset $\operatorname{CB}(X, X)$ can be of $B(X, X)$ is not clear. The most definitive result in this direction is due to Oikhberg [154]. He exhibits an operator space structure on Hilbert space such that the completely bounded maps from the space back to itself are just the scalar multiples of the identity plus the compacts. On the other hand it is quite easy to show that, independent of the operator space structure on Hilbert space, every map that is the sum of a scalar multiple of the identity and a trace class operator is necessarily completely bounded. These results are in sharp contrast to the Banach space setting, where it is still unknown if any Banach space X exists such that every operator in $B(X, X)$ is a scalar multiple of the identity plus a compact.

Hilbert C-Modules*

We close this chapter by studying some objects that play an important role in the study of C^*-algebras. To this end, let $\mathcal{A}$ be a unital C^*-algebra and let V be a vector space that is also a *left $\mathcal{A}$-module*. Thus, in particular, we assume $a_1 \cdot (v_1 + v_2) = a_1 \cdot v_1 + a_1 \cdot v_2$, $(a_1 + a_2) \cdot v_1 = a_1 \cdot v_1 + a_2 \cdot v_2$, $(a_1 a_2) \cdot v_1 = a_1 \cdot (a_2 \cdot v_1)$, and $(\lambda a_1) \cdot v_1 = \lambda(a_1 \cdot v_1)$, for $a_1, a_2 \in \mathcal{A}$ and v_1, $v_2 \in V$, and $\lambda \in \mathbb{C}$. We also assume that $1 \cdot v = v$ for every v in V. Sometimes modules satisfying this latter condition are called *unitary* $\mathcal{A}$-modules, but we find this terminology can be confusing.

We call V a *Hilbert C^*-module* provided it is equipped with a map $\langle\, ,\, \rangle\colon V \times V \to \mathcal{A}$ satisfying:

(i) $\langle v_1 + v_2, v_3\rangle = \langle v_1, v_3\rangle + \langle v_2, v_3\rangle$,
(ii) $\langle av_1, v_2\rangle = a\langle v_1, v_2\rangle$,
(iii) $\langle v_1, av_2\rangle = \langle v_1, v_2\rangle a^*$,
(iv) $\langle v_1, v_2\rangle = \langle v_2, v_1\rangle^*$,
(v) $\langle v_1, v_1\rangle \geq 0$ and $\langle v_1, v_1\rangle = 0$ if and only if $v_1 = 0$,

for every $v_1, v_2, v_3 \in V$ and $a \in \mathcal{A}$. We call such a map an *$\mathcal{A}$-valued inner product.*

We note that we have made no attempt to present a minimal set of axioms. For example, (ii) and (iv) taken together imply (iii).

Thus, a Hilbert C^*-module is like a Hilbert space, where the role of $\mathbb{C}$ is played by $\mathcal{A}$.

Perhaps the simplest example of a Hilbert C^*-module over $\mathcal{A}$ is to let $V = \{(a_1, \ldots, a_n)\colon a_i \in \mathcal{A}\}$ with module action given by $a(a_1, \ldots, a_n) = (aa_1, \ldots, aa_n)$ and an $\mathcal{A}$-valued inner product given by

$$\langle (a_1, \ldots, a_n), (b_1, \ldots, b_n)\rangle = a_1 b_1^* + \cdots + a_n b_n^*.$$

Note that if we identify $V = M_{1,n}(\mathcal{A})$, then $V^* = M_{n,1}(\mathcal{A})$ and for $v_1 = (a_1, \ldots, a_n)$ and $v_2 = (b_1, \ldots, b_n)$ in V we have that

$$\langle v_1, v_2\rangle = (a_1, \ldots, a_n) \cdot \begin{pmatrix} b_1^* \\ \vdots \\ b_n^* \end{pmatrix} = v_1 v_2^*.$$

Thus, we see that V is identified with an operator space $M_{1,n}(\mathcal{A})$ in such a way that the $\mathcal{A}$-valued inner product becomes an actual product between V and V^*. It is a well-known result in the theory of Hilbert C^*-modules that such a representation is always possible. Thus, in particular, every Hilbert C^*-module has a "natural" operator space structure.

Using Ruan's theorem, it is possible to give an intrinsic description of this operator space structure on a Hilbert C^*-module. Clearly, if we want $\langle v_1, v_2\rangle = v_1 v_2^*$, then for any $X = (v_{ij}) \in M_{p,q}(V)$ we should have that $XX^* \in M_{p,p}(\mathcal{A})$ is the element given by

$$XX^* = \left(\sum_{\ell=1}^{q} v_{i\ell} v_{j\ell}^*\right) = \left(\sum_{\ell=1}^{q} \langle v_{i\ell}, v_{j\ell}\rangle\right).$$

Thus, we need that this latter matrix should be a positive element of $M_p(\mathcal{A})$ and $\|X\|^2 = \|(\sum_{\ell=1}^{q} \langle v_{i\ell}, v_{j\ell}\rangle)\|$. This hopefully motivates the following theorem.

Theorem 14.10. *Let V be a Hilbert C^*-module over $\mathcal{A}$. Then for every p, q and $X = (v_{ij}) \in M_{p,q}(V)$ the matrix $(\sum_{\ell=1}^{q} \langle v_{i\ell}, v_{j\ell} \rangle) \in M_p(\mathcal{A})^+$. If we define $\|X\|_{p,q} = \|(\sum_{\ell=1}^{q} \langle v_{i\ell}, v_{j\ell} \rangle)\|^{1/2}$, then V endowed with this family of matrix norms is an operator space.*

Proof. To verify that the matrix $(\sum_{\ell=1}^{q} \langle v_{i\ell}, v_{j\ell} \rangle)$ is a positive element of $M_p(\mathcal{A})$, by Exercise 3.18 it is sufficient to prove that for every p-tuple $(a_1, \dots, a_p)$ of elements of $\mathcal{A}$, we have that $\sum_{i,j=1}^{p} \sum_{\ell=1}^{q} a_i \langle v_{i\ell}, v_{j\ell} \rangle a_j^*$ is a positive element of $\mathcal{A}$. Setting $x_\ell = \sum_{i=1}^{p} a_i v_{i\ell}$, we see that this latter sum is $\sum_{\ell=1}^{p} \langle x_\ell, x_\ell \rangle$, which is positive by property (v).

To see that $\|v\| = \|\langle v, v \rangle\|^{1/2}$ is a norm on V, we note that $\|\lambda v\| = |\lambda| \|v\|$ is clear and establish the triangle inequality. To this end, let $v, w \in V$ and set $X = \binom{v}{w} \in M_{2,1}(V)$. Since

$$XX^* = \begin{pmatrix} \langle v, v \rangle & \langle v, w \rangle \\ \langle w, v \rangle & \langle w, w \rangle \end{pmatrix} \in M_2(\mathcal{A})^+,$$

by Exercise 3.2 (ii) we have that

$$\begin{aligned} \|\langle v + w, v + w \rangle\| &= \|\langle v, v \rangle + \langle v, w \rangle + \langle w, v \rangle + \langle w, w \rangle\| \\ &\le \left(\|\langle v, v \rangle\|^{1/2} + \|\langle w, w \rangle\|^{1/2} \right)^2 \end{aligned}$$

and thus $\|v + w\| \le \|v\| + \|w\|$.

To see that the above quantity defines a norm on $M_{p,q}(V)$, simply note that the "natural" matrix product makes $M_{p,q}(V)$ a left $M_p(\mathcal{A})$-module. For $X = (v_{ij})$ and $Y = (w_{ij})$ in $M_{p,q}(V)$ set $\langle X, Y \rangle_{p,q} = (\sum_{k=1}^{q} \langle v_{ik}, w_{jk} \rangle) \in M_p(\mathcal{A})$. Since, formally, $\langle X, Y \rangle_{p,q} = X \cdot Y^*$, it is readily seen that $\langle \cdot, \cdot \rangle_{p,q}$ is an $M_p(\mathcal{A})$-valued inner product on $M_{p,q}(V)$. Thus, by the last paragraph, we know that setting

$$\|X\|_{p,q} = \|\langle X, X \rangle_{p,q}\|^{1/2} = \left\| \left(\sum_{k=1}^{q} \langle v_{ik}, v_{jk} \rangle \right) \right\|_{M_p(\mathcal{A})}^{1/2}$$

defines a norm on $M_{p,q}(\mathcal{A})$.

The fact that this system of norms satisfies Ruan's axioms is straightforward. In particular, notice that the L^∞ condition follows from the fact that for $X \in M_{p,q}(V)$, and $Y \in M_{r,s}(V)$ we have that

$$\begin{aligned} \|X \oplus Y\|_{p+r,q+s} &= \|\langle X \oplus Y, X \oplus Y \rangle_{p+r,q+s}\|^{1/2} \\ &= \|\langle X, X \rangle_{p,q} \oplus \langle Y, Y \rangle_{r,s}\|^{1/2} = \max\{\|X\|, \|Y\|\}. \end{aligned}$$

□

Notes

The theory of operator spaces is currently a very exciting and rapidly expanding area. See [88] and [193] for two recent texts in this field.

The constant α was first introduced and studied in [163], where many of the early estimates of its values were obtained. Some small improvements can be found in [164], and the key improvement on the lower bound for $\alpha(\ell_n^2)$ appears in [193].

The geometry of finite-dimensional operator spaces now plays an important role in many questions in the general theory of C^*-algebras. One of the most exciting breakthroughs was its use by Junge and Pisier [127] to prove that the maximal and minimal C^*-tensor norms of $B(\mathcal{H})$ with itself are different, thus settling an old conjecture.

Exercises

14.1 Let V be a normed space, X an operator space, and let $\varphi: X \to \text{MIN}(V)$, $\psi: \text{MAX}(V) \to X$ be linear maps. Prove that $\|\varphi\| = \|\varphi\|_{\text{cb}}$ and $\|\psi\| = \|\psi\|_{\text{cb}}$. Deduce that $\text{MIN}(V)$ and $\text{MAX}(V)$ are homogeneous.

14.2 Let V be a normed space, X an operator space, and let $\varphi: \text{MIN}(V) \to X$. Prove that $\|\varphi\|_{\text{cb}} \le \alpha(V)\|\varphi\|$.

14.3 (Zhang) Let $\mathbb{F}_m$ denote the free group on m generators

$$\{u_1^{(m)}, \ldots, u_m^{(m)}\}.$$

Prove that the maps $\varphi: \text{MAX}(\ell_n^1) \to C_u^*(\mathbb{F}_n)$ and $\psi: \text{MAX}(\ell_n^1) \to C_u^*(\mathbb{F}_{n-1})$ given by $\varphi((\lambda_1, \ldots, \lambda_n)) = \lambda_1 u_1^n + \cdots + \lambda_n u_n^n$ and

$$\psi((\lambda_1, \ldots, \lambda_n)) = \lambda_1 u_1^{(n-1)} + \lambda_2 u_2^{(n-1)} + \cdots + \lambda_{n-1} u_{n-1}^{(n-1)} + \lambda_n I$$

are complete isometries.

[Hint: Prove that the map sending $u_i^{(n-1)} \to u_n^{(n)*} u_i^{(n)}$ induces a $*$-isomorphism of $C_u^*(\mathbb{F}_{n-1})$ into $C_u^*(\mathbb{F}_n)$.]

14.4 Prove that the maps $\varphi: \text{MIN}(\ell_n^1) \to C(\mathbb{T}^n)$ and $\psi: \text{MIN}(\ell_n^1) \to C(\mathbb{T}^{n-1})$ given by $\varphi((\lambda_1, \ldots, \lambda_n)) = \lambda_1 z_1 + \cdots + \lambda_n z_n$ and $\psi((\lambda_1, \ldots, \lambda_n)) = \lambda_1 z_1 + \cdots + \lambda_{n-1} z_{n-1} + \lambda_n$ are complete isometries.

14.5 Prove Proposition 14.7.

14.6 (Blecher) Let V be a normed space. Prove that $\text{MIN}(V)^* = \text{MAX}(V^*)$ and $\text{MAX}(V)^* = \text{MIN}(V^*)$, completely isometrically. Use these identifications to prove that $\alpha(V) = \alpha(V^*)$.

14.7 Recall that for $\mathcal{H}, \mathcal{K}$ Hilbert spaces, $B(\mathcal{H}, \mathcal{K})$ is an operator space. The map $\varphi: B(\mathbb{C}, \mathcal{H}) \to \mathcal{H}$ via $\varphi(T) = T(1)$ for $T \in B(\mathbb{C}, \mathcal{H})$ is an isometry.

Prove that $B(\mathbb{C}, \ell_n^2)$ is completely isometrically isomorphic to C_n. Show similarly, that $B(\ell_n^2, \mathbb{C})$ is completely isometrically isomorphic to R_n. *We set $\mathcal{H}_c = B(\mathbb{C}, \mathcal{H})$ and $\mathcal{H}_r = B(\mathcal{H}, \mathbb{C})$, and call these operator spaces Hilbert column space and Hilbert row space, respectively.*

14.8 Let $\gamma\colon C_n \to R_n$ be the map $\gamma(\sum_{i=1}^n \lambda_i E_{i1}) = \sum_{i=1}^n \lambda_i E_{1i}$. Prove that γ is an isometry and $\|\gamma\|_{\text{cb}} = \|\gamma^{-1}\|_{\text{cb}} = \sqrt{n}$.

14.9 Let V be a Hilbert C^*-module over $\mathcal{A}$. Prove the following analogue of the Cauchy–Schwarz inequality: $\langle v, w\rangle\langle w, v\rangle \le \|\langle w, w\rangle\|\langle v, v\rangle$.

14.10 Prove the analogue of Theorem 14.6 with $H^\infty(G)$ replaced by $\mathcal{P}(G)$. Prove that the inclusion of $\mathcal{P}(G)$ into $H^\infty(G)$ induces a completely isometric isomorphism between the quotient spaces $\mathcal{P}_0(G)/\mathcal{P}_{00}(G)$ and the quotient space $H_0^\infty(G)/H_{00}^\infty(G)$.

Chapter 15
Injective Envelopes

In this chapter we take a closer look at injectivity and introduce injective envelopes and C^*-envelopes of operator systems, operator algebras, and operator spaces. Loosely speaking, the injective envelope of an object is a "minimal" injective object that contains the original object. The C^*-envelope of an operator algebra is a generalization of the Silov boundary of a uniform algebra. The C^*-envelope of an operator algebra $\mathcal{A}$ is the "smallest" C^*-algebra that contains $\mathcal{A}$ as a subalgebra, up to completely isometric isomorphism. These ideas will be made precise in this chapter. Many of the ideas of this chapter are derived from the work of M. Hamana [112].

Injectivity is really a categorical concept. Suppose that we are given some category $\mathcal{C}$ consisting of objects and morphisms. Then an object I is called *injective in* $\mathcal{C}$ provided that for every pair of objects $E \subseteq F$ and every morphism $\varphi\colon E \to I$, there exists a morphism $\psi\colon F \to I$ that extends φ, i.e., such that $\psi(e) = \varphi(e)$ for every e in E.

If we let $\mathfrak{S}$ denote the collection of operator systems and define the morphisms between operator systems to be the completely positive maps, then since the composition of completely positive maps is again completely positive, we shall have a category, which we call the *category of operator systems*. Arveson's extension theorem (Theorem 7.5) for completely positive maps is equivalent to the statement that $B(\mathcal{H})$ is injective in the category $\mathfrak{S}$. The earlier definition of injectivity for C^*-algebras that was given in Chapter 7 was really injectivity in the category $\mathfrak{S}$.

Similarly, we let $\mathcal{O}$ denote the category whose objects consist of operator spaces and whose morphisms are the completely bounded maps. Wittstock's extension theorem (Theorem 8.2) for completely bounded maps shows that $B(\mathcal{H})$ is also injective in the category $\mathcal{O}$. However, note that for it to be injective in $\mathcal{O}$, we only need that every completely bounded map has a completely bounded extension, but not necessarily of the same cb norm! To capture the full force of

Wittstock's extension theorem, we need a slightly different category. We let $\mathcal{O}_1$ denote the category whose objects are operator spaces and whose morphisms are the completely contractive maps. Now by a simple scaling argument, it is easy to see (Exercise 15.1) that I is injective in $\mathcal{O}_1$ if and only if every completely bounded map into I has a completely bounded extension of the same cb norm. Thus, Wittstock's extension theorem is equivalent to the statement that $B(\mathcal{H})$ is injective in $\mathcal{O}_1$.

In this chapter we focus on injectivity in $\mathcal{O}_1$.

For operator systems it might seem natural to also study injectivity in $\mathfrak{S}$ or in the category $\mathfrak{S}_1$ consisting of operator systems and unital completely positive maps, but it turns out that for operator systems injectivity in these three categories is equivalent (Proposition 15.1).

We begin with an elementary characterization of injectivity for operator systems.

Proposition 15.1. *Let $S \subseteq B(\mathcal{H})$ be an operator system. Then the following are equivalent:*

(i) S is injective in $\mathcal{O}_1$,
(ii) S is injective in $\mathfrak{S}$,
(iii) S is injective in $\mathfrak{S}_1$,
(iv) there exists a completely positive projection ϕ: $B(\mathcal{H}) \to S$ onto S.

Proof. We begin by proving that (i) is equivalent to (iv). Assuming (i), then the identity map from S to S extends to a completely contractive map ϕ: $B(\mathcal{H}) \to S$. Since ϕ extends the identity map, ϕ is a projection onto S. Since $\phi(1) = 1$, ϕ must be completely positive.

Conversely, assume (iv), and suppose we are given operator spaces $E \subseteq F$ and a completely contractive map, γ: $E \to S$. Then γ has a completely contractive extension ψ: $F \to B(\mathcal{H})$, and $\phi \circ \psi$: $F \to S$ is the desired completely contractive extension of γ into S.

The proof of the equivalence of (ii) and (iii) to (iv) is similar. □

Thus, for operator systems we have that injectivity in $\mathfrak{S}$, in $\mathfrak{S}_1$, and in $\mathcal{O}_1$ are all equivalent. For this reason we shall continue to simply call such systems *injective*. However, it is important to remark that there is an example due to Huruya [125] of a C^*-algebra that he proves is not injective, but is easily seen [102] to be injective in $\mathcal{O}$. Currently, there is little that is known about the C^*-algebras that are injective in $\mathcal{O}$. Every C^*-algebra that is completely boundedly linearly isomorphic to a C^*-algebra that is injective in $\mathcal{O}_1$ can be easily seen to be injective in $\mathcal{O}$, but it is not known if this characterizes the C^*-algebras that are injective in $\mathcal{O}$.

The following result shows that every injective operator system is in an appropriate sense a C^*-algebra.

Theorem 15.2 (Choi–Effros). *Let $S \subseteq B(\mathcal{H})$ be an injective operator system, and let $\phi\colon B(\mathcal{H}) \to S$ be a completely positive projection onto S. Then setting $a \circ b = \phi(a \cdot b)$ defines a multiplication on S, and S together with this multiplication and its usual $*$-operation is a C^*-algebra. Moreover, the identity map from S to the C^*-algebra $(S, \circ)$ is a unital complete order isomorphism.*

Proof. Clearly, for a, b in S, $a \circ b$ is in S. Distributivity is clear, and clearly $a \circ 1 = 1 \circ a = a$. To show that $\circ$ defines a multiplication on S it remains to show associativity, $a \circ (b \circ c) = (a \circ b) \circ c$, i.e., that $\phi(a\phi(bc)) = \phi(\phi(ab)c)$.

We claim for any x in $B(\mathcal{H})$ and a in S that $\phi(\phi(x)a) = \phi(xa)$ and $\phi(a\phi(x)) = \phi(ax)$. Assuming the claim, we have that

$$\phi(a\phi(bc)) = \phi(abc) = \phi(\phi(ab)c),$$

and thus associativity follows.

To prove the claim recall the Schwarz inequality, $\psi(y^*y) - \psi(y)^*\psi(y) \geq 0$ for any unital completely positive map ψ. Applying this to $\psi = \phi^{(2)}$ and $y = \left(\begin{smallmatrix} a^* & x \\ 0 & 0 \end{smallmatrix}\right)$ yields

$$\begin{pmatrix} \phi(aa^*) & \phi(ax) \\ \phi(x^*a^*) & \phi(x^*x) \end{pmatrix} - \begin{pmatrix} aa^* & a\phi(x) \\ \phi(x)^*a^* & \phi(x)^*\phi(x) \end{pmatrix} \geq 0.$$

Applying $\phi^{(2)}$ to this inequality yields

$$\begin{pmatrix} 0 & \phi(ax) - \phi(a\phi(x)) \\ \phi(x^*a^*) - \phi(\phi(x)^*a^*) & \phi(x^*x) - \phi(\phi(x)^*\phi(x)) \end{pmatrix} \geq 0.$$

The positivity of this matrix forces $\phi(ax) - \phi(a\phi(x)) = 0$, and hence, since ϕ is self-adjoint, the claim follows.

Next we verify the C^* condition, $\|a^* \circ a\| = \|a\|^2$. Clearly, $\|a^* \circ a\| = \|\phi(a^*a)\| \leq \|a^*a\| = \|a\|^2$. But, again by the Schwarz inequality, $\phi(a^*a) \geq \phi(a)^*\phi(a) = a^*a$ and hence $\|a^* \circ a\| = \|\phi(a^*a)\| \geq \|a^*a\| = \|a\|^2$, and the proof that $(S, \circ)$ is a C^*-algebra is complete.

Clearly the identity map from S to $(S, \circ)$ is an isometry. We now argue that it is a complete isometry. To this end consider $M_n(S) \subseteq M_n(B(\mathcal{H})) = B(\mathcal{H}^{(n)})$ and $\phi^{(n)}\colon B(\mathcal{H}^{(n)}) \to M_n(S)$. By the above, $M_n(S)$ is a C^*-algebra with product

$A \circ_n B = \phi^{(n)}(A \cdot B)$, and this C^*-algebra is isometrically isomorphic to the operator system $M_n(S)$. But for $A = (a_{ij})$, $B = (b_{ij})$ we have

$$A \circ_n B = \phi^{(n)}\left(\sum_{k=1}^{n} a_{ik}b_{kj}\right) = \left(\sum_{k=1}^{n} \phi(a_{ik}b_{kj})\right) = \left(\sum_{k=1}^{n} a_{ik} \circ b_{kj}\right).$$

Thus, $(M_n(S), \circ_n)$ is the C^*-algebra tensor product of M_n and $(S, \circ)$, and hence, by the uniqueness of the C^*-norm on $M_n((S, \circ))$, the identity map from S to $(S, \circ)$ is an n-isometry for all n. Since the identity map is a unital complete isometry, it is a complete order isomorphism as claimed. □

We now turn our attention to operator spaces. Since in this chapter we are mainly concerned with injectivity in $\mathcal{O}_1$, we shall also simply refer to an operator space as *injective* when it is injective in $\mathcal{O}_1$. This is also consistent with the terminology used in most of the operator space literature. Arguing as in the proof of Proposition 15.1, it is easily seen that an operator space $E \subseteq B(\mathcal{H})$ is injective if and only if there exists a completely contractive projection of $B(\mathcal{H})$ onto E (Exercise 15.2).

We wish to construct the injective envelope of an operator space. There are many equivalent ways to define these objects. We prefer to take a definition with the fewest hypotheses and deduce the other properties as consequences.

Given an operator space F, we say that (E, κ) is *an injective envelope of* F provided that

(i) E is injective in $\mathcal{O}_1$,
(ii) $\kappa\colon F \to E$ is a complete isometry,
(iii) if E_1 is injective with $\kappa(F) \subseteq E_1 \subseteq E$, then $E_1 = E$.

Any (E, κ) satisfying (ii) is called an *extension of F.* When (E, κ) satisfies (i) and (ii), then it is called an *injective extension of F.* Identifying F with $\kappa(F)$, we often simply regard E as containing F, with the understanding that now κ is simply the inclusion of F into E. Thus, an injective envelope of F is, loosely, a minimal injective containing F.

To prove the existence of such an object, one would like to simply invoke Zorn's lemma. But to use this approach it would be necessary to prove that if $\{E_\lambda\}$ is a decreasing chain of injectives with $F \subseteq E_\lambda$, then $\bigcap_\lambda E_\lambda$ is also injective, and this is not clear. The use of minimal F-seminorms, which we define below, is a way to finesse this problem.

Assume that $F \subseteq B(\mathcal{H})$. We call a map $\varphi\colon B(\mathcal{H}) \to B(\mathcal{H})$ an F*-map* provided that φ is completely contractive and $\varphi(x) = x$ for all x in F. An F-map φ such that $\varphi \circ \varphi = \varphi$ is called an F*-projection*. Thus, an F-projection φ is a completely contractive projection onto $E = \varphi(B(H))$, with $F \subseteq E$, but we do not demand $F = E$.

We define a partial order on F-projections by setting $\psi \prec \varphi$ provided that $\psi \circ \varphi = \psi = \varphi \circ \psi$.

Given an F-map φ, we define a F*-seminorm* p_φ on $B(\mathcal{H})$ by setting $p_\varphi(x) = \|\varphi(x)\|$. There is a natural partial order on seminorms, defined by $p \leq q$ if and only if $p(x) \leq q(x)$ for all x.

Proposition 15.3. *Let $F \subseteq B(\mathcal{H})$ be an operator space. Then there exist minimal F-seminorms on $B(\mathcal{H})$.*

Proof. Let $\varphi_\lambda \colon B(\mathcal{H}) \to B(\mathcal{H})$ be F-maps such that p_{φ_λ} is a decreasing chain of F-seminorms. Recall the BW topology of Chapter 5. Since $CB_1(B(\mathcal{H}), B(\mathcal{H}))$ is BW-compact, $\{\varphi_\lambda\}$ has a subnet $\{\varphi_{\lambda_\mu}\}$ converging to, say, φ. Clearly, φ is an F-map, and since $|\langle \varphi(x)h, k\rangle| = \lim_\mu |\langle \varphi_{\lambda_\mu}(x)h, k\rangle| \leq \liminf_\mu \|\varphi_{\lambda_\mu}(x)\| \|h\| \|k\|$, it follows that $p_\varphi \leq p_{\varphi_\lambda}$ for all λ. Thus, every decreasing chain of F-seminorms has a lower bound, and it follows by Zorn's lemma that minimal F-seminorms exist. □

Theorem 15.4. *Let $F \subseteq B(\mathcal{H})$ be an operator space. If $\varphi \colon B(\mathcal{H}) \to B(\mathcal{H})$ is a F-map such that p_φ is a minimal F-seminorm, then φ is a minimal F-projection and the range $\varphi(B(\mathcal{H}))$ of φ is an injective envelope of F.*

Proof. We begin by proving that φ is a F-projection. Since $\varphi \circ \varphi$ is also an F-map and $\|\varphi(\varphi(x))\| \leq \|\varphi(x)\|$, we must have that $\|\varphi \circ \varphi(x)\| = \|\varphi(x)\|$ for all $x \in B(\mathcal{H})$. Set $\varphi^{(k+1)} = \varphi^{(k)} \circ \varphi$, then $\|\varphi^{(k)}(x)\| = \|\varphi(x)\|$ for all $k \geq 1$. Set $\psi_n(x) = [\varphi(x) + \cdots + \varphi^{(n)}(x)]/n$; then $\|\psi_n(x)\| \leq \|\varphi(x)\|$, and so $\|\psi_n(x)\| = \|\varphi(x)\|$, too. Hence,

$$\begin{aligned}\|\varphi(x) - \varphi \circ \varphi(x)\| &= \|\varphi(x - \varphi(x))\| = \|\psi_n(x - \varphi(x))\| \\ &= \left\| \frac{\varphi(x) + \cdots + \varphi^{(n)}(x)}{n} - \frac{\varphi^{(2)}(x) + \cdots + \varphi^{(n+1)}(x)}{n} \right\| \\ &\leq \frac{2\|\varphi(x)\|}{n} \to 0.\end{aligned}$$

Thus, $\|\varphi(x) - \varphi \circ \varphi(x)\| = 0$, and it follows that φ is an F-projection.

Now suppose that ψ is a F-projection with $\psi \prec \varphi$, so that $\psi \circ \varphi = \psi = \varphi \circ \psi$. Since $\|\psi(x)\| = \|\psi(\varphi(x))\| \leq \|\varphi(x)\|$, we have that $\|\psi(x)\| = \|\varphi(x)\|$ for all x. Finally, $\|\varphi(x) - \psi(x)\| = \|\varphi(\varphi(x) - \psi(x))\| = \|\psi(\varphi(x) - \psi(x))\| = \|\psi(x) - \psi(x)\| = 0$, and hence $\varphi(x) = \psi(x)$ for all x. Thus, φ is a minimal F-projection.

Since $B(\mathcal{H})$ is injective in $\mathcal{O}_1$, and φ is a completely contractive projection, it follows readily that $\varphi(B(\mathcal{H}))$ is injective in $\mathcal{O}_1$ as well. Now assume that $F \subseteq E_1 \subseteq \varphi(B(\mathcal{H}))$ with E_1 injective in $\mathcal{O}_1$. Then the identity map from E_1 to E_1 extends to a completely contractive projection γ from $B(\mathcal{H})$ to E_1. Since $\gamma \circ \varphi$ is a F-map and $\|\gamma(\varphi(x))\| \leq \|\varphi(x)\|$, we have that $\|\gamma \circ \varphi(x)\| = \|\varphi(x)\|$ by minimality of the seminorm p_φ. Since γ is an isometry on $\varphi(B(H))$ and since $\gamma(\varphi(x) - \gamma \circ \varphi(x)) = 0$, we have $\varphi(x) = \gamma(\varphi(x))$, and it follows that $E_1 = \varphi(B(\mathcal{H}))$. Hence, $\varphi(B(\mathcal{H}))$ is an injective envelope of F. □

Note that we do not assert that if φ is a minimal F-projection then p_φ is a minimal F-seminorm.

Now that we know the existence of an injective envelope, we turn our attention to its uniqueness and further properties.

Lemma 15.5. *Let $F \subseteq B(\mathcal{H})$ be an operator space, with φ: $B(\mathcal{H}) \to B(\mathcal{H})$ a F-map such that p_φ is a minimal F-seminorm. If γ: $\varphi(B(\mathcal{H})) \to \varphi(B(\mathcal{H}))$ is completely contractive and $\gamma(x) = x$ for all x in F, then $\gamma(\varphi(x)) = \varphi(x)$ for all x in $B(\mathcal{H})$.*

Proof. Since $\|\gamma(\varphi(x))\| \leq \|\varphi(x)\|$, we have that $\|\gamma(\varphi(x))\| = \|\varphi(x)\|$ by minimality of p_φ. Hence γ is an isometry and, in particular, one-to-one. Since $p_{\gamma\circ\varphi} = p_\varphi$ is a minimal F-seminorm, Theorem 15.4 implies that $\gamma \circ \varphi$ is a projection. Hence $\gamma \circ \varphi = \gamma \circ \varphi \circ \gamma \circ \varphi = \gamma \circ \gamma \circ \varphi$, because φ is the identity on the range of $\gamma \circ \varphi$. Thus, $\gamma \circ (\varphi - \gamma \circ \varphi) = 0$. But since γ is one-to-one, $\varphi = \gamma \circ \varphi$ and the result follows. □

Theorem 15.6. *Let (E_1, κ_1) and (E_2, κ_2) be two injective envelopes of the operator space F. Then the map i: $\kappa_1(F) \to \kappa_2(F)$ given by $i(\kappa_1(m)) = \kappa_2(m)$ extends uniquely to a completely isometric isomorphism of E_1 onto E_2.*

Proof. Let $F \subseteq B(\mathcal{H})$ with φ: $B(\mathcal{H}) \to B(\mathcal{H})$ a F-map such that p_φ is a minimal F-seminorm. If we can prove that the map κ_1: $F \to E_1$ has a unique extension to a completely isometric isomorphism γ_1: $\varphi(B(H)) \to E_1$, then the result will follow, since $\gamma_2^{-1} \circ \gamma_1$ will then yield the desired map between E_1 and E_2.

By injectivity of E_1, a completely contractive map γ_1: $\varphi(B(\mathcal{H})) \to E_1$ extending κ_1 exists, and since $\varphi(B(\mathcal{H}))$ is injective, there is a completely contractive map β: $E_1 \to \varphi(B(\mathcal{H}))$ with $\beta(\kappa_1(x)) = x$ for all x in F. Now $\beta \circ \gamma_1$: $\varphi(B(\mathcal{H})) \to \varphi(B(\mathcal{H}))$ is completely contractive and fixes, F so by Lemma 15.5 it is the identity on $\varphi(B(\mathcal{H}))$.

Since β and γ_1 are both completely contractive, it follows that γ_1 must be a complete isometry. But the range of γ_1 will then be an injective operator subspace of E_1, and so, by the minimality of E_1, γ_1 must be onto. Hence, γ_1 is a completely isometric isomorphism. □

The above result shows that the injective envelope really only depends on the operator space up to complete isometry. Thus, if $F \subseteq B(\mathcal{H})$, $\kappa\colon F \to B(\mathcal{H}_1)$ is a complete isometry, and $\psi\colon B(\mathcal{H}_1) \to B(\mathcal{H}_1)$ is a $\kappa(F)$-map such that p_ψ is a minimal $\kappa(F)$-seminorm, then $(\psi(B(\mathcal{H}_1)), \kappa)$ is an injective envelope of F and so completely isometrically isomorphic to $\varphi(B(\mathcal{H}))$ for any F-map φ with p_φ a minimal F-seminorm.

Corollary 15.7 (Rigidity). *Let (E, κ) be an injective envelope of F, and let $\psi\colon E \to E$ be completely contractive with $\psi(\kappa(x)) = \kappa(x)$ for all x in F. Then $\psi(e) = e$ for all e in E.*

Proof. By Lemma 15.5, this is true for $\varphi(B(\mathcal{H}))$, but by Theorem 15.6, $\varphi(B(\mathcal{H}))$ and E are completely isometrically isomorphic. Clearly, this property is preserved by completely isometric isomorphisms. □

This rigidity property can be used to characterize injective envelopes. Call an extension (E, κ) of F a *rigid extension* if whenever $\varphi\colon E \to E$ is completely contractive and $\varphi(\kappa(x)) = \kappa(x)$ for all X in F, then the map φ is the identity on E.

Rigid extensions are related to essential extensions. Let (E, κ) be an extension of F, let M be any operator space, and let $\varphi\colon E \to M$ be completely contractive. If $\varphi \circ \kappa\colon F \to M$ a complete isometry implies that φ is a complete isometry, then we call (E, κ) an *essential extension of F.*

Formally, these definitions of rigidity and essentiality both depend on the category $\mathcal{O}_1$, but since we will not be studying these concepts in any other categories, we have omitted this dependence from the definition.

Finally, we say that (E, κ) is the *maximal extension* of a space F with some property, if (E, κ) is an extension of F and whenever (M, β) is an extension of E and $(M, \beta \circ \kappa)$ has the property, then $\beta(E) = M$. Similarly, we call (E, κ) a *minimal extension* of a space F with some property, if (E, κ) is an extension of F and whenever $\kappa(F) \subset M \subset E$ and M has the property, then $M = E$.

Theorem 15.8. *Let F be an operator space, and let (E, κ) be an extension of F. Then the following are equivalent:*

(i) (E, κ) *is an injective envelope of* F*, i.e., a minimal injective extension of* F*,*
(ii) (E, κ) *is a maximal rigid extension of* F*,*
(iii) (E, κ) *is a maximal essential extension of* F*,*
(iv) E *is injective and* (E, κ) *is a rigid extension of* F*,*
(v) E *is injective and* (E, κ) *is an essential extension of* F*.*

Proof. We only prove the equivalence of (i) and (iii); the remaining implications are left as Exercise 15.4.

Assuming (i), let (M, β) be an extension of E, and assume that $(M, \beta \circ \kappa)$ is an essential extension of F. Since E is injective, the inverse of β extends to a completely contractive projection from M onto E. This map is a complete isometry on $\beta \circ \kappa(F)$ and so must be a complete isometry on M. Hence, $\beta(E) = M$.

Conversely, assume that (E, κ) is a maximal essential extension of F, and let (M, β) be an injective envelope of F. By the proof that (i) implies (iii), we know that (M, β) is an essential extension of F. The map $\beta \circ \kappa^{-1}$ from $\kappa(F)$ to $\beta(F)$ must extend to a completely contractive map γ of E into M. But since E is an essential extension of F, γ must be a complete isometry on E. Thus (M, γ) is an extension of E and an essential extension of F. Hence, by maximality, $\gamma(E) = M$, and so (E, κ) is an injective envelope of F. □

We let $I(F)$ denote the (essentially unique) injective envelope in $\mathcal{O}_1$ of an operator space F.

Some comments are now in order on injective envelopes of operator systems. If S is an operator system, then it is also an operator space. Thus, it makes sense to seek either a minimal injective in $\mathcal{O}_1$, a minimal injective in $\mathcal{S}_1$, or a minimal injective in $\mathcal{S}$. To construct such an object in $\mathcal{S}$, one could take $1 \in S \subseteq B(\mathcal{H})$ and consider completely positive S-projections and mimic the proofs of 15.3–15.7. But since these maps are necessarily unital, they are completely contractive, and we are back to the situation of an operator space. Thus, we see that there is nothing to be gained by reproving theorems about injective envelopes of operator systems in $\mathcal{S}$ or $\mathcal{S}_1$. We do record one observation.

Proposition 15.9. *Let* S *be an operator system. Then the injective envelope* $I(S)$ *of* S *in* $\mathcal{O}_1$ *is an injective operator system and hence is completely order-isomorphic to a* C^**-algebra.*

Proof. Let $1 \in S \subseteq B(\mathcal{H})$. Then $I(S) = \varphi(B(\mathcal{H}))$ for some S-projection φ that is completely contractive and hence completely positive. Thus, $\varphi(B(\mathcal{H}))$ is an operator system, and hence $I(S)$ is a C^*-algebra under the product $a \circ b = \varphi(ab)$ by Theorem 15.2. □

One consequence of the above result is that if one "forgets" the order structure on S and only remembers the operator space structure, and embeds S completely isometrically via $\psi: S \to B(\mathcal{H})$ with say $\psi(1) \neq 1$, then $I(\psi(S))$ is completely isometric to $I(S)$ and so is still completely isometrically isomorphic to a C^*-algebra.

The next few results record how injective envelopes respect algebraic and C^*-structures.

Proposition 15.10. *Let $1 \in \mathcal{A} \subseteq S \subseteq B(\mathcal{H})$ with S an operator system and $\mathcal{A}$ a subalgebra of $B(\mathcal{H})$. Then the inclusion of $\mathcal{A}$ into the C^*-algebra $I(S)$ is a completely isometric *-isomorphism.*

Proof. Let $\varphi: B(\mathcal{H}) \to B(\mathcal{H})$ be a S-projection such that $\varphi(B(\mathcal{H}))$ is a copy of $I(S)$. Then the multiplication on $I(S)$ is given by $x \circ y = \varphi(xy)$. Hence for a, b in $\mathcal{A}$, $a \circ b = \varphi(a \cdot b) = a \cdot b$. □

In the case of a unital C^*-algebra $\mathcal{A}$, if we set $S = \mathcal{A}$, then the above result implies that $\mathcal{A}$ is a C^*-subalgebra of $I(\mathcal{A})$. The next result explains the case of nonunital C^*-algebras.

Proposition 15.11. *Let $\mathcal{A} \subseteq B(\mathcal{H})$ be a nonunital C^*-algebra such that $\mathcal{A}\mathcal{H}$ is dense in $\mathcal{H}$, and let $\mathcal{A}_1$ denote the algebra obtained by adjoining the unit to $\mathcal{A}$. If $\varphi: B(\mathcal{H}) \to B(\mathcal{H})$ is an $\mathcal{A}$-map, then φ is an $\mathcal{A}_1$-map. Consequently, the inclusion of $\mathcal{A}$ into $\mathcal{A}_1$ extends uniquely to define a complete isometry of $I(\mathcal{A})$ onto $I(\mathcal{A}_1)$. Thus, $I(\mathcal{A})$ is a unital C^*-algebra in the $\circ$-product with $\mathcal{A}_1$ a C^*-subalgebra of $I(\mathcal{A})$.*

Proof. If every $\mathcal{A}$-map is an $\mathcal{A}_1$-map, then any minimal $\mathcal{A}$-seminorm is a minimal $\mathcal{A}_1$-seminorm and the last claim follows from the characterization of the injective envelope in terms of minimal seminorms. Thus, it is enough to prove that if φ is an $\mathcal{A}$-map, then $\varphi(I) = I$.

Let $\{e_\alpha\}$ be a contractive, positive approximate identity for $\mathcal{A}$. Since $\mathcal{A}\mathcal{H}$ is dense in $\mathcal{H}$, we have $e_\alpha \to I$ strongly. Since $\{e_\alpha^n\}$ is also a contractive approximate identity for $\mathcal{A}$, we have $e_\alpha^n \to I$ strongly for all n.

By considering the power series for e^x, we see that for t real

$$\varphi(e^{ite_\alpha}) = \varphi(I) + \sum_{n=1}^{\infty} \frac{(ite_\alpha)^n}{n!} \to \varphi(I) + (e^{it} - 1)I$$

in the strong operator topology.

Since e^{ite_α} is unitary and φ is contractive, we have that $\|(\varphi(I) - I) + e^{it}I\| \leq 1$ for all t. Now it is easy to check that $\|X + e^{it}I\| \leq 1$ for all t implies that $X = 0$. Hence $\varphi(I) = I$ and the proof is complete. □

Next we examine how the injective envelope respects module actions.

First suppose that $1 \in \mathcal{A} \subseteq \mathcal{S} \subseteq B(\mathcal{H})$ with $\mathcal{A}$ a C^*-subalgebra of $B(\mathcal{H})$ and $\mathcal{S}$ a concrete operator system such that $\mathcal{A} \cdot \mathcal{S} \subseteq \mathcal{S}$. In this case, we also have that $\mathcal{S} \cdot \mathcal{A} = \mathcal{S}^* \cdot \mathcal{A}^* = (\mathcal{A}\mathcal{S})^* \subseteq \mathcal{S}^* = \mathcal{S}$, so that $\mathcal{S}$ is an $\mathcal{A}$-bimodule. Clearly, for any $(a_{ij}) \in M_n(\mathcal{A})$ and $(s_{ij}) \in M_n(\mathcal{S})^+$ we have that $(a_{ij}) \cdot (s_{ij}) \cdot (a_{ij})^* \in M_n(\mathcal{S})^+$. We wish to characterize such situations abstractly.

Assume we are given an abstract operator system $\mathcal{S}$, with unit denoted by e, and a unital C^*-algebra $\mathcal{A}$ such that $\mathcal{S}$ is an $\mathcal{A}$-bimodule. We denote the bimodule action and multiplication by $\cdot$, so that $(a_1 \cdot a_2) \cdot s = a_1 \cdot (a_2 \cdot s)$, and we assume $1 \cdot s = s$.

We call such an $\mathcal{S}$ an *operator $\mathcal{A}$-system* provided that, in addition, $a \cdot e = e \cdot a$ and for all n and for any $(a_{ij}) \in M_n(A)$, $(s_{ij}) \in M_n(\mathcal{S})^+$ we have that

$$(a_{ij}) \cdot (s_{ij}) \cdot (a_{ij})^* = \left(\sum_{k,\ell=1}^{n} a_{ik} \cdot s_{k\ell} \cdot a_{j\ell}^* \right) \in M_n(\mathcal{S})^+.$$

Theorem 15.12. *Let $\mathcal{A}$ be a unital C^*-algebra, let $\mathcal{S}$ be an operator $\mathcal{A}$-system, and let $(I(\mathcal{S}), \circ)$ denote the injective envelope of $\mathcal{S}$, with $\circ$ denoting the multiplication that makes $I(\mathcal{S})$ into a C^*-algebra. Then the map $\pi\colon (\mathcal{A}, \cdot) \to (I(\mathcal{S}), \circ)$, $\pi(a) = a \cdot e$, is a $*$-homomorphism.*

Proof. We must show that $(a_1 a_2) \cdot e = (a_1 \cdot e) \circ (a_2 \cdot e)$. Because the unitaries span $\mathcal{A}$, it will be enough to verify the above identity when a_1 is unitary.

Fix a unitary $u \in \mathcal{A}$, and define $\Phi\colon M_2(\mathcal{S}) \to M_2(\mathcal{S})$ via

$$\Phi\left(\begin{pmatrix} s_1 & s_2 \\ s_3 & s_4 \end{pmatrix}\right) = \begin{pmatrix} u \cdot s_1 \cdot u^* & u \cdot s_2 \\ s_3 \cdot u^* & s_4 \end{pmatrix} = \begin{pmatrix} u & 0 \\ 0 & 1 \end{pmatrix} \begin{pmatrix} s_1 & s_2 \\ s_3 & s_4 \end{pmatrix} \begin{pmatrix} u^* & 0 \\ 0 & 1 \end{pmatrix}.$$

Since $\mathcal{S}$ is an operator $\mathcal{A}$-system, Φ is completely positive. Thus, we may extend Φ to a completely positive map on $I(M_2(\mathcal{S})) = M_2(I(\mathcal{S}))$, which we still denote by Φ. Since Φ fixes

$$\mathbb{C} \oplus \mathbb{C} = \left\{ \begin{pmatrix} \lambda e & 0 \\ 0 & \mu e \end{pmatrix} : \lambda, \mu \in \mathbb{C} \right\},$$

it is a bimodule map over this C^*-algebra by Corollary 3.20. Using the fact that Φ is a bimodule map, or arguing as in the proof of Theorem 8.3, one finds that there exist maps $\varphi_i \colon I(\mathcal{S}) \to I(\mathcal{S})$, $i = 1, 2, 3, 4$, such that

$$\Phi\left(\begin{pmatrix} x_1 & x_2 \\ x_3 & x_4 \end{pmatrix}\right) = \begin{pmatrix} \varphi_1(x_1) & \varphi_2(x_2) \\ \varphi_3(x_3) & \varphi_4(x_4) \end{pmatrix}.$$

Clearly, φ_1 and φ_4 are completely positive.

Since $\varphi_4(s_4) = s_4$ for all s_4 in $\mathcal{S}$, by rigidity, $\varphi_4 = \mathrm{id}_{I(\mathcal{S})}$. Thus, Φ fixes

$$\mathbb{C} \oplus I(\mathcal{S}) = \left\{ \begin{pmatrix} \lambda e & 0 \\ 0 & x \end{pmatrix} : \lambda \in \mathbb{C}, x \in I(\mathcal{S}) \right\}$$

and so it is a bimodule map over this C^*-algebra. Hence,

$$\begin{pmatrix} 0 & \varphi_2(x) \\ 0 & 0 \end{pmatrix} = \Phi\left(\begin{pmatrix} 0 & x \\ 0 & 0 \end{pmatrix}\right) = \Phi\left(\begin{pmatrix} 0 & e \\ 0 & 0 \end{pmatrix} \circ \begin{pmatrix} 0 & 0 \\ 0 & x \end{pmatrix}\right)$$
$$= \Phi\left(\begin{pmatrix} 0 & e \\ 0 & 0 \end{pmatrix}\right) \circ \begin{pmatrix} 0 & 0 \\ 0 & x \end{pmatrix} = \begin{pmatrix} 0 & \varphi_2(e) \circ x \\ 0 & 0 \end{pmatrix}.$$

Setting $x = a_2 \cdot e$, we have that

$$(u \cdot a_2) \cdot e = u \cdot (a_2 \cdot e) = \varphi_2(a_2 \cdot e) = \varphi_2(e) \circ (a_2 \cdot e) = (u \cdot e) \circ (a_2 \cdot e),$$

and the proof is complete. □

Corollary 15.13. *Let $\mathcal{A}$ be a unital C^*-algebra, and let $\mathcal{S}$ be an operator $\mathcal{A}$-system with unit e. Then there exists a Hilbert space $\mathcal{H}$, a complete order isomorphism $\gamma \colon \mathcal{S} \to B(\mathcal{H})$ with $\gamma(e) = I_H$, and a unital $*$-homomorphism $\pi \colon \mathcal{A} \to B(\mathcal{H})$ such that $\pi(a)\gamma(s) = \gamma(a \cdot s)$.*

Proof. Represent the C^*-algebra $(I(\mathcal{S}), \circ)$ $*$-isomorphically on a Hilbert space, and apply Theorem 15.12. □

We call (π, γ) a *representation* of the operator $\mathcal{A}$-system.

The next representation theorem concerns operator space bimodules. Suppose that $\mathcal{A} \subseteq B(\mathcal{H})$, $\mathcal{B} \subseteq B(\mathcal{K})$ are unital C^*-subalgebras and $X \subseteq B(\mathcal{K}, \mathcal{H})$ is a subspace. If $\mathcal{A} \cdot X \cdot \mathcal{B} \subseteq X$, then X is an $\mathcal{A}$-$\mathcal{B}$-bimodule. Given $(a_{ij}) \in$

$M_n(\mathcal{A})$, $(x_{ij}) \in M_n(X)$, and $(b_{ij}) \in M_n(\mathcal{B})$, we have that

$$\|(a_{ij}) \cdot (x_{ij}) \cdot (b_{ij})\| \le \|(a_{ij})\| \|(x_{ij})\| \|(b_{ij})\|,$$

where

$$(a_{ij}) \cdot (x_{ij}) \cdot (b_{ij}) = \left(\sum_{k,\ell=1}^{n} a_{ik} x_{k\ell} b_{\ell j} \right).$$

Such a module action is called *completely contractive*. We wish to characterize this situation abstractly.

So let $\mathcal{A}, \mathcal{B}$ be unital C^*-algebras, and let X be an operator space that is also an $\mathcal{A}$-$\mathcal{B}$-bimodule. We call X an *operator $\mathcal{A}$-$\mathcal{B}$-bimodule* provided that the module action is completely contractive.

Theorem 15.14 (Christensen–Effros–Sinclair representation theorem). *Let $\mathcal{A}, \mathcal{B}$ be unital C^*-algebras, and let X be an operator $\mathcal{A}$-$\mathcal{B}$-bimodule. Then there exist Hilbert spaces $\mathcal{H}$ and $\mathcal{K}$, unital $*$-isomorphisms $\pi\colon \mathcal{A} \to B(\mathcal{H})$, $\rho\colon \mathcal{B} \to B(\mathcal{K})$, and a complete isometry $\gamma\colon X \to B(\mathcal{K}, \mathcal{H})$ such that $\gamma(a \cdot x \cdot b) = \pi(a)\gamma(x)\rho(b)$.*

Proof. We imitate the proof of Ruan's theorem with $\mathcal{A}$ and $\mathcal{B}$ playing the role of the scalars. To this end let

$$\mathcal{S} = \left\{ \begin{pmatrix} a & x \\ y^* & b \end{pmatrix} : a \in \mathcal{A},\ b \in \mathcal{B},\ x, y \in X \right\}.$$

Setting $b \cdot y^* \cdot a = (a^* \cdot y \cdot b^*)^*$ makes X^* an operator $\mathcal{B}$-$\mathcal{A}$-bimodule.

We wish to make $\mathcal{S}$ into a matrix ordered $*$-vector space with Archimedean order unit

$$e = \begin{pmatrix} 1_A & 0 \\ 0 & 1_B \end{pmatrix}.$$

Let

$$\begin{pmatrix} a & x \\ y^* & b \end{pmatrix}^* = \begin{pmatrix} a^* & y \\ x^* & b^* \end{pmatrix},$$

and define $\mathcal{C}_n \subseteq M_n(\mathcal{S})_h$ by saying

$$\begin{pmatrix} (a_{ij}) & (x_{ij}) \\ (y_{ij}^*) & (b_{ij}) \end{pmatrix} \in \mathcal{C}_n$$

if and only if $(y_{ij}^*) = (x_{ij})^*$ and

$$\left\| ((a_{ij}) + \varepsilon I_n)^{-1/2} (x_{ij}) ((b_{ij}) + \varepsilon I_n)^{-1/2} \right\| \le 1$$

for all $\epsilon > 0$.

If one copies the proof of Ruan's theorem *mutatis mutandis*, then one finds that all the axioms for an abstract operator system are satisfied by $\mathcal{S}$ (Exercise 15.11). The identification of $\mathcal{A} \oplus \mathcal{B}$ with $\{(\begin{smallmatrix} a & 0 \\ 0 & b \end{smallmatrix}) : a \in \mathcal{A},\ b \in \mathcal{B}\}$ makes $\mathcal{S}$ an operator $\mathcal{A} \oplus \mathcal{B}$-system. That is, define

$$(a_1 \oplus b_1) \cdot \begin{pmatrix} a & x \\ y^* & b \end{pmatrix} \cdot (a_2 \oplus b_2) = \begin{pmatrix} a_1 a a_2 & a_1 x b_2 \\ b_1 y^* a_2 & b_1 b b_2 \end{pmatrix}.$$

Now take any representation (θ, γ) of $\mathcal{S}$ on a Hilbert space. The $*$-homomorphism θ of $\mathcal{A} \oplus \mathcal{B}$ decomposes the Hilbert space as $\mathcal{H} \oplus \mathcal{K}$ so that

$$\theta(a \oplus b) = \begin{pmatrix} \pi(a) & 0 \\ 0 & \rho(b) \end{pmatrix},$$

where $\pi \colon \mathcal{A} \to B(\mathcal{H})$ and $\rho \colon \mathcal{B} \to B(\mathcal{K})$. Since

$$\begin{aligned} \gamma\left(\begin{pmatrix} 0 & x \\ 0 & 0 \end{pmatrix}\right) &= \gamma\left(\begin{pmatrix} 1 & 0 \\ 0 & 0 \end{pmatrix} \cdot \begin{pmatrix} 0 & x \\ 0 & 0 \end{pmatrix} \cdot \begin{pmatrix} 0 & 0 \\ 0 & 1 \end{pmatrix}\right) \\ &= \begin{pmatrix} \pi(1) & 0 \\ 0 & 0 \end{pmatrix} \gamma\left(\begin{pmatrix} 0 & x \\ 0 & 0 \end{pmatrix}\right) \begin{pmatrix} 0 & 0 \\ 0 & \rho(1) \end{pmatrix}, \end{aligned}$$

we see that

$$\gamma\left(\begin{pmatrix} 0 & x \\ 0 & 0 \end{pmatrix}\right) = \begin{pmatrix} 0 & \gamma_1(x) \\ 0 & 0 \end{pmatrix} \qquad \text{where} \quad \gamma_1 \colon X \to B(\mathcal{K}, \mathcal{H}).$$

It is readily checked that since γ is a complete order isomorphism, γ_1 is a complete isometry, and that $\pi(a)\gamma_1(x)\rho(b) = \gamma_1(axb)$. Note that π and ρ are one-to-one, since $\mathcal{A} \oplus \mathcal{B}$ is a subset of $\mathcal{S}$. □

We call (π, ρ, γ) a *representation of the operator $\mathcal{A}$-$\mathcal{B}$-bimodule X*.

We turn our attention now to Hamana's construction of the C^*-envelope of an operator algebra. To motivate this material we begin with the example of the disk algebra, $A(\mathbb{D})$. Clearly, $A(\mathbb{D})$ can be regarded, up to completely isometric isomorphism, as either the uniformly closed subalgebra of $C(\mathbb{D}^-)$, or that of $C(\mathbb{T})$, generated by the coordinate function z. In both cases $A(\mathbb{D})$ separate the points on the underlying spaces, and so the C^*-algebra generated by $A(\mathbb{D})$ will be $C(\mathbb{D}^-)$ in the first case and $C(\mathbb{T})$ in the second case. Thus, the C^*-algebra generated by an operator algebra can depend on the representation.

Restricting a function in $C(\mathbb{D}^-)$ to $\partial\mathbb{D}^- = \mathbb{T}$ defines a $*$-homomorphism from $C(\mathbb{D}^-)$ onto $C(\mathbb{T})$, which is the identity map on $A(\mathbb{D})$. Thus, $C(\mathbb{T})$ is a "smaller" representation of $A(\mathbb{D})$ in the sense that it is a C^*-quotient of $C(\mathbb{D}^-)$.

We shall prove that among all the possible C^*-algebras that completely isometric representations of an operator algebra can generate, there exists a "smallest" such C^*-algebra in the sense that it is a universal quotient. To be more precise, given an operator algebra $\mathcal{A}$, we shall prove that there exists a completely isometric representation $\gamma\colon \mathcal{A} \to B(\mathcal{H})$ such that if $\rho\colon \mathcal{A} \to B(\mathcal{K})$ is any other completely isometric representation, then there exists an onto $*$-homomorphism $\pi\colon C^*(\rho(\mathcal{A})) \to C^*(\gamma(\mathcal{A}))$ with $\pi(\rho(a)) = \gamma(a)$.

It is easy to see that if such a representation γ exists, then $C^*(\gamma(\mathcal{A}))$ is uniquely determined up to $*$-isomorphism. That is, if γ_1 and γ_2 were two representations with the above property, then there would exist onto $*$-homomorphisms $\pi_1\colon C^*(\gamma_1(\mathcal{A})) \to C^*(\gamma_2(\mathcal{A}))$ with $\pi_1(\gamma_1(a)) = \gamma_2(a)$ and $\pi_2\colon C^*(\gamma_2(\mathcal{A})) \to C^*(\gamma_1(\mathcal{A}))$ with $\pi_2(\gamma_2(a)) = \gamma_1(a)$, but these two equations force $\pi_2 \circ \pi_1 = \mathrm{id}_{C^*(\gamma_1(\mathcal{A}))}$ and $\pi_1 \circ \pi_2 = \mathrm{id}_{C^*(\gamma_2(\mathcal{A}))}$, and so π_1 and π_2 are $*$-isomorphisms.

Proposition 15.15. *Let $1 \in \mathcal{A} \subseteq B(\mathcal{H})$ be an operator algebra. Then $I(\mathcal{A}) = I(\mathcal{A} + \mathcal{A}^*)$ is a C^*-algebra, and the inclusion of $\mathcal{A}$ into the C^*-algebra $I(\mathcal{A})$ is a completely isometric isomorphism.*

Proof. Since $1 \in \mathcal{A}$, any completely contractive $\varphi\colon B(\mathcal{H}) \to B(\mathcal{H})$ that fixes $\mathcal{A}$ is completely positive, and hence for a and b in $\mathcal{A}$

$$\varphi(a + b^*) = \varphi(a) + \varphi(b)^* = a + b^*.$$

Thus, every $\mathcal{A}$-map is actually an $(\mathcal{A} + \mathcal{A}^*)$-map, and so $I(\mathcal{A}) = I(\mathcal{A} + \mathcal{A}^*)$. The fact that the inclusion is a completely isometric isomorphism follows from Proposition 15.10. □

It is worth noticing that if $\rho\colon \mathcal{A} \to B(\mathcal{H})$ is a completely isometric isomorphism, then the operator systems $\mathcal{A} + \mathcal{A}^*$ and $\rho(\mathcal{A}) + \rho(\mathcal{A})^*$ are completely order-isomorphic via the map $a + b^* \to \rho(a) + \rho(b)^*$. Thus, the operator system $\mathcal{A} + \mathcal{A}^*$ is really well defined independent of the particular representation of $\mathcal{A}$. This is not the case for the operator system $\mathcal{A}^*\mathcal{A} = \mathrm{span}\{a^*b\colon a, b \in \mathcal{A}\}$, as can be seen by regarding $A(\mathbb{D}) \subseteq C(\mathbb{T})$ versus $A(\mathbb{D}) \subseteq C(\mathbb{D}^-)$.

Given an operator algebra $\mathcal{A}$, we let $C_e^*(\mathcal{A})$ denote the C^*-subalgebra of $I(\mathcal{A})$ generated by $\mathcal{A}$. We call $C_e^*(\mathcal{A})$ the *C^*-envelope of $\mathcal{A}$*.

Theorem 15.16 (Hamana's theorem). *Let $\mathcal{A}$ be an operator algebra with unit, let $\rho\colon \mathcal{A} \to B(\mathcal{H})$ be a completely isometric homomorphism, and let $C^*(\rho(\mathcal{A}))$ denote the C^*-subalgebra of $B(\mathcal{H})$ generated by $\rho(\mathcal{A})$. Then there exists a* *unique onto $*$-homomorphism $\pi\colon C^*(\rho(\mathcal{A})) \to C_e^*(\mathcal{A})$ with $\pi(\rho(a)) = a$ for all a in $\mathcal{A}$.*

Proof. Once we show the existence of π, the fact that it is unique and onto follows from the fact that the C^*-algebras are generated by $\rho(\mathcal{A})$ and $\mathcal{A}$, respectively.

By injectivity of $I(\mathcal{A})$, there is a unital completely contractive (and hence completely positive) map $\varphi\colon B(\mathcal{H}) \to I(\mathcal{A})$ with $\varphi(\rho(a)) = a$ for a in $\mathcal{A}$. Similarly, there is a completely positive map $\psi\colon I(\mathcal{A}) \to B(\mathcal{H})$ with $\psi(a) = \rho(a)$ for a in $\mathcal{A}$. Since $\varphi \circ \psi(a) = a$ for a in $\mathcal{A}$, by rigidity $\varphi \circ \psi = \mathrm{id}_{I(\mathcal{A})}$.

We claim that $\varphi(\rho(a)^*\rho(a)) = a^*a$ and $\varphi(\rho(a)\rho(a)^*) = aa^*$. Indeed, by the Schwarz inequality, $\varphi(\rho(a)^*\rho(a)) \geq \varphi(\rho(a))^*\varphi(\rho(a)) = a^*a$, while $\psi(a^*a) \geq \psi(a)^*\psi(a) = \rho(a)^*\rho(a)$. Applying φ to the second inequality yields

$$a^*a = \varphi \circ \psi(a^*a) \geq \varphi(\rho(a)^*\rho(a)) \geq a^*a$$

and so $\varphi(\rho(a)^*\rho(a)) = a^*a$. Similarly, $\varphi(\rho(a)\rho(a)^*) = aa^*$.

Recalling Theorem 3.18 (iii), on the multiplicative domains of completely positive maps, these equalities imply that φ is a $*$-homomorphism when restricted to $C^*(\rho(\mathcal{A}))$. □

Thus, by the above theorem we see that $C_e^*(\mathcal{A})$ is a universal quotient for the C^*-algebra generated by any completely isometric representation of $\mathcal{A}$.

Some examples are helpful. We begin by showing how the classic theory of the Silov boundary of a uniform algebra follows from Hamana's theorem. Recall that if X is a compact Hausdorff space, then a closed subalgebra, $\mathcal{A} \subseteq C(X)$ is called a *uniform algebra on* X provided that $1 \in \mathcal{A}$ and $\mathcal{A}$ separates points, i.e., for every $x \neq y$ there exists $f \in \mathcal{A}$ with $f(x) \neq f(y)$.

Corollary 15.17 (Silov). *Let $\mathcal{A} \subseteq C(X)$ be a uniform algebra on X. Then there exists a unique compact subset $Y \subseteq X$ such that:*

(i) the restriction homomorphism $r\colon C(X) \to C(Y)$, $r(f) = f|_Y$ is isometric on $\mathcal{A}$,
(ii) if $W \subseteq X$ is any other set such that (i) holds, then $Y \subseteq W$.

Moreover, if $\rho\colon \mathcal{A} \to C(X_1)$ is any isometric isomorphism such that $\rho(\mathcal{A})$ is a uniform algebra on X_1, and $Y_1 \subseteq X_1$ is the corresponding set for $\rho(\mathcal{A})$, then there exists a homeomorphism $h\colon Y_1 \to Y$ such that $\rho(f)|_{Y_1} = f \circ h$.

Proof. Since the range of ρ is a commutative C^*-algebra, any isometric map is automatically completely isometric.

By Hamana's theorem, there exists a $*$-homomorphism $\pi: C(X) \to C_e^*(A)$, onto. Hence, $C_e^*(A) \cong C(Z)$ for some compact Hausdorff space Z. By the characterization of $*$-homomorphisms between commutative C^*-algebras (see [76]), there exists a one-to-one continuous function $h: Z \to X$ such that $\pi(g) = g \circ h$. Let $Y = h(Z)$, and we have (i).

Now suppose W satisfies (ii). Regarding $\mathcal{A} \subseteq C(W)$, via the restriction map, we obtain $\pi_1: C(W) \to C(Z)$ and $h_1: Z \to W$ such that $\pi_1(g) = g \circ h_1$. Let $r: C(X) \to C(W)$ denote the restriction map, so that $r(g) = g \circ i$, where $i: W \to X$ denotes the inclusion. Since $\pi = \pi_1 \circ r$, it follows that $h = i \circ h_1$ and so $Y = h(Z) = i(h_1(Z))$ must be a subset of W.

The remainder of the proof follows similar arguments. □

The set Y, which only depends on $\mathcal{A}$ up to homeomorphism, is called the *Silov boundary of* $\mathcal{A}$ and denoted $\partial_S \mathcal{A}$.

It can be easily shown that $\partial_S A(\mathbb{D}) = \mathbb{T}$ (Exercise 15.8).

Corollary 15.18. *Let $\mathcal{A}$ be a uniform algebra. Then $C_e^*(\mathcal{A}) = C(\partial_S \mathcal{A})$.*

By the above result we see that any time we represent a uniform algebra completely isometrically as an algebra of operators on a Hilbert space, the C^*-algebra generated by the image has a commutative C^*-algebra as a quotient.

The following is another case where it is easy to recognize $C_e^*(\mathcal{A})$. Recall that a C^*-algebra is *simple* if it contains no nontrivial two-sided ideals.

Proposition 15.19. *Let $\mathcal{A} \subseteq B(\mathcal{H})$ be an operator algebra. If $C^*(\mathcal{A})$ is simple, then $C_e^*(\mathcal{A}) = C^*(\mathcal{A})$.*

Proof. There exists a $*$-homomorphism from $C^*(\mathcal{A})$ onto $C_e^*(\mathcal{A})$, but because $C^*(\mathcal{A})$ is simple, this map must be one-to-one. □

Thus, for example if $\mathcal{A} \subseteq M_n$ is the algebra of upper triangular matrices, then $C_e^*(\mathcal{A}) = M_n$.

The existence of the C^*-envelope of an operator algebra with the properties of Hamana's theorem was conjectured by Arveson [6], and he attempted to prove its existence by developing the theory of what are called *boundary representations*. Surprisingly, Hamana's theorem does not guarantee the existence of boundary representations, and so many questions remain unanswered. We discuss these issues now.

Assume $\mathcal{A}$ is an operator algebra and $\mathcal{B} = C^*(\mathcal{A})$. Any irreducible representation of $C_e^*(\mathcal{A})$ on a Hilbert space would give rise to an irreducible representation of $\mathcal{B}$. Since to determine the structure of $C_e^*(\mathcal{A})$ it is enough to know all its irreducible representations, we would like a condition that guarantees an irreducible representation of $\mathcal{B}$ corresponds to an irreducible representation of $C_e^*(\mathcal{A})$. We call an irreducible representation $\pi: \mathcal{B} \to B(\mathcal{H})$ a *boundary representation* provided that if $\Phi: \mathcal{B} \to B(\mathcal{H})$ is a completely positive map such that $\Phi(a) = \pi(a)$ for all a in $\mathcal{A}$, then $\pi = \Phi$.

Assume that π is a boundary representation; then $\pi|_{\mathcal{A}}$ will have a completely positive extension $\gamma: C_e^*(\mathcal{A}) \to B(\mathcal{H})$. Let $\rho: \mathcal{B} \to C_e^*(\mathcal{A})$ be the onto $*$-homomorphism, so that $\gamma \circ \rho: \mathcal{B} \to B(\mathcal{H})$ is a completely positive map that extends $\pi|_{\mathcal{A}}$. Hence, $\gamma \circ \rho = \pi$, and it follows that γ must be a homomorphism and also be unique. Thus, every boundary representation of $\mathcal{B}$ gives rise to an irreducible representation of $C_e^*(\mathcal{A})$ that is also a boundary representation. Thus, one would hope that there exist enough boundary representations so that if one took the direct sum over all boundary representations, then that would yield a faithful, one-to-one representation of $C_e^*(\mathcal{A})$. Whether or not sufficiently many boundary representations exist is still unknown.

To illustrate the difficulties, we consider the compact Hausdorff space $X = \mathbb{D}^- \times [0, 1]$ and the uniform algebra $\mathcal{A}$ on X given by

$$\mathcal{A} = \{f \in C(X): f(z, 0) \in A(\mathbb{D})\}.$$

It can be shown that $X = \partial_S \mathcal{A}$, so that $C_e^*(\mathcal{A}) = C(X)$. Since $C_e^*(\mathcal{A})$ is commutative, the irreducible representations are all one-dimensional and are given by $\pi_x(g) = g(x)$ for points x in X. The boundary representations can be shown to correspond to evaluation at points of the form (z, t), $t > 0$, and $(z, 0)$, $|z| = 1$, but none of the points $(z, 0)$, $|z| < 1$, correspond to boundary representations. To see this last fact recall the Poisson kernel for $|z| < 1$,

$$P_z(\theta) = \frac{1}{1 - e^{-i\theta} z} + \frac{1}{1 - e^{+i\theta} \bar{z}} - 1.$$

We have that $P_z(\theta) \geq 0$ and for $f \in A(\mathbb{D})$,

$$f(z) = \frac{1}{2\pi} \int_0^{2\pi} f(e^{i\theta}) P_z(\theta)\, d\theta.$$

Thus, the map $\Phi_z: C(X) \to \mathbb{C}$ defined by $\Phi_z(g) = \frac{1}{2\pi} \int_0^{2\pi} g(e^{i\theta}, 0) P_z(\theta)\, d\theta$ is a completely positive map such that $\Phi_z(f) = \pi_z(f) = f(z, 0)$ for $f \in \mathcal{A}$, but $\Phi_z \neq \pi_{(z,0)}$. Thus, $\pi_{(z,0)}$, $|z| < 1$, is not a boundary representation.

For a general uniform algebra $\mathcal{A}$, Arveson's boundary representations can be shown to correspond to evaluations at what are called *peak points*, and these points are known to form a dense subset of $\partial_S \mathcal{A}$ called the *Choquet boundary* (see [103] for definitions and results). Thus, a successful theory of boundary representations would lead to some noncommutative analogues of the fact that peak points are dense in the Silov boundary.

Notes

The theory of the injective envelope of an operator system, in the category $\mathfrak{S}$, was developed in the work of M. Hamana [112]. Hamana not only recognized their importance, but used them to prove the existence of the C^*-envelope. The analogue of Theorem 15.8 for operator systems also appears in that work. These ideas were subsequently extended to the setting of operator spaces by Z.J. Ruan [204]. We will take a deeper look at these topics in Chapter 16.

The original proof of the Christensen–Effros–Sinclair representation theorem used a representation theorem for multilinear completely bounded maps, which we will examine in Chapter 17. The proof that we present here seems to be new.

There is still much to be understood about injectivity in the category $\mathcal{O}$. For an operator space $E \subset B(\mathcal{H})$ to be injective in this category it is necessary and sufficient that there exist a completely bounded projection of $B(\mathcal{H})$ onto E (compare with Exercise 15.2). As mentioned earlier, there exist C^*-algebras that are not injective in the usual sense (i.e., in $\mathcal{O}_1$) but are injective in $\mathcal{O}$. However, it is known that for von Neumann algebras injectivity in $\mathcal{O}_1$ and injectivity in $\mathcal{O}$ are equivalent [61, 182]. A natural question that still remains open is whether or not every C^*-algebra that is injective in $\mathcal{O}$ is completely boundedly isomorphic to a C^*-algebra that is injective in $\mathcal{O}_1$. All the known examples of C^*-algebras that are injective in $\mathcal{O}$ have this property.

M. Frank and the author [102] pursue the study of injectivity in the category whose objects are left operator $\mathcal{A}$-modules and whose morphisms are the completely bounded left $\mathcal{A}$-module maps for a fixed C^*-algebra $\mathcal{A}$. In contrast to the above results, if $\mathcal{A} \subset B(\mathcal{H})$ and there exists a completely bounded projection of $B(\mathcal{H})$ onto $\mathcal{A}$ that is also a left $\mathcal{A}$-module map, then $\mathcal{A}$ is injective in the usual sense. It is shown that the usual injective envelope, $I(\mathcal{A})$, is still the injective envelope in this new category. The authors also show that the theory of multipliers follows from properties of the injective envelope. In particular, if J is any essential ideal in $\mathcal{A}$, then every multiplier of J is given as multiplication by a unique element of $I(\mathcal{A})$. See [173] for the definitions of multipliers and essential ideals.

M. Hamana [113], C. Zhang [247, 249], and D.P. Blecher [23] have developed the analogue of Hamana's C^*-envelope of an operator algebra for operator spaces.

Exercises

15.1 Verify that an operator space I is injective in $\mathcal{O}_1$ if and only if every completely bounded map into I has a completely bounded extension of the same completely bounded norm.

15.2 Let $E \subseteq B(\mathcal{H})$ be an operator space. Prove that E is injective if and only if there exists a completely contractive map $\varphi: B(\mathcal{H}) \to E$ such that $\varphi(e) = e$ for all $e \in E$.

15.3 Let $M \subseteq B(\mathcal{H})$ be an operator space, and let $\varphi: B(\mathcal{H}) \to B(\mathcal{H})$ be an M-projection with p_φ a minimal M-seminorm. Assume that $u \in B(\mathcal{H})$ is a unitary that commutes with M. Prove that if $\gamma(x) = u^*\varphi(x)u$, then p_γ is a minimal M-seminorm and γ is a projection onto $u^*\varphi(B(\mathcal{H}))u$. Prove that if $\psi(x) = \varphi(u^*xu)$, then p_ψ is a minimal M-seminorm and ψ is a (possibly different) projection onto $\varphi(B(\mathcal{H}))$.

15.4 Complete the proof of Theorem 15.8.

15.5 Let $\varphi: B(\mathcal{H}) \to B(\mathcal{H})$ be a unital completely positive map that fixes the compacts. Prove that φ is necessarily the identity map. [Hint: Let $0 \le P$, and set $\mathcal{P} = \{0 \le K \le P: K \text{ compact}\}$. Prove that if $K \le Q$ for all $K \in \mathcal{P}$, then $P \le Q$.]

15.6 Let $\mathcal{S} \subseteq B(\mathcal{H})$ be an operator system that contains the compacts. Prove that $I(\mathcal{S}) = B(\mathcal{H})$.

15.7 Let $\mathcal{A} \subseteq B(\ell^2)$ be the algebra of upper triangular operators. Prove that $C_e^*(\mathcal{A}) = B(\ell^2)$.

15.8 Prove that $\partial_S A(\mathbb{D}) = \mathbb{T}$.

15.9 Prove that if $\mathcal{A} \subseteq M_n$ then $C_e^*(\mathcal{A})$ is finite-dimensional.

15.10 Let $\mathcal{A}$ and $\mathcal{B}$ be unital C^*-algebras, and let $\varphi: \mathcal{A} \to \mathcal{B}$ be a complete order isomorphism with $\varphi(1) = 1$. Prove that there exists a $*$-homomorphism $\pi: C^*(\varphi(\mathcal{A})) \to \mathcal{A}$ with $\pi(\varphi(a)) = a$ for all a in $\mathcal{A}$.

15.11 Verify that the set $\mathcal{S}$ defined in the proof of Theorem 15.14 is a matrix-ordered *-vector space with complete Archimedean order unit e.

15.12 Let $\{u_1, \dots, u_n\}$ denote the unitaries that generate $C_u^*(\mathbb{F}^n)$, and let M denote the $(n+1)$-dimensional subspace generated by these unitaries and the identity I. If $\mathcal{A} \subseteq M_2(C_u^*(\mathbb{F}^n))$ denotes the $(n+3)$-dimensional operator algebra

$$\mathcal{A} = \left\{ \begin{pmatrix} \lambda I & x \\ 0 & \mu I \end{pmatrix} : \lambda \in \mathbb{C},\ \mu \in \mathbb{C},\ x \in M \right\},$$

then prove that $C_e^*(\mathcal{A}) = M_2(C_u^*(\mathbb{F}^n))$. Thus, a finite-dimensional operator algebra can have an infinite-dimensional C^*-envelope.

Chapter 16

Abstract Operator Algebras

In previous chapters we have given abstract characterizations of operator systems and operator spaces. In this chapter we wish to give an abstract characterization of operator algebras. That is, given an algebra $\mathcal{A}$ and norms on $M_{m,n}(\mathcal{A})$, we desire necessary and sufficient conditions for there to exist a Hilbert space $\mathcal{H}$ and a completely isometric isomorphism $\pi\colon \mathcal{A} \to B(\mathcal{H})$. We present a theorem of Blecher, Ruan, and Sinclair (BRS) [31] that gives such necessary and sufficient conditions. Thus, the BRS theorem plays exactly the same role for operator algebras that the celebrated Gelfand–Naimark–Segal theorem plays for C^*-algebras.

Moreover, like those in the GNS theorem, the conditions in the BRS theorem are natural and quite often easy to verify. Briefly, the BRS theorem says that the matrix norms on $\mathcal{A}$ must satisfy Ruan's axioms for operator spaces (otherwise there wouldn't even exist a linear complete isometry) and, for all n, the induced multiplication on $M_n(\mathcal{A})$ must be contractive. That is, for $A = (a_{ij})$, $B = (b_{ij})$ in $M_n(\mathcal{A})$ one has

$$\left\| \left(\sum_{k=1}^{n} a_{ik} b_{kj} \right) \right\|_n \leq \|A\|_n \|B\|_n .$$

This last condition is often summarized by saying that the multiplication on $\mathcal{A}$ is *completely contractive.*

It is important not to confuse these results with other work on algebras of operators, where no matrix norms are assumed. In particular, Varopoulos [237] gives necessary and sufficient conditions for a Banach algebra to have an isometric representation as an algebra of operators on a Hilbert space. This result has not been as widely used, perhaps because of the difficulty of verifying the conditions.

There is still one outstanding problem in the isometric theory of operator algebras. If a unital Banach algebra $\mathcal{B}$ has the property that there exists a Hilbert space $\mathcal{H}$ and a unital isometric isomorphism $\pi\colon \mathcal{B} \to B(\mathcal{H})$, then by von Neumann's inequality, for every $b \in \mathcal{B}$ with $\|b\| \leq 1$, and for every polynomial p we have $\|p(b)\| \leq \|p\|_\infty$, where the supremum is over the unit disk. A Banach algebra with this latter property is said to *satisfy von Neumann's inequality*. Thus, Banach algebras that are isometrically isomorphic to an algebra of operators on a Hilbert space must satisfy von Neumann's inequality. The problem is whether or not the converse is true. That is, if $\mathcal{B}$ is a Banach algebra that satisfies von Neumann's inequality, then is $\mathcal{B}$ isometrically isomorphic to an algebra of operators on a Hilbert space? See [74] and [168] for two recent contributions to this problem.

The key to the proof of the BRS theorem that we present here is a better understanding of the injective envelope of the "canonical" operator system associated to an operator space.

Recall that if $X \subseteq B(\mathcal{K}, \mathcal{H})$ is a (concrete) operator space, then we may form the (concrete) operator system, in $B(\mathcal{H} \oplus \mathcal{K})$,

$$\mathcal{S}_X = \left\{ \left(\begin{array}{c|c} \lambda I_{\mathcal{H}} & x \\ \hline y^* & \mu I_{\mathcal{K}} \end{array} \right) : \lambda, \mu \in \mathbb{C},\ x, y \in X \right\}.$$

Given a complete isometry $\varphi\colon X \to B(\mathcal{K}_1, \mathcal{H}_1)$, the operator system

$$\mathcal{S}_{\varphi(X)} = \left\{ \begin{pmatrix} \lambda I_{\mathcal{H}_1} & \varphi(x) \\ \varphi(y)^* & \mu I_{\mathcal{K}_1} \end{pmatrix} : \lambda, \mu \in \mathbb{C},\ x, y \in X \right\}$$

is completely order-isomorphic to $\mathcal{S}_X$ via the map $\Phi\colon \mathcal{S}_X \to \mathcal{S}_{\varphi(X)}$ defined by

$$\Phi\left(\begin{pmatrix} \lambda I_{\mathcal{H}} & x \\ y^* & \mu I_{\mathcal{K}} \end{pmatrix} \right) = \begin{pmatrix} \lambda I_{\mathcal{H}_1} & \varphi(x) \\ \varphi(y)^* & \mu I_{\mathcal{K}_1} \end{pmatrix}.$$

One way to see that Φ is a complete order isomorphism is to recall that by Lemma 7.1, since φ and φ^{-1} are completely contractive, Φ and Φ^{-1} are completely positive. Thus, the operator system $\mathcal{S}_X$ only depends on the operator space structure of X and not on any particular representation of X. This can also be seen in the proof of Ruan's theorem, where for an arbitrary abstract operator space X we constructed this same operator system, abstractly, with only reference to the operator space structure of X.

Since

$$\mathbb{C} \oplus \mathbb{C} \cong \left\{ \begin{pmatrix} \lambda I_{\mathcal{H}} & 0 \\ 0 & \mu I_{\mathcal{K}} \end{pmatrix} : \lambda, \mu \in \mathbb{C} \right\}$$

is a C^*-subalgebra of $\mathcal{S}_X$, by Proposition 15.10 $\mathbb{C} \oplus \mathbb{C}$ will still be a C^*-subalgebra of the C^*-algebra $I(\mathcal{S}_X)$, with

$$\begin{pmatrix} I_{\mathcal{H}} & 0 \\ 0 & 0 \end{pmatrix} \text{ and } \begin{pmatrix} 0 & 0 \\ 0 & I_{\mathcal{K}} \end{pmatrix}$$

corresponding to projections e_1 and e_2, respectively, in the C^*-algebra $I(\mathcal{S}_X)$. We have that $e_1 + e_2$ is equal to the identity and $e_1 \cdot e_2 = 0$.

A few words on such a situation are in order. Let $\mathcal{A}$ be any unital C^*-algebra with projections e_1 and e_2 satisfying $e_1 + e_2 = 1, e_1 \cdot e_2 = 0$, and let $\pi\colon \mathcal{A} \to B(\mathcal{H})$ be a one-to-one unital $*$-homomorphism. Setting $\mathcal{H}_1 = \pi(e_1)\mathcal{H}, \mathcal{H}_2 = \pi(e_2)\mathcal{H}$, we have that $\mathcal{H} = \mathcal{H}_1 \oplus \mathcal{H}_2$, and relative to this decomposition every $T \in B(\mathcal{H})$ has the form $T = (T_{ij})$ where $T_{ij} \in B(\mathcal{H}_j, \mathcal{H}_i)$. In particular, identifying $\mathcal{A}$ with $\pi(\mathcal{A})$, we have that

$$\mathcal{A} = \left\{ \begin{pmatrix} a_{11} & a_{12} \\ a_{21} & a_{22} \end{pmatrix} : a_{ij} \in \mathcal{A}_{ij} \right\},$$

where $\mathcal{A}_{ij} = e_i \mathcal{A} e_j$, with $\mathcal{A}_{ii} \subseteq B(\mathcal{H}_i)$ unital C^*-subalgebras and $\mathcal{A}_{21} = \mathcal{A}_{12}^*$. The operator space $\mathcal{A}_{12} \subseteq B(\mathcal{H}_2, \mathcal{H}_1)$ will be referred to as a *corner of* $\mathcal{A}$. Note that $\mathcal{A}_{11} \cdot \mathcal{A}_{12} \cdot \mathcal{A}_{22} \subseteq \mathcal{A}_{12}$, so that $\mathcal{A}_{12}$ is an $\mathcal{A}_{11}$–$\mathcal{A}_{12}$-bimodule.

Returning to $I(\mathcal{S}_X)$, relative to e_1 and e_2, we wish to identify each of these four subspaces. Note that $X \subseteq e_1 I(\mathcal{S}_X) e_2 = I(\mathcal{S}_X)_{12}$.

Theorem 16.1. *Let X be an operator space, and $\mathcal{S}_X, I(\mathcal{S}_X), e_1$, and e_2 be as above. Then $I(\mathcal{S}_X)_{11}, I(\mathcal{S}_X)_{22}$ are injective C^*-algebras, and $I(\mathcal{S}_X)_{12}$ is the injective envelope of X.*

Proof. Since the maps $\gamma_{ij}\colon I(\mathcal{S}_X) \to I(\mathcal{S}_X)_{ij}$ defined by $\gamma_{ij}(a) = e_i a e_j$ are completely contractive projections, it follows that $I(\mathcal{S}_X)_{ij}, i, j = 1, 2$, are injective operator spaces and so $I(\mathcal{S}_X)_{11}$ and $I(\mathcal{S}_X)_{22}$ are injective C^*-algebras.

To prove the final claim it will be enough to prove that $I(\mathcal{S}_X)_{12}$ is a minimal injective extension of X.

To this end, let $\varphi\colon I(\mathcal{S}_X)_{12} \to I(\mathcal{S}_X)_{12}$ be completely contractive, and suppose $\varphi(x) = x$ for all $x \in X$. Let

$$\mathcal{R} = \left\{ \begin{pmatrix} \lambda & z \\ w^* & \mu \end{pmatrix} : \lambda, \mu \in \mathbb{C},\ z, w \in I(\mathcal{S}_X)_{12} \right\} \subseteq I(\mathcal{S}_X),$$

and let $\Phi\colon \mathcal{R} \to \mathcal{R}$ be defined by

$$\Phi\left(\begin{pmatrix} \lambda & z \\ w^* & \mu \end{pmatrix} \right) = \begin{pmatrix} \lambda & \varphi(z) \\ \varphi(w)^* & \mu \end{pmatrix}.$$

Then Φ is completely positive and so extends to a completely positive map, which we still denote by Φ, from $I(\mathcal{S}_X)$ to $I(\mathcal{S}_X)$. Since Φ is the identity on $\mathcal{S}_X$, by rigidity of the injective envelope of $\mathcal{S}_X$, Φ is the identity on $I(\mathcal{S}_X)$ and hence φ must be the identity on $I(\mathcal{S}_X)_{12}$. If $X \subseteq Y \subseteq I(\mathcal{S}_X)_{12}$ with Y injective, then there would exist a projection φ of $I(\mathcal{S}_X)_{12}$ onto Y. By the above, φ would be the identity on $I(\mathcal{S}_X)_{12}$ and hence $Y = I(\mathcal{S}_X)_{12}$. Thus, $I(\mathcal{S}_X)_{12}$ is a minimal injective extension of X, i.e., an injective envelope. □

Using the uniqueness of the injective envelope, we see that if $X \subseteq I$ with I a minimal injective, then there exists a completely isometric isomorphism between I and $I(\mathcal{S}_X)_{12}$ that, restricted to X, is the identity. Thus, we may identify $I(\mathcal{S}_X)_{12}$ with $I(X)$, and under this identification, $I(X)$ is a corner of the C^*-algebra $I(\mathcal{S}_X)$.

We define $I_{11}(X) = I(\mathcal{S}_X)_{11}$ and $I_{22}(X) = I(\mathcal{S}_X)_{22}$. Thus we have the following picture of the C^*-algebra $I(\mathcal{S}_X)$:

$$I(\mathcal{S}_X) = \left\{ \begin{pmatrix} a & z \\ w^* & b \end{pmatrix} : a \in I_{11}(X),\ b \in I_{22}(X),\ z, w \in I(X) \right\}$$

where $I_{11}(X)$ and $I_{22}(X)$ are injective C^*-algebras and $I(X)$ is an operator $I_{11}(X)$-$I_{22}(X)$-bimodule. Moreover, the fact that $I(\mathcal{S}_X)$ is a C^*-algebra means that for $z, w \in I(X)$,

$$\begin{pmatrix} 0 & z \\ w^* & 0 \end{pmatrix} \begin{pmatrix} 0 & z \\ w^* & 0 \end{pmatrix} = \begin{pmatrix} zw^* & 0 \\ 0 & w^*z \end{pmatrix},$$

and consequently there are natural products $z \cdot w^* \in I_{11}(X)$, $w^* \cdot z \in I_{22}(X)$.

It is interesting to note that setting $\langle z, w \rangle = zw^*$ defines an $I_{11}(X)$-valued inner product that makes $I(X)$ a Hilbert C^*-module over $I_{11}(X)$, but we shall not use this additional structure.

This rather complete picture of $I(\mathcal{S}_X)$ has some immediate consequences.

Corollary 16.2 (Youngson). *An operator space is injective if and only if it is completely isometrically isomorphic to a corner of an injective C^*-algebra.*

Proof. If X is an injective operator space, then $X = I(X)$, which is a corner of the injective C^*-algebra $I(\mathcal{S}_X)$.

Conversely, every corner of an injective C^*-algebra is the image of a completely contractive projection and hence is injective. □

If X is an operator space, then it is clear that all of the spaces $M_{m,n}(X)$ are themselves operator spaces. Technically, the matrix-norm structure on $M_{m,n}(X)$ comes from the identification $M_{p,q}(M_{m,n}(X)) = M_{pm,qn}(X)$. We set $C_m(X) = M_{m,1}(X)$ and $R_n(X) = M_{1,n}(X)$ and call these the *column* and *row* operator spaces of X, respectively. In the following results, we characterize the maps from X to itself that are given as multiplication by elements of the corners of $I(\mathcal{S}_X)$. All products in the following results are defined by the picture of $I(\mathcal{S}_X)$ outlined above.

Lemma 16.3. *Let X be an operator space and let $a \in I_{11}(X)$. If $a \cdot x = 0$ for all $x \in X$, then $a = 0$. Similarly, if $b \in I_{22}(X)$ and $xb = 0$ for all $x \in X$, then $b = 0$.*

Proof. Without loss of generality we can assume that $\|a\| \leq 1$. Let $p = a^*a$ so that $0 \leq p \leq 1$, where 1 denotes the unit of $I_{11}(X)$. Define $\varphi\colon I(X) \to I(X)$ by $\varphi(z) = (1-p)z$. Since $\|1-p\| \leq 1$, φ is a completely contractive map. But for $x \in X$ we have $\varphi(x) = x$, and hence by rigidity $\varphi(z) = z$ for all $z \in I(X)$. Hence, $pI(X) = 0$.

Now let K denote the two-sided ideal in $I(\mathcal{S}_X)$ generated by $\left(\begin{smallmatrix} p & 0 \\ 0 & 0 \end{smallmatrix}\right)$. Since

$$\begin{pmatrix} p & 0 \\ 0 & 0 \end{pmatrix} I(\mathcal{S}_X) = \begin{pmatrix} pI_{11}(X) & 0 \\ 0 & 0 \end{pmatrix} \quad \text{and} \quad I(\mathcal{S}_X)\begin{pmatrix} p & 0 \\ 0 & 0 \end{pmatrix} = \begin{pmatrix} I_{11}(X)p & 0 \\ 0 & 0 \end{pmatrix},$$

we see that

$$K = \left\{ \begin{pmatrix} k & 0 \\ 0 & 0 \end{pmatrix} : k \in K_1 \right\},$$

where K_1 is the two-sided ideal in $I_{11}(X)$ generated by p.

For $a_1 p a_2 \in K_1$ and $z \in I(X)$ we have that,

$$(a_1 p a_2)z = a_1 p(a_2 z) = 0,$$

since $a_2 z \in I(X)$. Thus, $K_1 \neq I_{11}(X)$, and the quotient C^*-algebra $I(\mathcal{S}_X)/K$ is isomorphic to

$$\begin{pmatrix} I_{11}(X)/K_1 & I(X) \\ I(X)^* & I_{22}(X) \end{pmatrix}.$$

From this it is clear that the quotient map $\pi\colon I(\mathcal{S}_X) \to I(\mathcal{S}_X)/K$ is a complete order isomorphism on $\mathcal{S}_X$. Hence, there is a complete isometry $\gamma\colon \pi(\mathcal{S}_X) \to I(\mathcal{S}_X)$ such that $\gamma \circ \pi$ is the identity on $\mathcal{S}_X$.

By the injectivity of $I(\mathcal{S}_X)$ we may extend γ to a completely contractive map, which we still denote by γ, from $I(\mathcal{S}_X)/K$ to $I(\mathcal{S}_X)$.

However, since $\gamma \circ \pi$ is the identity on $\mathcal{S}_X$, by rigidity it must be the identity on all of $I(\mathcal{S}_X)$. But this implies that π is one-to-one and hence $K = (0)$. Therefore, $p = 0$ and so $a = 0$.

The proof for $b \in I_{22}(X)$ is similar. $\square$

Theorem 16.4. *Let X be an operator space, let $\varphi\colon X \to X$ be completely bounded, and define $\tau\colon C_2(X) \to C_2(X)$ via*

$$\tau\left(\begin{pmatrix} x_1 \\ x_2 \end{pmatrix}\right) = \begin{pmatrix} \varphi(x_1) \\ \|\varphi\|_{\mathrm{cb}} x_2 \end{pmatrix}.$$

Then $\|\tau\|_{\mathrm{cb}} = \|\varphi\|_{\mathrm{cb}}$ if and only if there exists a unique $a \in I_{11}(X)$, $\|a\| = \|\varphi\|_{\mathrm{cb}}$, such that $\varphi(x) = ax$ for all $x \in X$.

Proof. By scaling it is enough to consider the case $\|\varphi\|_{\mathrm{cb}} = 1$. If such an element a exists, then it must be unique by Lemma 16.3. In this case

$$\tau\left(\begin{pmatrix} x_1 \\ x_2 \end{pmatrix}\right) = \begin{pmatrix} a & 0 \\ 0 & 1 \end{pmatrix}\begin{pmatrix} x_1 \\ x_2 \end{pmatrix},$$

and since $\|\left(\begin{smallmatrix} a & 0 \\ 0 & 1 \end{smallmatrix}\right)\| \leq 1$ in $M_2(I_{11}(X))$, it follows easily that τ is completely contractive.

So assume that τ is completely contractive, and consider

$$\mathcal{B} = \begin{pmatrix} I_{11}(X) & I_{11}(X) & I(X) \\ I_{11}(X) & I_{11}(X) & I(X) \\ I(X)^* & I(X)^* & I_{22}(X) \end{pmatrix}.$$

Since

$$M_2(I(\mathcal{S}_X)) \cong \begin{pmatrix} I_{11}(X) & I_{11}(X) & I(X) & I(X) \\ I_{11}(X) & I_{11}(X) & I(X) & I(X) \\ I(X)^* & I(X)^* & I_{22}(X) & I_{22}(X) \\ I(X)^* & I(X)^* & I_{22}(X) & I_{22}(X) \end{pmatrix},$$

we see that $\mathcal{B}$ is the upper 3×3 corner of this injective C^*-algebra. Hence $\mathcal{B}$ is also an injective C^*-algebra, since it is the image of a completely contractive projection applied to an injective C^*-algebra .

Let $\mathcal{S} \subseteq \mathcal{B}$ be the operator system

$$\mathcal{S} = \begin{pmatrix} \mathbb{C} & 0 & X \\ 0 & \mathbb{C} & X \\ X^* & X^* & \mathbb{C} \end{pmatrix}.$$

It is also useful to think of $\mathcal{S}$ as

$$\begin{pmatrix} \mathcal{D}_2 & C_2(X) \\ C_2(X)^* & \mathbb{C} \end{pmatrix}, \qquad \text{where} \quad \mathcal{D}_2 = \mathbb{C} \oplus \mathbb{C},$$

and to regard $C_2(X)$ as a $\mathcal{D}_2$-$\mathbb{C}$-bimodule. Here the left action is given by

$$\begin{pmatrix} \lambda_1 & 0 \\ 0 & \lambda_2 \end{pmatrix} \begin{pmatrix} x_1 \\ x_2 \end{pmatrix} = \begin{pmatrix} \lambda_1 \cdot x_1 \\ \lambda_2 \cdot x_2 \end{pmatrix},$$

and the map τ is a $\mathcal{D}_2$-$\mathbb{C}$-bimodule map.

If we define $\Phi\colon \mathcal{S} \to \mathcal{S}$ via

$$\Phi\left(\begin{pmatrix} \lambda_1 & 0 & x_1 \\ 0 & \lambda_2 & x_2 \\ y_1^* & y_2^* & \lambda_3 \end{pmatrix}\right) = \begin{pmatrix} \lambda_1 & 0 & \varphi(x_1) \\ 0 & \lambda_2 & x_2 \\ \varphi(y_1)^* & y_2^* & \lambda_3 \end{pmatrix},$$

then since τ is a completely contractive $\mathcal{D}_2$-$\mathbb{C}$-bimodule map, Φ is completely positive.

By injectivity we may extend Φ to a completely positive map from $\mathcal{B}$ to $\mathcal{B}$, which we still denote by Φ. Since Φ fixes the diagonal C^*-subalgebra $\mathcal{D}_3 = \mathbb{C} \oplus \mathbb{C} \oplus \mathbb{C}$ of $\mathcal{B}$, Φ must be a $\mathcal{D}_3$-bimodule map. This implies that there exist maps $\varphi_{i,j}$, $i, j = 1, 2, 3$, such that $\Phi((z_{ij})) = (\varphi_{ij}(z_{ij}))$, where the 3×3 matrix (z_{ij}) represents a typical element of $\mathcal{B}$.

In particular, $\varphi_{13}\colon I(X) \to I(X)$ is an extension of $\varphi\colon X \to X$, and $\varphi_{23}\colon I(X) \to I(X)$ is an extension of the identity map on X.

Now the map

$$\begin{pmatrix} \varphi_{22} & \varphi_{23} \\ \varphi_{32} & \varphi_{33} \end{pmatrix} : \begin{pmatrix} I_{11}(X) & I(X) \\ I(X)^* & I_{22}(X) \end{pmatrix} \to \begin{pmatrix} I_{11}(X) & I(X) \\ I(X)^* & I_{22}(X) \end{pmatrix}$$

is completely positive and is the identity on $\left(\begin{smallmatrix} \mathbb{C} & X \\ X^* & \mathbb{C} \end{smallmatrix}\right) = \mathcal{S}_X$. Thus, by rigidity it must be the identity on

$$\begin{pmatrix} I_{11}(X) & I(X) \\ I(X)^* & I_{22}(X) \end{pmatrix} = I(\mathcal{S}_X).$$

This means that Φ is the identity on the C^*-subalgebra

$$\begin{pmatrix} \mathbb{C} & 0 & 0 \\ 0 & I_{11}(X) & I(X) \\ 0 & I(X)^* & I_{22}(X) \end{pmatrix}$$

and hence a bimodule map over this C^*-algebra.

Finally, for $x \in X$ we have that

$$\begin{aligned}\begin{pmatrix} 0 & 0 & \varphi(x) \\ 0 & 0 & 0 \\ 0 & 0 & 0 \end{pmatrix} &= \Phi\left(\begin{pmatrix} 0 & 1 & 0 \\ 0 & 0 & 0 \\ 0 & 0 & 0 \end{pmatrix}\begin{pmatrix} 0 & 0 & 0 \\ 0 & 0 & x \\ 0 & 0 & 0 \end{pmatrix}\right) \\ &= \Phi\left(\begin{pmatrix} 0 & 1 & 0 \\ 0 & 0 & 0 \\ 0 & 0 & 0 \end{pmatrix}\right)\begin{pmatrix} 0 & 0 & 0 \\ 0 & 0 & x \\ 0 & 0 & 0 \end{pmatrix} \\ &= \begin{pmatrix} 0 & \varphi_{12}(1) & 0 \\ 0 & 0 & 0 \\ 0 & 0 & 0 \end{pmatrix}\begin{pmatrix} 0 & 0 & 0 \\ 0 & 0 & x \\ 0 & 0 & 0 \end{pmatrix} \\ &= \begin{pmatrix} 0 & 0 & \varphi_{12}(1)x \\ 0 & 0 & 0 \\ 0 & 0 & 0 \end{pmatrix},\end{aligned}$$

and we see that necessarily φ is left multiplication by the element $a = \varphi_{12}(1) \in I_{11}(X)$. Since Φ is unital and completely positive, $\|\varphi_{12}(1)\| \leq 1$, which completes the proof. □

Corollary 16.5. *Let X be an operator space, let $\varphi\colon X \to X$ be completely bounded, and define $\gamma\colon R_2(X) \to R_2(X)$ via*

$$\gamma((x_1, x_2)) = (\varphi(x_1), \|\varphi\|_{\text{cb}} x_2).$$

Then $\|\gamma\|_{\text{cb}} = \|\varphi\|_{\text{cb}}$ if and only if there exists a unique $b \in I_{22}(X)$, $\|b\| = \|\varphi\|_{\text{cb}}$, such that $\varphi(x) = xb$ for all $x \in X$.

Proof. The proof follows similarly, or by taking adjoints. □

We are now in a position to prove the BRS theorem. If $\mathcal{A}$ is an algebra, then $M_n(\mathcal{A})$ is also an algebra with product given by $(a_{ij})(b_{ij}) = (\sum_k a_{ik}b_{kj})$. When $\mathcal{A}$ is a normed algebra, we call the product *contractive* provided that $\|ab\| \leq \|a\|\|b\|$ for every a and b in $\mathcal{A}$. If $\mathcal{A}$ is an algebra and an operator space, we call the product *completely contractive* provided that the product on $M_n(\mathcal{A})$

is contractive for every n, and we call $\mathcal{A}$ an *abstract operator algebra*. If $\mathcal{A}$ also has a unit e and $\|e\| = 1$, then we call $\mathcal{A}$ an *abstract unital operator algebra*. The following theorem shows that every abstract unital operator algebra is in fact, a concrete operator algebra.

Theorem 16.6. *Let $\mathcal{A}$ be an abstract unital operator algebra. Then there exists a unital completely isometric homomorphism $\pi\colon \mathcal{A} \to I_{11}(\mathcal{A})$.*

Proof. Let $a \in \mathcal{A}$, and let $\varphi_a\colon \mathcal{A} \to \mathcal{A}$ be left multiplication by a, i.e., $\varphi_a(x) = ax$. Since multiplication is completely contractive, $\|\varphi\|_{\text{cb}} \le \|a\|$, but $\|a\| = \|\varphi(e)\| \le \|\varphi\|_{\text{cb}}$ and so $\|\varphi\|_{\text{cb}} = \|a\|$. Consider the map $\tau\colon C_2(\mathcal{A}) \to C_2(\mathcal{A})$,

$$\tau\left(\begin{pmatrix} x_1 \\ x_2 \end{pmatrix}\right) = \begin{pmatrix} \varphi_a(x_1) \\ \|a\|x_2 \end{pmatrix} = \begin{pmatrix} a & 0 \\ 0 & \|a\|e \end{pmatrix}\begin{pmatrix} x_1 \\ x_2 \end{pmatrix}.$$

Since

$$\left\|\begin{pmatrix} a & 0 \\ 0 & \|a\|e \end{pmatrix}\right\| = \|a\|$$

and the multiplication on $\mathcal{A}$ is completely contractive, it follows that $\|\tau\|_{\text{cb}} = \|a\| = \|\varphi\|_{\text{cb}}$. Thus, by Theorem 16.4, there is a unique element $\pi(a) \in I_{11}(\mathcal{A})$, $\|\pi(a)\| = \|a\|$, such that $ax = \pi(a) \circ x$, where we let $\circ$ denote the action of $I_{11}(\mathcal{A})$ on $I(\mathcal{A})$.

Thus, there is a well-defined map $\pi\colon \mathcal{A} \to I_{11}(\mathcal{A})$ satisfying $\|\pi(a)\| = \|a\|$ and $\pi(a) \circ x = ax$ for all x. By Lemma 16.3, it follows that $\pi(a+b) = \pi(a) + \pi(b)$, $\pi(ab) = \pi(a)\pi(b)$, and $\pi(e) = 1$, since their differences annihilate $\mathcal{A}$. Hence π is an isometric homomorphism.

It remains to show that π is a complete isometry. To see that π is n-contractive, apply the same reasoning to obtain a contractive map

$$\beta\colon M_n(\mathcal{A}) \to I_{11}(M_n(\mathcal{A})) = M_n(I_{11}(\mathcal{A}))$$

satisfying $\beta((a_{ij})) \circ (x_{ij}) = (a_{ij})(x_{ij})$. Then, again by Lemma 16.3, $\beta((a_{ij})) = (\pi(a_{ij}))$ and hence π is n-contractive, for all n. On the other hand,

$$\|(\pi(a_{ij}))\| \ge \left\|(\pi(a_{ij})) \circ \begin{pmatrix} e & & 0 \\ & \ddots & \\ 0 & & e \end{pmatrix}\right\| = \|(a_{ij})\|,$$

which shows that π is a complete isometry. □

Corollary 16.7 (Blecher–Ruan–Sinclair Theorem). *Let $\mathcal{A}$ be an abstract unital operator algebra. Then there exists a Hilbert space $\mathcal{H}$ and a unital completely isometric homomorphism $\pi\colon \mathcal{A} \to B(\mathcal{H})$.*

Proof. Since $I_{11}(\mathcal{A})$ is a unital C^*-algebra, by the Gelfand–Naimark–Segal theorem there exists a Hilbert space $\mathcal{H}$ and a unital *-isomorphism of $I_{11}(\mathcal{A})$ into $B(\mathcal{H})$. Since *-isomorphisms are complete isometries, the result follows. □

We are now in a position to generalize the Christensen–Effros–Sinclair representation theorem. Let X be an operator space, and let $\mathcal{A}$ and $\mathcal{B}$ be unital operator algebras. We call X an *operator $\mathcal{A}$-$\mathcal{B}$-bimodule* provided that X is an $\mathcal{A}$-$\mathcal{B}$-bimodule and

$$\|(a_{ij})(x_{ij})(b_{ij})\| \leq \|(a_{ij})\|\,\|(x_{ij})\|\,\|(b_{ij})\|$$

for every m, n, p, q, and every $(a_{ij}) \in M_{m,n}(\mathcal{A})$, $(x_{ij}) \in M_{np}(X)$, and $(b_{ij}) \in M_{pq}(\mathcal{B})$.

When $\mathcal{B} = \mathbb{C}$, we call X a *left operator $\mathcal{A}$-module*, and when $\mathcal{A} = \mathbb{C}$, we call X a *right operator $\mathcal{B}$-module*. It is important to recall that, generally, there is a difference between X being an $\mathcal{A}$-$\mathcal{B}$-bimodule and X being both a left $\mathcal{A}$-module and a right $\mathcal{B}$-module. The difference is that a bimodule action is assumed to be associative, i.e., $(a \circ x) \circ b = a \circ (x \circ b)$, whereas if X is only a left $\mathcal{A}$-module and a right $\mathcal{B}$-module, then no associativity is assumed. Thus, to be an $\mathcal{A}$-$\mathcal{B}$-bimodule is, generally, a stronger condition. However, as we shall see, in this situation it is all the same.

Theorem 16.8. *Let X be an operator space, with $\mathcal{A}$ and $\mathcal{B}$ unital operator algebras. If X is a left operator $\mathcal{A}$-module (respectively, right operator $\mathcal{B}$-module), then there is a unique completely contractive homomorphism $\pi\colon \mathcal{A} \to I_{11}(X)(\rho\colon \mathcal{B} \to I_{22}(X))$ such that $\pi(a) \cdot x = a \circ x (x \cdot \rho(b) = x \circ b)$, where $\circ$ denotes the module action and $\cdot$ denotes the product in $I(\mathcal{S}_X)$.*

Proof. As in the proof of Theorem 16.6, for $a \in \mathcal{A}$ the map

$$\tau\left(\begin{pmatrix} x_1 \\ x_2 \end{pmatrix}\right) = \begin{pmatrix} a \circ x_1 \\ \|a\| x_2 \end{pmatrix} = \begin{pmatrix} a & 0 \\ 0 & \|a\| e \end{pmatrix} \circ \begin{pmatrix} x_1 \\ x_2 \end{pmatrix}$$

is completely bounded with $\|\tau\|_{\mathrm{cb}} = \|a\|$ and so is given as multiplication by a unique element of $I_{11}(X)$. This defines $\pi\colon \mathcal{A} \to I_{11}(X)$.

To define ρ one uses similar arguments on $\gamma((x_1, x_2)) = (x_1 \circ b, \|b\| x_2)$ and invokes Corollary 16.5. □

One consequence of the above theorem is that the associativity condition for bimodule actions is automatic.

Corollary 16.9. *Let X be an operator space, and let $\mathcal{A}$ and $\mathcal{B}$ be unital operator algebras. If X is a left operator $\mathcal{A}$-module and a right operator $\mathcal{B}$-module, then X is an operator $\mathcal{A}$-$\mathcal{B}$-bimodule.*

Proof. Let π and ρ be as above. Then for any $a \in \mathcal{A}$, $b \in \mathcal{B}$, and $x \in X$, we have

$$a \circ (x \circ b) = \pi(a)(x\rho(b)) = (\pi(a)x)\rho(b) = (a \circ x) \circ b. \qquad \square$$

The following result generalizes the Christensen–Effros–Sinclair representation theorem (Theorem 15.14).

Corollary 16.10. *Let $\mathcal{A}$ and $\mathcal{B}$ be unital operator algebras, and let X be an operator $\mathcal{A}$-$\mathcal{B}$-bimodule. Then there exist Hilbert spaces $\mathcal{H}$ and $\mathcal{K}$, unital completely contractive homomorphisms $\pi\colon \mathcal{A} \to B(\mathcal{H})$, $\rho\colon \mathcal{B} \to B(\mathcal{K})$, and a complete isometry $\varphi\colon X \to B(\mathcal{K}, \mathcal{H})$ such that $\pi(a)\varphi(x)\rho(b) = \varphi(axb)$.*

Proof. Represent the C^*-algebra $I(\mathcal{S}_X)$ as a C^*-subalgebra of a Hilbert space $\mathcal{L}$, set $\mathcal{H} = e_1\mathcal{L}$, $\mathcal{K} = e_2\mathcal{L}$, and apply Theorem 16.8. $\square$

A representation of an operator $\mathcal{A}$-$\mathcal{B}$-bimodule of the form given by Corollary 16.10 is called a *Christensen–Effros–Sinclair representation*. Theorem 16.8 is much stronger than Corollary 16.10; it shows that every $\mathcal{A}$-$\mathcal{B}$-bimodule action of operator algebras on an operator space X factors through the actions of $I_{11}(X)$ and $I_{22}(X)$ on X. To be more precise, let

$$M_\ell(X) = \{a \in I_{11}(X) \colon aX \subseteq X\}$$

and

$$M_r(X) = \{b \in I_{22}(X) \colon Xb \subseteq X\};$$

then these sets are easily seen to be unital subalgebras of the C^*-algebras $I_{11}(X)$ and $I_{22}(X)$, respectively. We call these operator algebras, respectively, the *left* and *right multiplier algebras of X*. We have that X is an operator $M_\ell(X)$-$M_r(X)$-bimodule. By Theorem 16.8, if X is an operator $\mathcal{A}$-$\mathcal{B}$-bimodule, then there exist completely contractive homomorphisms $\pi\colon \mathcal{A} \to M_\ell(X)$ and $\rho\colon \mathcal{B} \to M_r(X)$

such that $a \cdot x \cdot b = \pi(a)x\rho(b)$. Thus, in this sense, $M_\ell(X)$ and $M_r(X)$ are the "universal" operator algebras that can act on X.

It is also possible to give an "extrinsic" characterization of multipliers. Following Blecher [22], we call a linear map $\varphi: X \to X$ a *left multiplier* provided that there exists a Hilbert space $\mathcal{H}$ an operator T in $B(\mathcal{H})$ and a completely isometric map $\psi: X \to B(\mathcal{H})$ such that $\psi(\varphi(x)) = T\psi(x)$ for every x in X. When this is the case, we say that (ψ, T) is an *implementing pair* for φ. Thus, a map is a left multiplier exactly when, after a suitable representation of the operator space X, the map is given as multiplication by an operator. If φ is a left multiplier, then we define its *multiplier norm* $\|\varphi\|_m$ to be the infimum of $\|T\|$ over all such representations. The next result shows that this infimum is always attained.

Proposition 16.11. *Let X be an operator space, and let φ be a left multiplier of X. Then there exists a Hilbert space $\mathcal{H}$, a complete isometry $\psi: X \to B(\mathcal{H})$, and an operator T in $B(\mathcal{H})$ such that (ψ, T) implements φ and $\|\varphi\|_m = \|T\|$.*

Proof. Let (ψ_n, T_n) be a sequence of implementing pairs on Hilbert spaces $\mathcal{H}_n$ for φ with $\|\varphi\|_m = \lim_n \|T_n\|$. Let $\mathcal{F}$ be the vector space of bounded sequences of vectors $h = (h_1, h_2, \dots)$ with h_n in $\mathcal{H}_n$. Fix a Banach generalized limit glim, and define a sesquilinear form on $\mathcal{F}$ by setting $\langle h, k\rangle = \operatorname{glim}_n \langle h_n, k_n\rangle$ for $h = (h_1, h_2, \dots)$ and $k = (k_1, k_2, \dots)$ in $\mathcal{F}$. Let $\mathcal{H}$ denote the Hilbert space obtained by forming the quotient of $\mathcal{F}$ by the null space of this sesquilinear form and completing.

If we define $\psi: X \to B(\mathcal{H})$ and define T in $B(\mathcal{H})$ by setting $\langle Th, k\rangle = \operatorname{glim}_n \langle T_n h_n, k_n\rangle$ and $\langle \psi(x)h, k\rangle = \operatorname{glim}_n \langle \psi_n(x)h_n, k_n\rangle$, then it is easily checked that (ψ, T) is an implementing pair for φ with $\|\varphi\|_m = \|T\|$. □

These "extrinsically" defined maps have the following "intrinsic" characterizations.

Theorem 16.12 (Blecher–Effros–Zarikian Theorem). *Let X be an operator space, and let $\varphi: X \to X$ be a linear map. Then φ is a left multiplier with $\|\varphi\|_m \le 1$ if and only if the map $\tau: C_2(X) \to C_2(X)$ defined by*

$$\tau\left(\begin{pmatrix} x_1 \\ x_2 \end{pmatrix}\right) = \begin{pmatrix} \varphi(x_1) \\ x_2 \end{pmatrix}$$

is a complete contraction.

Proof. If τ is a complete contraction, then by Theorem 16.4 there exists an element $a \in I_{11}(X)$ such that $\varphi(x) = ax$ with $\|a\| = 1$. Representing $I(\mathcal{S}_X)$ as operators on a Hilbert space, letting ψ be the induced mapping of X, and letting T be the image of a yields the desired implementing pair.

Conversely, if $\|\varphi\|_m \leq 1$, then by using an implementing pair (ψ, T) with $\|\varphi\|_m = \|T\|$ it is easy to see that τ is completely contractive. □

Combining Theorem 16.4 with the above result completes the description of multipliers.

Corollary 16.13. *Let X be an operator space, and let $\varphi: X \to X$ be a linear map. Then φ is a left multiplier if and only if there exists a unique $a \in I_{11}(X)$ such that $\varphi(x) = ax$. Moreover, in this case $\|\varphi\|_m = \|a\|$.*

Thus, we see that "extrinsically" defined left multipliers are exactly the maps given as left multiplications by the elements of $M_\ell(X)$. There is a parallel theory for right multipliers.

Notes

The original proofs of the BRS theorem and the representation theorem for operator bimodules (i.e., the generalization of the Christensen–Effros–Sinclair representation theorem) made use of a representation theory for completely bounded multilinear maps, which we shall study in the next chapter.

In [102] it was shown that the left multiplier algebra of an ideal J of a C^*-algebra $\mathcal{A}$ can be represented as the subalgebra $\{x \in I(\mathcal{A}): xJ \subset J\}$.

Blecher [23] developed the study of multipliers of operator spaces. Many of these ideas were developed, independently, by Werner [250]. The techniques of [102] were then extended to multipliers of operator spaces in [30], where Theorem 16.8 appeared. But this earlier proof of Theorem 16.8 still used the representation theory for completely bounded multilinear maps in an essential way.

Blecher, Effros, and Zarikian [25], in their proof of Theorem 16.12, introduced the use of the map τ. Theorem 16.4 is implied by Theorem 16.12 together with Theorem 16.8, which both appeared earlier.

The proof of Theorem 16.4 given here is new and avoids reliance on the representation theory for completely bounded multilinear maps. The key ingredients are the τ-trick of [25] and the methods from [102].

The important role played by $I(\mathcal{S}_X)$ for many results about operator spaces was recognized earlier in [204], [113], and [249].

Exercises

16.1 Let C_n denote n-dimensional column Hilbert space. Prove that $I(\mathcal{S}_{C_n}) \cong M_{n+1}$ in such a way that

$$I_{11}(C_n) = M_\ell(C_n) \cong M_n$$

and

$$I_{22}(C_n) = M_r(C_n) \cong \mathbb{C}.$$

16.2 Prove, by showing that the BRS axioms are met, that the matrix-normed algebra $(\mathcal{P}_n, \|\cdot\|_{u,k})$, introduced in Chapter 5, is a unital operator algebra.

16.3 Let $\mathcal{A}$ be a unital operator algebra, and let J be a nontrivial two-sided ideal in $\mathcal{A}$. Prove that the algebra $\mathcal{A}/J$ equipped with the quotient operator space structure is an operator algebra.

Chapter 17

Completely Bounded Multilinear Maps and the Haagerup Tensor Norm

In many parts of analysis an important role is played by multilinear maps. Recall that if E, F, and Z are vector spaces, then a map $\gamma\colon E \times F \to Z$ is *bilinear* provided that it is linear in each variable, i.e., $\gamma(e_1 + e_2, f_1) = \gamma(e_1, f_1) + \gamma(e_2, f_1)$, $\gamma(e_1, f_1 + f_2) = \gamma(e_1, f_1) + \gamma(e_1, f_2)$, and $\gamma(\lambda e_1, f_1) = \gamma(e_1, \lambda f_1) = \lambda\gamma(e_1, f_1)$ for any e_1 and e_2 in E, for any f_1 and f_2 in F, and for any λ in $\mathbb{C}$. If one forms the algebraic tensor product $E \otimes F$ of E and F, then there is a one-to-one correspondence between linear maps $\Gamma\colon E \otimes F \to Z$ and bilinear maps $\gamma\colon E \times F \to Z$ given by setting $\Gamma(e \otimes f) = \gamma(e, f)$.

Consequently, if one endows $E \otimes F$ with a matrix norm, then the completely bounded linear maps from $E \otimes F$ to another matrix-normed space Z correspond to a family of bilinear maps from $E \times F$ to Z that one would like to regard as the "completely bounded" bilinear maps. In this fashion, one often arrives at an important family of bilinear maps to study. Conversely, if one starts with a family of bilinear maps into Z that one chooses to regard as the "completely bounded" bilinear maps, then often there is a corresponding matrix-normed structure on $E \otimes F$ such that these maps are precisely the completely bounded linear maps from $E \otimes F$ to Z, and in this fashion one often arrives at an important tensor norm.

There are several important cases in the literature where operator space norms on tensor products of operator spaces and the corresponding families of bilinear maps are examined in detail. Among these are the *spatial tensor norm*, also called the *operator space injective tensor norm*, the *operator space projective tensor norm*, and the *Haagerup tensor norm*. We touched on some of these tensor norms in Chapter 12.

For the remainder of this book only the Haagerup tensor norm and its corresponding family of "completely bounded" bilinear maps play a crucial role. So we will develop the properties of this norm and the corresponding maps in some depth. We refer the interested reader to [88] and [193] for a more in-depth study of the tensor theory of operator spaces.

The original proofs of the Blecher–Ruan–Sinclair characterization of operator algebras, and the various characterizations of operator bimodules, such as the Christensen–Effros–Sinclair theorem, used the Haagerup tensor theory. In particular, these proofs used a Stinespring-like representation theorem for the corresponding family of bilinear maps, which we shall prove later in this chapter. The Haagerup tensor theory is also closely connected with the theory of *free products of C^*-algebras.*

Consequently, we begin this chapter with an analysis of the Haagerup tensor norm followed by some discussion of free products. We then present alternate proofs of several of the results of Chapter 16. In particular, we present proofs of the BRS characterization theorem of operator algebras (Corollary 16.7) and of the characterization theorem for operator bimodules (Corollary 16.10) that use this Stinespring-like representation theorem for the corresponding family of completely bounded bilinear maps. The Haagerup tensor norm also will play a central role in Chapter 19.

In a certain sense the Haagerup tensor norm is the "universal norm for products." To make this clear we introduce a "faux" product operation, denoted $\odot$. Given an element e of E and an element f of F, we do not necessarily have a way to take their product, but the element $e \otimes f$ obeys the same bilinear relations as a product. Extending this idea, given $(e_{ij}) \in M_{m,k}(E)$ and $(f_{i,j}) \in M_{k,n}(F)$, we define an element $(e_{ij}) \odot (f_{ij})$ of $M_{m,n}(E \otimes F)$ by setting

$$(e_{ij}) \odot (f_{ij}) = \left(\sum_{\ell=1}^{k} e_{i\ell} \otimes f_{\ell j} \right).$$

For $(x_{ij}) \in M_{m,n}(E \otimes F)$ we define the *Haagerup norm* by setting

$$\|(x_{ij})\|_h = \inf\{\|(e_{ij})\| \|(f_{ij})\|: (x_{ij}) = (e_{ij}) \odot (f_{ij})\},$$

where the infimum is taken over all possible ways to represent $(x_{ij}) = (e_{ij}) \odot (f_{ij})$ with $(e_{ij}) \in M_{m,k}(E)$, $(f_{ij}) \in M_{k,n}(F)$, and k arbitrary. We let $E \otimes_h F$ denote $E \otimes F$ equipped with this set of norms. It is important to note that the above infimum is nonempty, that is, given (x_{ij}), there is always a k, a $(e_{ij}) \in M_{mk}(E)$, and a $(f_{ij}) \in M_{kn}(F)$ such that $(x_{ij}) = (e_{ij}) \odot (f_{ij})$. To see this, note that if $x = \sum_{\ell=1}^{k} e_\ell \otimes f_\ell$ in $E \otimes F$ then

$$x = (e_1, \ldots, e_k) \odot \begin{pmatrix} f_1 \\ \vdots \\ f_k \end{pmatrix},$$

while

$$\begin{pmatrix} e_1 \otimes f_1 & e_2 \otimes f_2 \\ e_3 \otimes f_3 & e_4 \otimes f_4 \end{pmatrix} = \begin{pmatrix} e_1 & e_2 & 0 & 0 \\ 0 & 0 & e_3 & 3_4 \end{pmatrix} \odot \begin{pmatrix} f_1 & 0 \\ 0 & f_2 \\ f_3 & 0 \\ 0 & f_4 \end{pmatrix}.$$

Combining these two tricks allows one to express an arbitrary (x_{ij}) as $(e_{ij}) \odot (f_{ij})$.

We wish to show that the Haagerup norm is indeed a norm on $E \otimes F$ and that $E \otimes F$ is an L^∞-matrix-normed space, i.e., an abstract operator space, when endowed with the family of Haagerup norms. To do this a few lemmas are useful.

Lemma 17.1. *Let $(e_{ij}) \in M_{m,k}(E)$, $(f_{ij}) \in M_{k,n}(F)$, and let A, B, and C be scalar matrices with $A = (a_{ij}) \in M_{m_1,m}$, $B \in M_{k,k}$, and $C \in M_{n,n_1}$. Then,*

(i) $A \cdot [(e_{ij}) \odot (f_{ij})] = [A(e_{ij})] \odot (f_{ij})$,
(ii) $(e_{ij})B \odot (f_{ij}) = (e_{ij}) \odot B(f_{ij})$,
(iii) $[(e_{ij}) \odot (f_{ij})] \cdot C = (e_{ij}) \odot [(f_{ij})C]$.

Proof. Straightforward. □

We shall call a family of norms on $M_{m,n}(V)$ satisfying all the axioms for an L^∞-matrix norm except that $\|x\| = 0$ implies $x = 0$ for $x \in V$ an *L^∞-matrix seminorm*. If we let $W = \{x\colon \|x\| = 0\}$, then these induce an L^∞-matrix norm on V/W and it becomes an abstract operator space (Exercise 17.1).

Proposition 17.2. *Let E and F be operator spaces. Then the Haagerup norm is an L^∞-matrix seminorm on $E \otimes F$.*

Proof. By Lemma 17.1 we have for $(x_{ij}) \in M_{m,n}(E \otimes F)$, $A \in M_{m_1,m}$, and $B \in M_{n,n_1}$ that $\|A(x_{ij})B\|_h \le \|A\| \|(x_{ij})\|_h \|B\|$.

Next we show the triangle inequality. Given (x_{ij}) and (y_{ij}) in

$$M_{m,n}(E \otimes F),$$

and $\epsilon > 0$, choose $(e_{ij}) \in M_{m,k_1}(E)$, $(f_{ij}) \in M_{k_1,n}(F)$, $(\hat{e}_{ij}) \in M_{m,k_2}(E)$, and $(\hat{f}_{ij}) \in M_{k_2,n}(F)$ such that $(x_{ij}) = (e_{ij}) \odot (f_{ij})$, $(y_{ij}) = (\hat{e}_{ij}) \odot (\hat{f}_{ij})$, and

$$\|(e_{ij})\| \|(f_{ij})\| \le \|(x_{ij})\|_h + \epsilon,$$
$$\|(\hat{e}_{ij})\| \|(\hat{f}_{ij})\| \le \|(y_{ij})\|_h + \epsilon.$$

Replacing $(e_{ij}), (f_{ij})$ by $r(e_{ij}), r^{-1}(f_{ij})$, we may assume that $\|(e_{ij})\| = \|(f_{ij})\|$ and similarly that $\|(\hat{e}_{ij})\| = \|(\hat{f}_{ij})\|$. Now consider

$$e = ((e_{ij}), (\hat{e}_{ij})) \in M_{m,k_1+k_2}(E) \quad \text{and} \quad f = \begin{pmatrix} (f_{ij}) \\ (\hat{f}_{ij}) \end{pmatrix} \in M_{k_1+k_2,n}(F).$$

We have that $(x_{ij}) + (y_{ij}) = e \odot f$ and hence

$$\begin{aligned} \|(x_{ij}) + (y_{ij})\|_h &\leq \|e\| \|f\| \leq \sqrt{\|(e_{ij})\|^2 + \|(\hat{e}_{ij})\|^2} \cdot \sqrt{\|(f_{ij})\|^2 + \|(\hat{f}_{ij})\|^2} \\ &\leq \|(x_{ij})\|_h + \|(y_{ij})\|_h + 2\epsilon. \end{aligned}$$

To see $\|e\|^2 \leq \|(e_{ij})\|^2 + \|(\hat{e}_{ij})\|^2$, it is perhaps useful to represent E as a concrete operator space. Since ϵ was arbitrary, the triangle inequality follows.

Finally, to see the L^∞ condition, note that if $(x_{ij}) = (e_{ij}) \odot (f_{ij})$ and $(y_{ij}) = (\hat{e}_{ij}) \odot (\hat{f}_{ij})$, then

$$\begin{pmatrix} (x_{ij}) & 0 \\ 0 & (y_{ij}) \end{pmatrix} = \begin{pmatrix} (e_{ij}) & 0 \\ 0 & (\hat{e}_{ij}) \end{pmatrix} \odot \begin{pmatrix} (f_{ij}) & 0 \\ 0 & (\hat{f}_{ij}) \end{pmatrix}. \qquad \square$$

The Haagerup tensor norm has many useful identifications. We introduce one for future use. Note that if E and F are operator spaces then $C_m(E) = M_{m,1}(E)$ is an operator space. Similarly, $R_n(E) = M_{1,n}(F)$ is an operator space. Hence, there is a Haagerup norm on $C_m(F) \otimes R_n(F)$.

Proposition 17.3. *Let E and F be operator spaces. Then the map*

$$\Gamma: C_m(E) \otimes_h R_n(F) \to M_{m,n}(E \otimes_h F)$$

given by

$$\Gamma\left(\begin{pmatrix} e_1 \\ \vdots \\ e_m \end{pmatrix} \otimes (f_1, \ldots, f_n)\right) = (e_i \otimes f_j)$$

is an isometry of the Haagerup norm on $C_m(E) \otimes R_n(F)$ with the Haagerup norm on the $m \times n$ matrices over $E \otimes F$.

Proof. One notes that to compute the Haagerup norm of an element of $C_m(E) \otimes R_n(F)$, one represents it as a row of, say, k columns, which is really an $m \times k$ matrix over E, "times" a column of k rows, which is a $k \times n$ matrix over F.

Now one checks that Γ preserves this identification. $\square$

Before preceding further in our analysis of the Haagerup norm, it is essential to make a few observations about representations of elements of a tensor product. First recall that if E and F are vector spaces and $\varphi\colon E \to \mathbb{C}$ is any linear functional, then there is a linear map $\varphi \otimes \mathrm{id}\colon E \otimes F \to F$ with $(e \otimes f) = \varphi(e) f$. Now suppose that

$$x = \sum_{i=1}^{n} e_i \otimes f_i = \sum_{\ell=1}^{m} \tilde{e}_\ell \otimes \tilde{f}_\ell$$

are two ways to express x and that $\{e_1, \ldots, e_n\}$ are linearly independent. Choosing functionals φ_i such that $\varphi_i(e_j) = \delta_{ij}$ we see that $f_i = \varphi_i \otimes \mathrm{id}(x) = \sum_{\ell=1}^{m} \varphi_i(\tilde{e}_\ell)\tilde{f}_\ell$. Hence, $\mathrm{span}\{f_1, \ldots, f_n\} \subseteq \mathrm{span}\{\tilde{f}_1, \ldots, \tilde{f}_m\}$. Thus, if all the sets are linearly independent in both representations, then $\mathrm{span}\{f_1, \ldots, f_n\} = \mathrm{span}\{\tilde{f}_1, \ldots, \tilde{f}_n\}$, $\mathrm{span}\{e_1, \ldots, e_n\} = \mathrm{span}\{\tilde{e}_1, \ldots, \tilde{e}_m\}$, and $n = m$.

Also note that one can always write $x = \sum_{i=1}^{n} e_i \otimes f_i$ with $\{e_1, \ldots, e_n\}$ and $\{f_1, \ldots, f_n\}$ linearly independent. For if, say, $e_n = \lambda_1 e_1 + \cdots + \lambda_{n-1} e_{n-1}$, then $x = \sum_{i=1}^{n-1} e_i \otimes (f_i + \lambda_i f_n)$ and we proceed inductively. The unique integer n such that $x = \sum_{i=1}^{n} e_i \otimes f_i$ with both $\{e_1, \ldots, e_n\}$ and $\{f_1, \ldots, f_n\}$ linearly independent is called the *rank of the tensor* x, denoted $\mathrm{rank}(x)$.

Those more familiar with Banach space approaches to tensor theory know that $x = \sum_{i=1}^{n} e_i \otimes f_i$ in $E \otimes F$ is often identified with a linear map $T_x\colon E^* \to F$ given by $T_x(\varphi) = \sum_{i=1}^{n} \varphi(e_i) f_i$ and that $\mathrm{rank}(x) = \dim \mathrm{range}(T_x)$. The fact that the map T_x is independent of the particular representation of x as a sum of elementary tensors is essentially equivalent to our observations above.

Theorem 17.4. *Let $E \subseteq E_1$ and $F \subseteq F_1$ be operator spaces. Then the inclusion of $E \otimes_h F$ into $E_1 \otimes_h F_1$ is a complete isometry.*

Proof. Let x be in $E \otimes F$. Write $\|x\|_0$ for its Haagerup norm as an element of $E \otimes_h F$, and $\|x\|_1$ for its Haagerup norm as an element of $E_1 \otimes_h F_1$. We wish to prove that $\|x\|_0 = \|x\|_1$. By the definition of the Haagerup norm as an infimum, we have that $\|x\|_1 \leq \|x\|_0$.

Now let $x = \sum_{i=1}^{n} e_i \otimes f_i$ with $e_i \in E_1$ and $f_i \in F_1$. To prove that $\|x\|_0 \leq \|x\|_1$, it will be enough to prove that

$$\|x\|_0 \leq \|(e_1, \ldots, e_n)\| \left\| \begin{pmatrix} f_1 \\ \vdots \\ f_n \end{pmatrix} \right\|.$$

Note that the rank of x must be the same whether we regard it as an element of $E \otimes F$ or of $E_1 \otimes F_1$. So let $\mathrm{rank}(x) = m \leq n$, and write $x = \sum_{i=1}^{m} \tilde{e}_i \otimes \tilde{f}_i$ with $\{\tilde{e}_1, \ldots, \tilde{e}_m\} \subseteq E$ and $\{\tilde{f}_1, \ldots, \tilde{f}_m\} \subseteq F$.

If $\{e_1, \ldots, e_n\}$ and $\{f_1, \ldots, f_n\}$ are linearly independent, then necessarily $m = n$, $\{e_1, \ldots, e_n\} \subseteq E$, $\{f_1, \ldots, f_n\} \subseteq F$, and hence

$$\|x\|_0 \leq \|(e_1, \ldots, e_n)\| \left\| \begin{pmatrix} f_1 \\ \vdots \\ f_n \end{pmatrix} \right\|.$$

Thus, we can assume that there are some linear dependences. Say $\{e_1, \ldots, e_k\}$ is a maximal linearly independent set of $\{e_1, \ldots, e_n\}$. The linear equations expressing $\{e_1, \ldots, e_n\}$ as linear combinations of $\{e_1, \ldots, e_k\}$ can be written as

$$(e_1, \ldots, e_n) = (e_1, \ldots, e_k)B$$

for some $k \times n$ scalar matrix B.

Polar-decompose $B = PW$ with W a $k \times n$ partial isometry satisfying $WW^* = I_k$. We have that

$$\begin{aligned} x &= (e_1, \ldots, e_k)PW \odot \begin{pmatrix} f_1 \\ \vdots \\ f_n \end{pmatrix} \\ &= (e_1, \ldots, e_k)P \odot W \begin{pmatrix} f_1 \\ \vdots \\ f_n \end{pmatrix} = (\hat{e}_1, \ldots, \hat{e}_k) \odot \begin{pmatrix} \hat{f}_1 \\ \vdots \\ \hat{f}_k \end{pmatrix}, \end{aligned}$$

where

$$(\hat{e}_1, \ldots, \hat{e}_k) = (e_1, \ldots, e_k)P \quad \text{and} \quad \begin{pmatrix} \hat{f}_1 \\ \vdots \\ \hat{f}_k \end{pmatrix} = W \begin{pmatrix} f_1 \\ \vdots \\ f_n \end{pmatrix}.$$

Note that $\|(\hat{e}_1, \ldots, \hat{e}_k)\| = \|(e_1, \ldots, e_n)\|$, and that

$$\left\| \begin{pmatrix} \hat{f}_1 \\ \vdots \\ \hat{f}_k \end{pmatrix} \right\| \leq \left\| \begin{pmatrix} f_1 \\ \vdots \\ f_n \end{pmatrix} \right\|.$$

Thus, we have that

$$\|(\hat{e}_1, \ldots, \hat{e}_k)\| \left\| \begin{pmatrix} \hat{f}_1 \\ \vdots \\ \hat{f}_k \end{pmatrix} \right\| \leq \|(e_1, \ldots, e_n)\| \left\| \begin{pmatrix} f_1 \\ \vdots \\ f_n \end{pmatrix} \right\|.$$

Now repeat this process if necessary on $\{\hat{f}_1, \ldots, \hat{f}_k\}$ until one achieves a situation where both sets $\{\hat{e}_1, \ldots, \hat{e}_k\}$ and $\{\hat{f}_1, \ldots, \hat{f}_k\}$ are linearly independent and hence lie in E and F, respectively. Since these elements lie in E and F, one has

$$\|x\|_0 \leq \|(\hat{e}_1, \ldots, \hat{e}_k)\| \left\| \begin{pmatrix} \hat{f}_1 \\ \vdots \\ \hat{f}_k \end{pmatrix} \right\| \leq \|(e_1, \ldots, e_n)\| \left\| \begin{pmatrix} f_1 \\ \vdots \\ f_n \end{pmatrix} \right\|$$

and hence $\|x\|_0 \leq \|x\|_1$.

This proves that the inclusion of $E \otimes_h F$ into $E_1 \otimes_h F_1$ is isometric. To prove that it is a complete isometry, one notes that $C_m(E) \subseteq C_m(E_1)$, $R_n(F) \subseteq R_n(F_1)$ and applies the result just proven to conclude that the inclusion of $C_m(E) \otimes_h R_n(F)$ into $C_m(E_1) \otimes_h R_n(F_1)$ is isometric. Now, applying Proposition 17.3, we conclude that

$$M_{m,n}(E \otimes_h F) = C_m(E) \otimes_h R_n(F) \subseteq C_m(E_1) \otimes_h R_n(F_1) = M_{m,n}(E_1 \otimes_h F_1)$$

isometrically.

Hence the inclusion of $E \otimes_h F$ into $E_1 \otimes_h F_1$ is a complete isometry. □

Corollary 17.5. *Let E and F be operator spaces, and let $x \in E \otimes F$ be a rank k tensor. Then there exists a representation $x = \sum_{i=1}^{k} e_i \otimes f_i$ such that*

$$\|x\|_h = \|(e_1, \ldots, e_k)\| \left\| \begin{pmatrix} f_1 \\ \vdots \\ f_k \end{pmatrix} \right\|.$$

In particular, $\|x\|_h \neq 0$, and so the Haagerup norm is indeed a norm.

Proof. By the proof of Theorem 17.4, the expression in the Haagerup norm is reduced by making the terms used in the expression of x as a sum of elementary tensors linearly independent. Thus, if $\text{rank}(x) = k$, then

$$\|x\|_h = \inf \left\{ \|(e_1, \ldots, e_k)\| \left\| \begin{pmatrix} f_1 \\ \vdots \\ f_k \end{pmatrix} \right\| : x = \sum_{i=1}^{k} e_i \otimes f_i \right\}.$$

But if

$$x = \sum_{i=1}^{k} e_i \otimes f_i = \sum_{i=1}^{k} \tilde{e}_i \otimes \tilde{f}_i,$$

then $\text{span}\{e_1, \ldots, e_k\} = \text{span}\{\tilde{e}_1, \ldots, \tilde{e}_k\} = E_1$ and $\text{span}\{f_1, \ldots, f_k\} = \text{span}\{\tilde{f}_1, \ldots, \tilde{f}_k\} = F_1$.

Thus, one can choose sequences of bases for $E_1, \{e_1^n, \dots, e_k^n\}$, and for $F_1, \{f_1^n, \dots, f_k^n\}$, such that $x = \sum_{i=1}^k e_i^n \otimes f_i^n$ and

$$\|(e_1^n, \dots, e_k^n)\| \leq \|x\|_h^{1/2} + \frac{1}{n}, \qquad \left\| \begin{pmatrix} f_1^n \\ \vdots \\ f_k^n \end{pmatrix} \right\| \leq \|x\|_h^{1/2} + \frac{1}{n}.$$

Since E_1 and F_1 are finite-dimensional, these bounded sequences will have a common convergent subsequence. Letting $e_1, \dots, e_k, f_1, \dots, f_k$ denote their limits, we have that

$$x = \sum_{i=1}^k e_i \otimes f_i \qquad \text{with} \quad \|(e_1, \dots, e_k)\| \left\| \begin{pmatrix} f_1 \\ \vdots \\ f_k \end{pmatrix} \right\| = \|x\|_h.$$

Since $\operatorname{rank}(x) = k$, the sets $\{e_1, \dots, e_k\}$ and $\{f_1, \dots, f_k\}$ are bases. Thus, $\|e_i\| \neq 0$, $\|f_j\| \neq 0$ for all i and j, and hence

$$\|x\|_h = \|(e_1, \dots, e_k)\| \left\| \begin{pmatrix} f_1 \\ \vdots \\ f_k \end{pmatrix} \right\| \neq 0. \qquad \square$$

Combining Proposition 17.2 with Corollary 17.5 establishes that the Haagerup tensor product of two operator spaces is again an operator space. We now wish to give a more concrete realization of the Haagerup tensor product. Since operator spaces E and F can be represented as subspaces of C^*-algebras $\mathcal{A}$ and $\mathcal{B}$, and since $E \otimes_h F \subseteq \mathcal{A} \otimes_h \mathcal{B}$ completely isometrically, it will be sufficient to understand $\mathcal{A} \otimes_h \mathcal{B}$ more clearly when $\mathcal{A}$ and $\mathcal{B}$ are C^*-algebras.

To this end we introduce the *free product* of algebras. Given two algebras $\mathcal{A}$ and $\mathcal{B}$, their free product [10] is another algebra, denoted $\mathcal{A} * \mathcal{B}$, that contains $\mathcal{A}$ and $\mathcal{B}$ as subalgebras and satisfies the following universal property: If $\mathcal{C}$ is any algebra, and $\pi\colon \mathcal{A} \to \mathcal{C}$ and $\rho\colon \mathcal{B} \to \mathcal{C}$ are algebra homomorphisms, then there exists a unique algebra homomorphism $\gamma\colon \mathcal{A} * \mathcal{B} \to \mathcal{C}$ with $\gamma(a) = \pi(a)$ and $\gamma(b) = \rho(b)$.

The map γ is denoted $\pi * \rho$.

Alternatively, one can regard $\mathcal{A} * \mathcal{B}$ as consisting of linear combinations of words in $\mathcal{A}$ and $\mathcal{B}$,

$$\mathcal{A} * \mathcal{B} = \operatorname{span}\{a_1, b_1, a_2 * b_2, b_3 * a_3, a_4 * b_4 * a_5, \dots\},$$

where the a_i's and b_i's are arbitrary elements of $\mathcal{A}$ and $\mathcal{B}$, respectively, subject to certain relations. Among these relations we would have that the product

w_1w_2, for a word w_1 ending in a different letter than w_2 begins with, is simply their concatenation, while, for example, the product of $w_1 = a_1 * b_1$ with $w_2 = b_2 * a_2$ is $w_1w_2 = a_1 * (b_1b_2) * a_2$. Also, these words behave multilinearly in each variable, e.g., $a_1 * b_1 * a_2 + a_1 * b_2 * a_2 = a_1 * (b_1 + b_2) * a_2$ and $a_1 * (\lambda b_1) * a_2 = (\lambda a_1) * b_1 * a_2 = a_1 * b_1 * (\lambda a_2)$ for any scalar λ.

When $\mathcal{A}$ and $\mathcal{B}$ both contain a common subalgebra $\mathcal{C}$, then one can also form the *free product amalgamated over* $\mathcal{C}$, denoted $\mathcal{A} *_{\mathcal{C}} \mathcal{B}$, which satisfies the following universal property: If $\mathcal{D}$ is any algebra, and $\pi\colon \mathcal{A} \to \mathcal{D}$ and $\rho\colon \mathcal{B} \to \mathcal{D}$ are algebra homomorphisms with $\pi(c) = \rho(c)$ for every c in $\mathcal{C}$, then there exists a unique algebra homomorphism $\gamma\colon \mathcal{A} *_{\mathcal{C}} \mathcal{B} \to \mathcal{D}$ with $\gamma(a) = \pi(a)$ and $\gamma(b) = \rho(b)$.

One can also give $\mathcal{A} *_{\mathcal{C}} \mathcal{B}$ a definition in terms of words where one allows elements of $\mathcal{C}$ to commute with concatenation – for example, $(ac) * b = a * (cb)$.

Now assume that $\mathcal{A}$ and $\mathcal{B}$ are unital C^*-algebras (we will consider operator algebras later) that both contain the algebra $\mathcal{C} = \mathbb{C} \cdot 1$. We endow $\mathcal{A} *_{\mathbb{C}} \mathcal{B}$ with a seminorm by setting

$$\|x\| = \sup \|\pi * \rho(x)\|,$$

where the supremum is over all Hilbert spaces $\mathcal{H}$ and all unital $*$-homomorphisms $\pi\colon \mathcal{A} \to B(\mathcal{H})$ and $\rho\colon \mathcal{B} \to B(\mathcal{H})$. Clearly, $\mathcal{A} *_{\mathbb{C}} \mathcal{B}$ is a $*$-algebra and $\|x^*x\| = \|x\|^2$. If $\mathcal{J} = \{x\colon \|x\| = 0\}$, then $\mathcal{J}$ is a two-sided $*$-ideal, and $(\mathcal{A} *_{\mathbb{C}} \mathcal{B})/\mathcal{J}$ is a pre-C^*-algebra. We denote its completion by $\mathcal{A} *_1 \mathcal{B}$. It turns out that $\mathcal{J} = \{0\}$, but we shall not need this harder fact here. See [10] for a proof of it and more on representations of $\mathcal{A} *_1 \mathcal{B}$.

This construction allows us to introduce another tensor (semi)norm on operator spaces. If $E \subseteq \mathcal{A}$ and $F \subseteq \mathcal{B}$ are concrete operator subspaces, then there is a linear map $\gamma\colon E \otimes F \to \mathcal{A} *_1 \mathcal{B}$ defined by setting $\gamma(e \otimes f) = e * f$ and extending linearly. For $(x_{ij}) \in M_{m,n}(E \otimes F)$ we define $\|(x_{ij})\|_1 = \|(\gamma(x_{ij}))\|_{M_{m,n}(\mathcal{A} *_1 \mathcal{B})}$ and shall refer to this as the *induced amalgamated free product norm.*

Without citing [10], it is only apparent that $\|\cdot\|_1$ is a seminorm. But we shall prove that it is equal to the Haagerup norm and, consequently, give an independent proof that it is a norm. More importantly, this will also prove that $\|\cdot\|_1$ is independent of the particular representations of E and F as subspaces of C^*-algebras.

There is a third way to define a tensor norm on operator spaces, which we wish to prove equals the Haagerup tensor norm. This definition was first introduced in [170], where it is called the *Brown tensor norm*, but has since come to be widely known as the *factorization norm.*

To define this tensor norm, first suppose that we are given three Hilbert spaces, $\mathcal{H}_1, \mathcal{H}_2$, and $\mathcal{H}_3$, and linear maps $\varphi: E \to B(\mathcal{H}_2, \mathcal{H}_3)$ and $\psi: F \to B(\mathcal{H}_1, \mathcal{H}_2)$. Then there is a linear map $\varphi \cdot \psi: E \otimes F \to B(\mathcal{H}_1, \mathcal{H}_3)$ defined by setting $\varphi \cdot \psi(e \otimes f) = \varphi(e)\psi(f)$ and extending linearly. Thus, we may define the *factorization (semi)norm* of (x_{ij}) in

$$M_{m,n}(E \otimes F)$$

by setting

$$\|(x_{ij})\|_f = \sup\|(\varphi \cdot \psi(x_{ij}))\|,$$

where the supremum is taken over all $\mathcal{H}_1, \mathcal{H}_2, \mathcal{H}_3$ and all pairs of completely contractive maps $\varphi: E \to B(\mathcal{H}_2, \mathcal{H}_3)$, $\psi: F \to B(\mathcal{H}_1, \mathcal{H}_2)$.

Since the C^*-algebra $\mathcal{A} *_1 \mathcal{B}$ can be represented on a Hilbert space and since the inclusions $E \subseteq \mathcal{A}$ and $F \subseteq \mathcal{B}$ yield completely contractive inclusions $\varphi: E \to \mathcal{A} *_1 \mathcal{B}$ and $\psi: F \to \mathcal{A} *_1 \mathcal{B}$, we see that the map $\gamma = \varphi \cdot \psi$ used to define the induced amalgamated free product (semi)norm is just one example of a map $\varphi \cdot \psi$. Hence, we have that

$$\|(x_{ij})\|_1 \le \|(x_{ij})\|_f$$

for all (x_{ij}) in $M_{m,n}(E \otimes f)$.

Now assume that φ and ψ are as in the definition of the factorization norm and that $(x_{ij}) \in M_{m,n}(E \otimes F)$ is represented as $(x_{ij}) = (e_{ij}) \odot (f_{ij})$ with $(e_{ij}) \in M_{m,k}(E)$, $(f_{ij}) \in M_{k,m}(F)$. Then we have that

$$\begin{aligned}(\varphi \cdot \psi(x_{ij})) &= \left(\varphi \cdot \psi\left(\sum_{\ell=1}^{k} e_{ik} \otimes f_{kj}\right)\right) = \left(\sum_{\ell=1}^{k} \varphi(e_{ik})\psi(f_{kj})\right) \\ &= (\varphi(e_{ij}))(\psi(f_{ij})).\end{aligned}$$

Hence, $\|(\varphi \cdot \psi(x_{ij}))\| \le \|(\varphi(e_{ij}))\|\|(\psi(f_{ij}))\| \le \|(e_{ij})\|\|(f_{ij})\|$. Taking the infimum over all factorizations and then the supremum over all φ and ψ yields

$$\|(x_{ij})\|_1 \le \|(x_{ij})\|_f \le \|(x_{ij})\|_h.$$

We now wish to prove that $\|(x_{ij})\|_h \le \|(x_{ij})\|_1$, so that all of these matrix norms are equal.

Theorem 17.6 (Christensen–Effros–Sinclair–Pisier). *Let $\mathcal{A}$ and $\mathcal{B}$ be unital C^*-algebras, and let $E \subseteq \mathcal{A}$, $F \subseteq \mathcal{B}$ be subspaces. Then the map $\gamma: E \otimes_h F \to \mathcal{A} *_1 \mathcal{B}$ given by $\gamma(e \otimes f) = e * f$ is a complete isometry.*

Proof. By Theorem 17.4, it will be sufficient to prove the case $E = \mathcal{A}$, $F = \mathcal{B}$. To this end we show that $\mathcal{A} \otimes_h \mathcal{B}$ is an operator $\mathcal{A}$-$\mathcal{B}$-bimodule.

Define an $\mathcal{A}$-$\mathcal{B}$-bimodule action on $\mathcal{A} \otimes_h \mathcal{B}$ by setting $a \cdot (c \otimes d) \cdot b = (ac) \otimes (db)$. We wish to show that this module action is completely contractive. To see this, note that if $(x_{ij}) \in M_{m,n}(\mathcal{A} \otimes \mathcal{B})$, $(a_{ij}) \in M_{m_1,m}(\mathcal{A})$, $(b_{ij}) \in M_{n,n}(\mathcal{B})$, and $(x_{ij}) = (c_{ij}) \odot (d_{ij})$, then

$$(a_{ij}) \cdot (x_{ij}) \cdot (b_{ij}) = [(a_{ij})(c_{ij})] \odot [(d_{ij})(b_{ij})].$$

Hence,

$$\begin{aligned}\|(a_{ij}) \cdot (x_{ij}) \cdot (b_{ij})\|_h &\leq \inf\{\|(a_{ij})(c_{ij})\| \cdot \|(d_{ij})(b_{ij})\| \colon (x_{ij}) = (c_{ij}) \odot (d_{ij})\} \\ &\leq \|(a_{ij})\| \|(b_{ij})\| \cdot \|(x_{ij})\|_h.\end{aligned}$$

Thus, by the Christensen–Effros–Sinclair representation theorem (Theorem 15.14 or Corollary 16.10), there exist Hilbert spaces $\mathcal{H}$ and $\mathcal{K}$, unital $*$-homomorphisms $\pi\colon \mathcal{A} \to B(\mathcal{H})$, $\rho\colon \mathcal{B} \to B(\mathcal{K})$, and a linear complete isometry $\psi\colon \mathcal{A} \otimes_h \mathcal{B} \to \mathcal{B}(\mathcal{K}, \mathcal{H})$ such that $\psi(a \cdot x \cdot b) = \pi(a)\psi(x)\rho(b)$.

Let $T = \psi(1 \otimes 1)$. Then $\|T\| = 1$, and for any $x = \sum_{i=1}^{k} a_i \otimes b_i$ we have that

$$\psi(x) = \sum_{i=1}^{k} \pi(a_i) T \rho(b_i).$$

Now assume that the spaces $\mathcal{H}$ and $\mathcal{K}$ are large enough so that there exist unital $*$-homomorphisms

$$\pi_1\colon \mathcal{A} \to \mathcal{B}(\mathcal{K}), \qquad \rho_1\colon \mathcal{B} \to \mathcal{B}(\mathcal{H}).$$

If this is not the case, then one can always replace (π, ρ, ψ) by a representation such that it is (Exercise 17.2). Let $\tilde{\pi} = \pi \oplus \pi_1$, $\tilde{\rho} = \rho_1 \oplus \rho$, and let

$$U = \begin{pmatrix} \sqrt{1 - TT^*} & T \\ -T^* & \sqrt{I - T^*T} \end{pmatrix},$$

so that U is unitary. We then have that

$$\sum_{i=1}^{k} \tilde{\pi}(a_i) U \tilde{\rho}(b_i) = \begin{pmatrix} * & \psi(x) \\ * & * \end{pmatrix}.$$

Let $\hat{\rho}(b) = U\tilde{\rho}(b)U^*$, so that

$$\|x\|_h = \|\psi(x)\| \le \left\| \sum_{i=1}^{k} \tilde{\pi}(a_i)U\tilde{\rho}(b_i) \right\| = \left\| \sum_{i=1}^{k} \tilde{\pi}(a_i)\hat{\rho}(b_i) \right\|$$

$$= \left\| \tilde{\pi} * \hat{\rho} \left(\sum_{i=1}^{k} a_i * b_i \right) \right\| \le \|x\|_1.$$

The same calculation applies for a matrix (x_{ij}), and so we have that γ is a complete isometry. □

A bilinear map φ is called *completely bounded* if and only if there exists a constant C such that

$$\left\| \left(\sum_{\ell=1}^{k} \varphi(e_{i\ell}, f_{\ell j}) \right) \right\| \le C\|(e_{ij})\| \|(f_{ij})\|$$

for every n, k, m and every $(e_{ij}) \in M_{nk}(E)$, $(f_{ij}) \in M_{km}(F)$. The least such constant C is denoted $\|\varphi\|_{\text{cb}}$.

Of course, as discussed at the beginning of this chapter, every linear map ψ on $E \otimes F$ corresponds to a bilinear map φ on $E \times F$ by simply setting $\psi(e \otimes f) = \varphi(e, f)$. It is easily checked (Exercise 17.3) that ψ is completely bounded on $E \otimes_h F$ if and only if φ is a completely bounded bilinear map, and that the norms coincide. Thus, completely bounded bilinear maps are precisely the family of bilinear maps that correspond to the Haagerup tensor norm in the sense discussed at the beginning of this chapter.

In texts that focus on additional norms on tensor products of operator spaces, the completely bounded bilinear maps corresponding to the Haagerup tensor product are sometimes called the *multiplicatively completely bounded maps* or the *completely bounded maps in the sense of Christensen and Sinclair*, in order to avoid confusion.

The above theorem leads to a representation theorem for these maps.

Corollary 17.7. *Let $\mathcal{A}$, $\mathcal{B}$ be unital C^*-algebras, let $E \subseteq \mathcal{A}$, $F \subseteq \mathcal{B}$ be subspaces, and let $\mathcal{H}$ be a Hilbert space. Then a bilinear map $\varphi\colon E \times F \to B(\mathcal{H})$ is completely bounded if and only if there exists a Hilbert space $\mathcal{K}$, unital $*$-homomorphisms $\pi\colon \mathcal{A} \to B(\mathcal{K})$, $\rho\colon \mathcal{B} \to B(\mathcal{K})$, and linear maps $V, W\colon \mathcal{H} \to \mathcal{K}$ such that $\varphi(e, f) = V^*\pi(e)\rho(f)W$. Moreover, such a representation can be chosen with $\|\varphi\|_{\text{cb}} = \|V\|\|W\|$.*

Proof. The linear map $\psi\colon E \otimes_h F \to B(\mathcal{H})$ with $\psi(e \otimes f) = \varphi(e, f)$ is completely bounded with $\|\psi\|_{\text{cb}} = \|\varphi\|_{\text{cb}}$. Regarding $E \otimes_h F \subseteq \mathcal{A} *_1 \mathcal{B}$ completely isometrically, we may extend ψ to a completely bounded map on $\mathcal{A} *_1 \mathcal{B}$ of the same cb norm.

The result now follows from the generalized Stinespring representation of completely bounded maps and the fact that $\pi * \rho$ is a $*$-homomorphism of $\mathcal{A} *_1 \mathcal{B}$. □

It is often convenient to think of the above result as a "factorization" theorem for bilinear maps that are completely bounded. The following result makes this more explicit.

Corollary 17.8. *Let E and F be operator spaces, and let $\mathcal{H}$ be a Hilbert space. Then a bilinear map $\gamma\colon E \times F \to B(\mathcal{H})$ is completely contractive if and only if there exists a Hilbert space $\mathcal{K}$ and completely contractive maps $\psi\colon F \to B(\mathcal{H}, \mathcal{K})$ and $\varphi\colon E \to B(\mathcal{K}, \mathcal{H})$ such that $\gamma(e, f) = \varphi(e)\psi(f)$ for every e in E and f in F.*

Proof. If γ is completely contractive, then by Corollary 17.7 we may write $\gamma(e, f) = V^*\pi(e)\rho(f)W$ with $\|V\| \le 1$ and $\|W\| \le 1$. Setting $\varphi(e) = V^*\pi(e)$ and $\psi(f) = \rho(f)W$ yields the desired factorization.

Conversely, if $\gamma(e, f) = \varphi(e)\psi(f)$ then

$$\left\|\left(\sum_k \gamma(e_{ik}, f_{kj})\right)\right\| = \left\|\left(\sum_k \varphi(e_{ik})\psi(f_{kj})\right)\right\|$$
$$= \|(\varphi(e_{ij}))\cdot(\psi(f_{ij}))\| \le \|(e_{ij})\|\|(f_{ij})\|$$

for arbitrary matrices (e_{ij}) in $M_{mp}(E)$ and (f_{ij}) in $M_{pn}(F)$. □

We are now prepared to present another proof of the abstract characterization of operator algebras. First we recall the construction of an *inverse limit of Hilbert spaces.*

Assume that we are given Hilbert spaces $\{\mathcal{M}_k\}_{k\ge1}$ and contraction operators $T_k\colon \mathcal{M}_{k+1} \to \mathcal{M}_k$. We call a sequence $(m_1, m_2, \dots)$ with m_k in $\mathcal{M}_k$ *coherent* provided $T_k(m_{k+1}) = m_k$, and *bounded* provided $\sup_k \|m_k\|$ is finite. The set $\mathcal{M}$ of bounded, coherent sequences with $\|(m_1, m_2, \dots)\| = \sup_k \|m_k\|$ is a Banach space, denoted $\mathcal{M} = \lim_k (\mathcal{M}_k, T_k)$. Since every T_k is a contraction, $\|m_k\| = \|T_k(m_{k+1})\| \le \|m_{k+1}\|$ and so $\sup_k \|m_k\| = \lim_k \|m_k\|$. We claim $\mathcal{M}$ is a Hilbert space. To see this one checks that the parallelogram identity holds.

Indeed, for $x = (x_1, \dots)$ and $y = (y_1, \dots)$ in $\mathcal{M}$,

$$\begin{aligned}\|x+y\|^2 + \|x-y\|^2 &= \lim_k (\|x_k + y_k\|^2 + \|x_k - y_k\|^2) \\ &= \lim_k (2\|x_k\|^2 + 2\|y_k\|^2) = 2\|x\|^2 + 2\|y\|^2.\end{aligned}$$

The following theorem is a restatement of Corollary 16.7. The proof borrows from ideas in [137].

Theorem 17.9 (Blecher–Ruan–Sinclair Theorem). *Let $\mathcal{A}$ be a unital algebra with unit e, and assume that $\mathcal{A}$ is an (abstract) operator space. If $\|e\| = 1$ and the bilinear map $m\colon \mathcal{A} \times \mathcal{A} \to \mathcal{A}$ given by $m(a, b) = ab$ is completely contractive, then there exists a Hilbert space $\mathcal{M}$ and a unital completely isometric algebra homomorphism $\rho\colon \mathcal{A} \to B(\mathcal{M})$.*

Proof. Let $\alpha_0\colon \mathcal{A} \to B(\mathcal{H}_0, \mathcal{H}_1)$ be some linear completely isometric map of the operator space $\mathcal{A}$. Applying Corollary 17.8, we have the completely contractive bilinear map $\gamma(a, b) = \alpha_0(ab) = \beta_1(a)\alpha_1(b)$ where $\alpha_1\colon \mathcal{A} \to B(\mathcal{H}_0, \mathcal{H}_2)$ and $\beta_1\colon \mathcal{A} \to B(\mathcal{H}_2, \mathcal{H}_0)$ are completely contractive. Since

$$\|(a_{ij})\| = \|(\alpha_0(a_{ij}e))\| = \|(\beta_1(a_{ij})\alpha_1(e))\| \le \|(\beta_1(a_{ij}))\|$$

we see that β_1, and similarly α_1, are complete isometries.

Inductively, we obtain complete isometries $\alpha_k\colon \mathcal{A} \to B(\mathcal{H}_0, \mathcal{H}_{k+1})$, $\beta_k\colon \mathcal{A} \to B(\mathcal{H}_{k+1}, \mathcal{H}_k)$ such that $\alpha_{k-1}(ab) = \beta_k(a)\alpha_k(b)$.

Now let $\mathcal{M}_k \subseteq \mathcal{H}_k$, $k \ge 1$, denote the closed linear span of vectors of the form $\alpha_{k-1}(a)h$ for a in $\mathcal{A}$ and h in $\mathcal{H}_0$. Since $\beta_k(e)\alpha_k(a)h = \alpha_{k-1}(a)h$, we have that $\beta_k(e)\colon \mathcal{M}_{k+1} \to \mathcal{M}_k$. Thus, we may form the Hilbert space $\mathcal{M} = \lim_k(\mathcal{M}_k, \beta_k(e))$.

If $(m_1, m_2, \dots)$ is a bounded, coherent sequence, then we claim that for any a in $\mathcal{A}$, $(\beta_1(a)m_2, \beta_2(a)m_3, \dots)$ is another bounded coherent sequence. Clearly, this sequence is bounded. To see that it is coherent note that if $m_{k+2} = \sum_i \alpha_{k+1}(b_i)h_i$, then $\beta_{k+1}(a)m_{k+2} = \sum_i \alpha_k(ab_i)h_i \in \mathcal{M}_{k+1}$ and so

$$\begin{aligned}\beta_k(e)[\beta_{k+1}m_{k+2}] &= \sum_i \beta_k(e)\alpha_k(ab_i)h_i = \sum_i \alpha_{k-1}(ab_i)h_i \\ &= \beta_k(a)\sum_i \alpha_k(b_i)h_i = \beta_k(a)\beta_{k+1}(e)m_{k+2} = \beta_k(a)m_{k+1}.\end{aligned}$$

Note that we have also shown that $\beta_k(e)\beta_{k+1}(a) = \beta_k(a)\beta_{k+1}(e)$ on $\mathcal{M}_{k+2}$.

Thus, there is a well-defined map $\rho\colon \mathcal{A} \to B(\mathcal{M})$ given by

$$\rho(b)(m_1, m_2, \dots) = (\beta_1(b)m_2, \beta_2(b)m_3, \dots).$$

Note that $\rho(e) = I_{\mathcal{M}}$ since $\beta_k(e)m_{k+1} = m_k$.

To verify $\rho(ab) = \rho(a)\rho(b)$ it is enough to compare the kth entry of any vector. If $m = (m_1, m_2, \dots)$ is in $\mathcal{M}$ and $m_{k+2} = \sum_i \alpha_{k+1}(c_i)h_i$, then the kth entry of $\rho(a)\rho(b)m$ is

$$\begin{aligned}\beta_k(a)\beta_{k+1}(b)m_{k+2} &= \beta_k(a)\sum_i \beta_{k+1}(b)\alpha_{k+1}(c_i)h_i = \beta_k(a)\sum_i \alpha_k(bc_i)h_i\\ &= \sum_i \alpha_{k-1}(abc_i)h_i = \sum_i \beta_k(ab)\alpha_k(c_i)h_i = \beta_k(ab)m_{k+1},\end{aligned}$$

which is the kth entry of $\rho(ab)m$.

To complete the proof, we must show that ρ is a complete isometry. Note that to compute $\|(\rho(a_{ij}))\|$ we must apply this matrix to a column of vectors from $\mathcal{M}$, but the norm of each such vector is obtained by taking its norm in $\mathcal{M}_k$ and taking a limit on k. Moreover these norms are increasing with k. Thus, if $v_j = (m_{1,j}, m_{2,j}, \dots)$, $j = 1, \dots, n$, are in $\mathcal{M}$ and $(a_{ij}) \in M_{m,n}(\mathcal{A})$, then

$$\begin{aligned}\left\| (\rho(a_{ij})) \begin{pmatrix} v_1 \\ \vdots \\ v_n \end{pmatrix} \right\|_{\mathcal{M}^{(m)}} &= \lim_k \left\| (\beta_k(a_{ij})) \begin{pmatrix} m_{k+1,1} \\ \vdots \\ m_{k+1,n} \end{pmatrix} \right\|_{\mathcal{M}_k^{(m)}}\\ &\le \lim_k \|(\beta_k(a_{ij}))\| \left\| \begin{pmatrix} m_{k+1,1} \\ \vdots \\ m_{k+1,n} \end{pmatrix} \right\|_{\mathcal{M}_k^{(m)}}\\ &= \|(a_{ij})\| \left\| \begin{pmatrix} v_1 \\ \vdots \\ v_n \end{pmatrix} \right\|.\end{aligned}$$

Hence, $\|(\rho(a_{ij}))\| \le \|(a_{ij})\|$, and ρ is completely contractive.

Finally, since α_0 is a complete isometry, given any $\epsilon > 0$ and $(a_{ij}) \in M_{m,n}(\mathcal{A})$, there exists $h_1, \dots, h_n$ in $\mathcal{H}_0$, with $\|h_1\|^2 + \dots + \|h_n\|^2 = 1$, such that

$$\|(a_{ij})\| - \epsilon \le \left\| (\alpha_0(\alpha_{ij})) \begin{pmatrix} h_1 \\ \vdots \\ h_n \end{pmatrix} \right\|.$$

Forming the coherent sequences $v_i = (\alpha_0(e)h_i, \alpha_1(e)h_i, \dots)$, we have that

$\|v_i\| \le \|h_i\|$ and

$$\|(\rho(a_{ij}))\| \ge \left\|(\rho(a_{ij}))\begin{pmatrix} v_1 \\ \vdots \\ v_n \end{pmatrix}\right\| = \lim_k \left\|(\beta_k(a_{ij}))\begin{pmatrix} \alpha_k(e)h_1 \\ \vdots \\ \alpha_k(e)h_n \end{pmatrix}\right\|$$
$$\ge \left\|(\beta_1(a_{ij}))\begin{pmatrix} \alpha_1(e)h_1 \\ \vdots \\ \alpha_1(e)h_n \end{pmatrix}\right\| = \left\|(\alpha_0(a_{ij}))\begin{pmatrix} h_1 \\ \vdots \\ h_n \end{pmatrix}\right\|.$$

Since $\epsilon > 0$ was arbitrary, $\|(\rho(a_i))\| \ge \|(a_{ij})\|$ and we have that ρ is a complete isometry. □

Note that by Exercise 17.3 the requirement that the multiplication map $m\colon \mathcal{A} \times \mathcal{A} \to \mathcal{A}$ be completely contractive is just the requirement that $\|(\sum_{\ell=1}^k a_{i\ell}b_{\ell j})\| \le \|(a_{ij})\|\|(b_{ij})\|$ for all $(a_{ij}) \in M_{m,k}(\mathcal{A})$, all $(b_{ij}) \in M_{k,n}(\mathcal{A})$, and all m, k, n. However, by filling matrices with 0's to make them square it is easily seen that it is sufficient to consider the cases $m = k = n$. In these cases, $\{M_n(\mathcal{A})\}_{n\ge 1}$ are all algebras, and the above condition is just the requirement that the product in these algebras satisfy $\|(a_{ij})(b_{ij})\| \le \|(a_{ij})\|\|(b_{ij})\|$. Thus, Theorem 17.9 is really just a restatement of Corollary 16.7, as claimed earlier.

The Haagerup tensor theory has a multilinear generalization. We discuss the trilinear case in enough detail that the multilinear cases should be transparent. Given three operator spaces E, F, G, introduce a "faux" product by setting

$$(e_{ij}) \odot (f_{ij}) \odot (g_{ij}) = \left(\sum_{k,\ell} e_{ik} \otimes f_{k\ell} \otimes g_{\ell j}\right)$$

whenever $(e_{ij}) \in M_{m,n_1}(E)$, $(f_{ij}) \in M_{n_1,n_2}(F)$, and $(g_{ij}) \in M_{n_2,n}(G)$. Using this operation, we define for each m, n a norm on $M_{m,n}(E \otimes F \otimes G)$ by setting

$$\|(x_{ij})\|_h = \inf\{\|(e_{ij})\|\|(f_{ij})\|\|(g_{ij})\|\colon (x_{ij}) = (e_{ij}) \odot (f_{ij}) \odot (g_{ij})\},$$

where the infimum is over all n_1, n_2 and all such representations of (x_{ij}). We let $E \otimes_h F \otimes_h G$ denote the resulting matrix-normed space.

On the other hand $E \otimes_h F$ and $F \otimes_h G$ are operator spaces, and so we can form $(E \otimes_h F) \otimes_h G$ and $E \otimes_h (F \otimes_h G)$.

Proposition 17.10. *Let E, F, and G be operator spaces. Then the linear isomorphisms $\varphi\colon (E \otimes_h F) \otimes_h G \to E \otimes_h F \otimes_h G$ and $\psi\colon E \otimes_h (F \otimes_h G) \to E \otimes_h F \otimes_h G$ given by $\varphi((e \otimes f) \otimes g) = e \otimes f \otimes g = \psi(e \otimes (f \otimes g))$ are complete isometries.*

Proof. Exercise 17.3. □

The above proposition is generally summarized by the statement that the *Haagerup tensor product is associative*, and given operator spaces $E_1, \ldots, E_N$ we write $E_1 \otimes_h \cdots \otimes_h E_N$ with no further explanation.

In a similar fashion the amalgamated free product is associative. That is, if $\mathcal{A}_1, \mathcal{A}_2$, and $\mathcal{A}_3$ are unital C^*-algebras, then one can define $\mathcal{A}_1 *_1 \mathcal{A}_2 *_1 \mathcal{A}_3$ and

$$(\mathcal{A}_1 *_1 \mathcal{A}_2) *_1 \mathcal{A}_3 = \mathcal{A}_1 *_1 \mathcal{A}_2 *_1 \mathcal{A}_3 = \mathcal{A}_1 *_1 (\mathcal{A}_2 *_1 \mathcal{A}_3)$$

$*$-isomorphically, via the natural identifications. This C^*-algebra is the universal C^*-algebra for triples of unital $*$-homomorphisms $\pi_i\colon \mathcal{A}_i \to B(\mathcal{H})$. Consequently, if $\mathcal{A}_1, \ldots, \mathcal{A}_N$ are unital C^*-algebras, we write $\mathcal{A}_1 *_1 \cdots *_1 \mathcal{A}_N$, unambiguously.

The following results summarize the multilinear theory. We leave the details of their proofs to the interested reader.

Theorem 17.11 (Christensen–Effros–Sinclair–Pisier). *Let $E_1, \ldots, E_N$ be operator spaces and $\mathcal{A}_1, \ldots, \mathcal{A}_N$ unital C^*-algebras, and assume that $E_1 \subseteq \mathcal{A}_1, \ldots, E_N \subseteq \mathcal{A}_N$. Then the linear map $\varphi\colon E_1 \otimes_h \cdots \otimes_h E_N \to \mathcal{A}_1 *_1 \cdots *_1 \mathcal{A}_N$ given by $\varphi(e_1 \otimes \cdots \otimes e_N) = e_1 * \cdots * e_N$ is a complete isometry.*

Theorem 17.12. *Let $E_1, \ldots, E_N$ be operator spaces and $\mathcal{A}_1, \ldots, \mathcal{A}_N$ unital C^*-algebras, and assume that $E_1 \subseteq \mathcal{A}_1, \ldots, E_N \subseteq \mathcal{A}_N$. If $\varphi\colon E_1 \otimes_h \ldots \otimes_h E_N \to B(\mathcal{H})$ is completely bounded, then there exists a Hilbert space $\mathcal{K}$, unital $*$-homomorphisms $\pi_i\colon \mathcal{A}_i \to B(\mathcal{K})$, and operators $S\colon \mathcal{K} \to \mathcal{H}$, $T\colon \mathcal{H} \to \mathcal{K}$ such that $\varphi(e_1 \otimes \cdots \otimes e_N) = S\pi_1(e_1) \cdots \pi_N(e_N)T$ with $\|\varphi\|_{\text{cb}} = \|S\|\|T\|$.*

Just as in the bilinear theory, one defines a multilinear map

$$\gamma\colon E_1 \times \cdots \times E_N \to B(\mathcal{H})$$

to be *completely bounded* if and only if it extends to give a completely bounded linear map $\varphi\colon E_1 \otimes_h \cdots \otimes_h E_N \to B(\mathcal{H})$ with $\varphi(e_1 \otimes \cdots \otimes e_N) = \gamma(e_1, \ldots, e_N)$, and one sets $\|\gamma\|_{\text{cb}} = \|\varphi\|_{\text{cb}}$. As in the bilinear case, the completely bounded norm of a multilinear map can be characterized by using "faux" products of matrices. The above theorem is generally referred to as the *representation* or *factorization theorem for completely bounded multilinear maps.*

We close this chapter with an application of the trilinear theory by proving a result that is equivalent to the generalized Christensen–Effros–Sinclair representation theorem (Corollary 16.10).

Theorem 17.13. *Let $\mathcal{A}$ and $\mathcal{B}$ be (abstract) unital operator algebras, and let X be an operator space that is also an $\mathcal{A}$-$\mathcal{B}$-bimodule with $e_A \cdot x \cdot e_B = x$ for all x in X.*

If the bimodule action $(a, x, b) \to axb$ is a completely contractive trilinear map, then there exist Hilbert spaces $\mathcal{H}$ and $\mathcal{K}$, unital completely isometric homomorphisms $\pi\colon \mathcal{A} \to B(\mathcal{H})$, $\rho\colon \mathcal{B} \to B(\mathcal{K})$, and a complete isometry $\alpha\colon X \to B(\mathcal{H}, \mathcal{K})$ such that $\pi(a)\alpha(x)\rho(b) = \alpha(a \cdot x \cdot b)$.

Proof. Consider the set

$$\mathcal{C} = \left\{ \begin{pmatrix} a & x \\ 0 & b \end{pmatrix} : a \in \mathcal{A},\ b \in \mathcal{B},\ x \in X \right\}.$$

If we define

$$\begin{pmatrix} a_1 & x_1 \\ 0 & b_1 \end{pmatrix} + \begin{pmatrix} a_2 & x_2 \\ 0 & b_2 \end{pmatrix} = \begin{pmatrix} a_1 + a_2 & x_1 + x_2 \\ 0 & b_1 + b_2 \end{pmatrix}$$

and

$$\begin{pmatrix} a_1 & x_1 \\ 0 & b_1 \end{pmatrix} \cdot \begin{pmatrix} a_2 & x_2 \\ 0 & b_2 \end{pmatrix} = \begin{pmatrix} a_1 a_2 & a_1 x_2 + x_1 b_2 \\ 0 & b_1 b_2 \end{pmatrix}$$

then $\mathcal{C}$ clearly becomes an algebra.

Suppose we can introduce an operator space structure on $\mathcal{C}$ such that each of the inclusions of $\mathcal{A}$, $\mathcal{B}$, and X into $\mathcal{C}$ are complete isometries and such that $\mathcal{C}$ satisfies the hypotheses of the BRS theorem. Then it is easily checked that any completely isometric isomorphism $\gamma\colon \mathcal{C} \to B(\mathcal{L})$ will lead to π, ρ, α as above, on setting

$$\mathcal{H} = \gamma\left(\begin{pmatrix} e_A & 0 \\ 0 & 0 \end{pmatrix}\right)\mathcal{L}, \qquad \mathcal{K} = \gamma\left(\begin{pmatrix} 0 & 0 \\ 0 & e_B \end{pmatrix}\right)\mathcal{L},$$

$$\pi(a) = P_{\mathcal{H}}\gamma\left(\begin{pmatrix} a & 0 \\ 0 & 0 \end{pmatrix}\right)\Big|_{\mathcal{H}}, \qquad \rho(b) = P_{\mathcal{K}}\gamma\left(\begin{pmatrix} 0 & 0 \\ 0 & b \end{pmatrix}\right)\Big|_{\mathcal{K}}, \quad \text{and}$$

$$\alpha(x) = P_{\mathcal{H}}\gamma\left(\begin{pmatrix} 0 & x \\ 0 & 0 \end{pmatrix}\right)\Big|_{\mathcal{K}}.$$

Thus, it remains to show that we can endow $\mathcal{C}$ with the desired operator algebra structure. To this end we embed $\mathcal{A} \subseteq \mathcal{C}_1$, $\mathcal{B} \subseteq \mathcal{C}_3$ as unital subalgebras and embed $X \subseteq \mathcal{C}_2$ as a subspace, where $\mathcal{C}_1$, $\mathcal{C}_2$, and $\mathcal{C}_3$ are unital C^*-algebras. This allows us to regard $\mathcal{A}$ and $\mathcal{B}$ as unital subalgebras of $\mathcal{C}_1 *_1 \mathcal{C}_2 *_1 \mathcal{C}_3$ and identify

$\mathcal{A} \otimes_h X \otimes_h \mathcal{B} \cong \text{span}\{a * x * b \colon a \in \mathcal{A}, x \in X, b \in \mathcal{B}\}$, completely isometrically.

The set

$$\tilde{\mathcal{C}} = \left\{ \begin{pmatrix} a & v \\ 0 & b \end{pmatrix} : a \in \mathcal{A},\ v \in \mathcal{A} \otimes_h X \otimes_h B,\ b \in \mathcal{B} \right\} \subseteq M_2(\mathcal{C}_1 *_1 \mathcal{C}_2 *_1 \mathcal{C}_3)$$

is readily seen to form a subalgebra of this C^*-algebra. Note that the map $\varphi \colon \tilde{\mathcal{C}} \to \mathcal{C}$ defined by

$$\varphi\left(\begin{pmatrix} a & \sum a_i * x_i * b_i \\ 0 & b \end{pmatrix} \right) = \begin{pmatrix} a & \sum a_i x_i b_i \\ 0 & b \end{pmatrix}$$

is a homomorphism. We endow $\mathcal{C}$ with the quotient operator space structure induced by this map, that is,

$$\left\| \begin{pmatrix} a & x \\ 0 & b \end{pmatrix} \right\| = \inf \left\{ \left\| \begin{pmatrix} a & v \\ 0 & b \end{pmatrix} \right\| : \varphi\left(\begin{pmatrix} a & v \\ 0 & b \end{pmatrix} \right) = \begin{pmatrix} a & x \\ 0 & b \end{pmatrix} \right\}.$$

It is now easily checked that $\mathcal{C}$ endowed with this norm satisfies the hypotheses of the BRS theorem and that each of the inclusions is a complete isometry. □

Note that when $\mathcal{A}$ and $\mathcal{B}$ are C^*-algebras, the above result reduces to the Christensen–Effros–Sinclair representation theorem.

Notes

A version of the representation theorem for multilinear completely bounded maps (Theorem 17.12) was first obtained in the case that $E_1, \ldots, E_N$ are C^*-algebras by Christensen and Sinclair [58]. Their proof used Wittstock's set-valued extension theorem [242]. Later this result was generalized to operator spaces in [170], and the proof was simplified to the extent that the use of the set-valued extension theorem was replaced by the ordinary extension theorem for linear completely bounded maps. This multilinear representation theorem implied an extension theorem for multilinear completely bounded maps (see Exercise 17.5), which was in turn equivalent to the injectivity of the Haagerup tensor norm (Theorem 17.4). One of the more difficult parts of the proof in [170] was establishing that $E_1 \otimes_h E_2$ had a representation as a concrete operator space, since Ruan's abstract characterization of operator spaces came later. Indeed, in some ways the proof of that fact in [170] anticipates Ruan's theorem.

The direct proof of the injectivity of the Haagerup tensor norm presented in Theorem 17.4 appeared later in [28].

The connection between the Haagerup tensor product and free products was first made by Christensen, Effros, and Sinclair [57], who proved that the Haagerup tensor product embedded into the unamalgamated free product of the C^*-algebras. Pisier [193] was the first to realize that the embedding of $E_1 \otimes_h E_2$ into the amalgamated free product is completely isometric (see [162]), which leads to the cleaner realization of the Haagerup tensor product that appears here (Theorem 17.6). However, both of these earlier proofs used either the Christensen–Sinclair representation theorem for completely bounded multilinear maps or its generalization to operator spaces.

The original proof of the Christensen–Effros–Sinclair representation theorem for operator bimodules over C^*-algebras (Theorem 15.14) also used the Christensen–Sinclair representation theorem for completely bounded bilinear maps.

The proofs given here are new in that we have used injective envelopes to directly obtain the Christensen–Effros–Sinclair representation theorem for operator bimodules and the embedding of $\mathcal{A}_1 \otimes_h \mathcal{A}_2$ into $\mathcal{A}_1 *_1 \mathcal{A}_2$, without appealing to the Christensen–Sinclair representation theorem for completely bounded bilinear maps. Using these results, we are then able to deduce the Christensen–Sinclair representation theorem for bilinear maps and its generalization as a consequence of this embedding.

The Haagerup tensor norm first appeared in some uncirculated work of Haagerup. Effros and Kishimoto [81] recognized its value and adopted the name. This tensor norm was clearly inspired by Haagerup's proof [107] of the Wittstock extension theorem for completely bounded linear maps and his results on Schur product maps. Indeed, Wittstock's extension theorem can be deduced from the injectivity of the Haagerup tensor norm, and this approach recaptures the essence of Haagerup's proof.

We outline how this can be done. One first shows that for an operator space E there is an isometric identification between $\mathrm{CB}(E, M_n)$ and the dual of $R_n \otimes_h E \otimes_h C_n$. Now, let $E \subset F$ be operator spaces, so that by the injectivity of the Haagerup tensor norm, $R_n \otimes_h E \otimes_h C_n \subset R_n \otimes_h F \otimes_h C_n$, isometrically. Thus, by applying the ordinary Hahn–Banach extension theorem for linear functionals to this latter pair of spaces, one may extend a completely bounded map from E into M_n to a completely bounded map from F into M_n. The details of this approach appear in [28]. See also [162, Theorem 3].

Exercise 17.9 is a special case of the results in [85].

Exercises

17.1 Let V be a vector space, and let $\|\cdot\|_{m,n}$ be a family of norms on $M_{m,n}(V)$ that are L^∞-matrix seminorms. Let $W = \{x \in V : |x\| = 0\}$. Prove that for $(x_{i,j}) \in M_{m,n}(V)$, we have $\|(x_{i,j})\|_{m,n} = 0$ if and only if $(x_{i,j}) \in M_{m,n}(W)$. Conclude that setting $\|(x_{i,j} + W)\|_{m,n} = \|(x_{i,j})\|_{m,n}$ gives a well-defined L^∞-matrix norm on V/W.

17.2 Verify the claim in the proof of Theorem 17.6.

17.3 Verify that a bilinear map $\varphi\colon E \times F \to Z$ is completely bounded if and only if the induced linear map $\psi\colon E \otimes_h F \to Z$ is completely bounded and that $\|\psi\|_{\text{cb}}$ is the least constant C satisfying

$$\left\| \left(\sum_{\ell=1}^{k} \varphi(e_{i\ell}, f_{\ell j}) \right) \right\| \leq C \|(e_{ij})\| \|(f_{ij})\|$$

for every n, k, m and every $(e_{ij}) \in M_{nk}(E)$, $(f_{ij}) \in M_{km}(F)$. Deduce that if $E \subset E_1$ and $F \subset F_1$ are operator spaces and $\varphi\colon E \times F \to B(\mathcal{H})$ is a completely bounded bilinear map, then φ can be extended to a completely bounded bilinear map $\psi\colon E_1 \times F_1 \to B(\mathcal{H})$ of the same completely bounded norm.

17.4 Prove the associativity of the Haagerup tensor product as asserted in Proposition 17.10.

17.5 Extend Exercise 17.3 to the multilinear case, and deduce an extension theorem for completely bounded multilinear maps into $B(\mathcal{H})$.

17.6 Let $\mathcal{A}$ and $\mathcal{B}$ be (abstract) unital operator algebras; let X be an operator space that is also an $\mathcal{A}$-$\mathcal{B}$-bimodule with $e_A \cdot x \cdot e_B = x$ for all x in X. Use Exercise 17.5 to show directly that the bimodule action $(a, x, b) \to axb$ is a completely contractive trilinear map if and only if X is an operator $\mathcal{A}$-$\mathcal{B}$-bimodule.

17.7 Show directly that $C_m \otimes_h R_n$ is completely isometrically isomorphic to $M_{m,n}$ and that this isomorphism carries the tensor product of the standard basis vectors $e_i \otimes e_j$ to the standard matrix units $E_{i,j}$.

17.8 Prove directly that $R_m \otimes_h C_n$ is isometrically isomorphic to the dual of $M_{m,n}$.

17.9 (Effros–Ruan) Let E and F be finite-dimensional operator spaces, and let E^* and F^* denote their duals. Prove that $(E \otimes_h F)^*$ is completely isometrically isomorphic to $E^* \otimes_h F^*$. Deduce Exercise 17.8 from this result.

Chapter 18
Universal Operator Algebras and Factorization

In this chapter we examine some consequences of the BRS characterization of operator algebras. The main theme of this chapter is that by using the BRS theorem, we are able to give intrinsic formulas, similar to the description of MAX(X), for certain extrinsically defined universal operator algebra norms. Among the results that we shall obtain as consequences of the theory are Nevanlinna's theorem characterizing the set of analytic functions that map the disk to the disk in terms of positive definite functions, and Agler's recent generalizations of Nevanlinna's theorem. In addition, we obtain formulas for the norm in the full C^*-algebra of a group.

We begin with the construction of the full operator algebra of a semigroup. By a *semigroup* we shall mean a nonempty set S, together with a product $S \times S \to S, (s_1, s_2) \to s_1 \cdot s_2$, and a unit element e satisfying $e \cdot s = s \cdot e = s$ for every s in S. By the *semigroup algebra* $\mathbb{C}[S]$ we mean the vector space of all finite linear combinations $\sum \lambda_i s_i, \lambda_i \in \mathbb{C}, s_i \in S$, equipped with the product

$$\left(\sum \lambda_i s_i\right)\left(\sum \mu_j t_j\right) = \sum (\lambda_i \mu_j)(s_i t_j).$$

Note that e is the unit of this algebra. By a *representation of S on $\mathcal{H}$* we mean any map $\pi\colon S \to B(\mathcal{H})$ satisfying $\pi(e) = I_{\mathcal{H}}$ and $\pi(s_1 \cdot s_2) = \pi(s_1)\pi(s_2)$ for all s_1 and s_2 in S. Clearly, every representation π induces a unital algebra homomorphism of $\mathbb{C}[S]$ into $B(\mathcal{H})$, which we still denote by π, on setting $\pi(\sum \lambda_i s_i) = \sum \lambda_i \pi(s_i)$.

We call π *bounded* if there is a constant C with $\|\pi(s)\| \le C$ for all $s \in S$, and *contractive* when $C \le 1$.

If $S = \mathbb{Z}^+$, the nonnegative integers with addition for the product and $e = 0$ for the unit, then a representation π of $\mathbb{Z}^+$ simply corresponds to an operator $T = \pi(1)$, since $\pi(n) = \pi(1 + \cdots + 1) = \pi(1)^n = T^n$. The representation is bounded when T is power-bounded, and contractive when $\|T\| \leq 1$. The semigroup algebra $\mathbb{C}[\mathbb{Z}^+]$ can be identified with the algebra $\mathcal{P}$ of polynomials in one variable. The map $\gamma\colon \mathbb{C}[\mathbb{Z}^+] \to \mathcal{P}$, $\gamma(\sum \lambda_n \cdot n) = \sum \lambda_n z^n$, is an algebra isomorphism between $\mathbb{C}[\mathbb{Z}^+]$ and the algebra of polynomials in one variable.

Similarly, if $(\mathbb{Z}^+)^N$ denotes the Cartesian product of N copies of $\mathbb{Z}^+$, then a contractive representation $\pi\colon (\mathbb{Z}^+)^N \to B(\mathcal{H})$ is determined by choosing an N-tuple of commuting contractions $\{T_1, \ldots, T_N\}$ and setting $\pi((k_1, \ldots, k_N)) = T_1^{k_1} \cdots T_N^{k_N}$. The semigroup algebra $\mathbb{C}[(\mathbb{Z}^+)^N]$ is isomorphic to $\mathcal{P}_N$, the algebra of polynomials in N variables, via the isomorphism $\gamma\colon \mathbb{C}[(\mathbb{Z}^+)^N] \to \mathcal{P}_N$ defined by $\gamma((k_1, \ldots, k_N)) = z_1^{k_1} \cdots z_N^{k_N}$.

In this chapter we focus primarily on contractive representations. Note that every semigroup possesses at least one contractive representation, given by setting $\pi(s) = I_{\mathcal{H}}$ for all s.

We now wish to endow $M_{m,n}(\mathbb{C}[S])$ with a seminorm by setting

$$\|(a_{ij})\| = \sup \|(\pi(a_{ij}))\|,$$

where the supremum is over all contractive representations $\pi\colon S \to B(\mathcal{H})$ and all Hilbert spaces.

It is easy to see that $J = \{a \in \mathbb{C}[S]\colon \|a\| = 0\}$ is a two-sided ideal and that the above formulas define a matrix norm on $\mathbb{C}[S]/J$ that gives $\mathbb{C}[S]/J$ the structure of an abstract operator algebra. We denote this algebra by $\mathrm{OA}(S)$. We call $\mathrm{OA}(S)$ the *full semigroup operator algebra* and let $\|\cdot\|_{\mathrm{OA}(S)}$ denote the norm on $\mathrm{OA}(S)$ induced by the seminorm on $\mathbb{C}[S]$. When $S = G$ is a group, then every contractive representation is a unitary representation, since $\pi(g)$ and $\pi(g)^{-1} = \pi(g^{-1})$ must both be contractions. So in this case the completion of $\mathrm{OA}(G)$ is the full group C^*-algebra, $C^*(G)$, of the discrete group G.

By the theorems of von Neumann and Ando, we have that $\mathrm{OA}(\mathbb{Z}^+) = \mathcal{P}$ is completely isometrically isomorphic to the space of polynomials equipped with the supremum norm over the unit disk, and $\mathrm{OA}((\mathbb{Z}^+)^2) = \mathcal{P}_2$ is completely isometrically isomorphic to the space of polynomials in two variables equipped with the supremum norm over the bidisk. However, for $N \geq 3$ we have that $\mathrm{OA}((\mathbb{Z}^+)^N) = \mathcal{P}_N$ is completely isometrically isomorphic to the algebra of polynomials in N variables equipped with the norm obtained by taking the supremum over all commuting N-tuples of contractions, and this norm is strictly larger than the supremum norm over $\mathbb{D}^N$. This is the universal operator algebra for N commuting contractions, $(\mathcal{P}_N, \|\cdot\|_u)$, which was first introduced in Chapter 5.

The definition of the operator algebra structure on $\mathrm{OA}(S)$ given above is *extrinsic* in the sense that it requires a supremum over all representations into another object. We now wish to give an *intrinsic* characterization of this norm, which can be achieved internally to OA(S).

Theorem 18.1. *Let S be a semigroup and let $(a_{ij}) \in M_{m,n}(\mathbb{C}[S])$. Then $\|(a_{ij})\|_{\mathrm{OA}(S)} = \inf\{\|C_1\| \cdots \|C_\ell\|\}$, where the infimum is taken over all ways to factor $(a_{ij}) = C_1 D_1 \cdots C_{\ell-1} D_{\ell-1} C_\ell$ as a product of arbitrarily many matrices over $\mathbb{C}[S]$ of arbitrary size, with $C_1, \ldots, C_\ell$ scalar matrices and $D_1, \ldots, D_{\ell-1}$ diagonal matrices with entries from S.*

Proof. Let $\|(a_{ij})\|_f$ denote the infimum on the right hand side of the above equation. Note that the set of such factorizations is nonempty. For example, if $a = \sum_{i=1}^{n} \lambda_i s_i$ is in $\mathbb{C}[S]$, then

$$a = (\lambda_1, \ldots, \lambda_n) \begin{pmatrix} s_1 & 0 & \ldots & 0 \\ 0 & \ddots & \ddots & \vdots \\ \vdots & \ddots & \ddots & 0 \\ 0 & \ldots & 0 & s_n \end{pmatrix} \begin{pmatrix} 1 \\ 1 \\ \vdots \\ 1 \end{pmatrix},$$

and so $\|a\|_f \le \sqrt{n}(|\lambda_1|^2 + \cdots + |\lambda_n|^2)^{1/2}$.

Given a diagonal matrix

$$D = \begin{pmatrix} s_1 & & \\ & \ddots & \\ & & s_n \end{pmatrix}$$

as above and a contractive representation π, set

$$\pi(D) = \begin{pmatrix} \pi(s_1) & & \\ & \ddots & \\ & & \pi(s_n) \end{pmatrix}$$

and note that $\|\pi(D)\| \le 1$.

Given a factorization of (a_{ij}) as in the statement of the theorem and a contractive representation π, observe that

$$(\pi(a_{ij})) = C_1 \pi(D_1) \cdots C_{\ell-1} \pi(D_{\ell-1}) C_\ell$$

and hence

$$\|(\pi(a_{ij}))\| \le \|C_1\| \cdots \|C_\ell\|.$$

Taking the supremum over all such π and the infimum over all such factorizations yields $\|(a_{ij})\|_{\mathrm{OA}(S)} \le \|(a_{ij})\|_f$.

Assume for the moment that we have proven that $(\mathbb{C}[S], \|\cdot\|_f)$ satisfies the BRS axioms to be an abstract operator algebra seminorm. Then there must exist a Hilbert space $\mathcal{H}$ and an algebra homomorphism $\pi\colon \mathbb{C}[S] \to B(\mathcal{H})$ with $\|(a_{ij})\|_f = \|(\pi(a_{ij}))\|$ for any m, n and (a_{ij}) in $M_{m,n}(\mathcal{A})$. Since $\|s\|_f \leq 1$ for $s \in S$, we have that π is a contractive representation of S, and hence $\|(\pi(a_{ij}))\| \leq \|(a_{ij})\|_{\mathrm{OA}(S)}$. Thus, $\|(a_{ij})\|_f \leq \|(a_{ij})\|_{\mathrm{OA}(S)}$, and the result follows.

Thus, it remains to check that the axioms of BRS are satisfied by the algebra $(\mathbb{C}[S], \|\cdot\|_f)$. Note that if L and M are scalar matrices of appropriate sizes and $(a_{ij}) = C_1 D_1 \cdots D_{\ell-1} C_\ell$ is any factorization of the desired type, then

$$L(a_{ij})M = (LC_1)D_1 \cdots D_{\ell-1}(C_\ell M)$$

and hence $\|L(a_{ij})M\|_f \leq \|L\|\|(a_{ij})\|_f\|M\|$.

If $(a_{ij}) = C_1 D_1 \cdots D_{\ell-1} C_\ell$ and $(b_{ij}) = C_1' D_1' \cdots D_{\ell'-1}' C_{\ell'}'$, then

$$(a_{ij})(b_{ij}) = C_1 D_1 \cdot \cdots \cdot D_{\ell-1}(C_\ell C_1')D_1' \cdots D_{\ell'-1}' C_{\ell'}'$$

and hence $\|(a_{ij})(b_{ij})\|_f \leq \|C_1\| \cdots \|C_\ell\|\|C_1'\| \cdots \|C_{\ell'}'\|$. Taking the infimum over all such factorizations yields $\|(a_{ij})(b_{ij})\|_f \leq \|(a_{ij})\|_f\|(b_{ij})\|_f$, and so the multiplication is completely contractive.

Finally, to see that the triangle inequality and L^∞ condition are met, note that a given factorization can always be made to contain more terms by inserting diagonal matrices of e's and scalar identity matrices, and the product of the norms remains the same. Thus, given (a_{ij}) and (b_{ij}) and $\epsilon > 0$, it is enough to assume that we have factorizations

$$(a_{ij}) = C_1 D_1 \cdots D_{\ell-1} C_\ell \quad \text{and} \quad (b_{ij}) = C_1' D_1' \cdots D_{\ell-1}' C_\ell'$$

with $\|C_1\| \cdots \|C_\ell\| < \|(a_{ij})\|_f + \epsilon$, $\|C_1'\| \cdots \|C_\ell'\| < \|(b_{ij})\|_f + \epsilon$. Replacing C_i by $r_i C_i$ with $r_1 \cdots r_\ell = 1$, we may assume that $\|C_1\| = \cdots = \|C_\ell\|$ and $\|C_1'\| = \cdots = \|C_\ell'\|$. Hence,

$$(a_{ij}) \oplus (b_{ij}) = (C_1 \oplus C_1')(D_1 \oplus D_1') \cdots (D_{\ell-1} \oplus D_{\ell-1}')(C_\ell \oplus C_\ell')$$

and so

$$\begin{aligned}\|(a_{ij}) \oplus (b_{ij})\|_f &\leq \|C_1 \oplus C_1'\| \cdots \|C_\ell \oplus C_\ell'\| \\ &\leq (\max\{\|C_1\|, \|C_1'\|\})^\ell \leq \max\{\|(a_{ij})\|_f, \|(b_{ij})\|_f\} + \epsilon,\end{aligned}$$

and the L^∞ condition follows.

Finally, assuming that (a_{ij}) and (b_{ij}) are the same size, to prove the triangle inequality we scale so that $\|C_1\|^2 = \|C_\ell\|^2 \leq \|(a_{ij})\|_f + \epsilon$, $\|C_1'\|^2 = \|C_\ell'\|^2 \leq$

$\|(b_{ij})\|_f + \epsilon$, and $\|C_2\| = \cdots = \|C_{\ell-1}\| = \|C_2'\| = \cdots = \|C_{\ell-1}'\| = 1$. Then $(a_{ij}) + (b_{ij}) = (C_1, C_1')(D_1 \oplus D_1')(C_2 \oplus C_2') \cdots (C_{\ell-1} \oplus C_{\ell-1}')(D_{\ell-1} \oplus D_{\ell-1}')$ $\begin{pmatrix} C_\ell \\ C_\ell' \end{pmatrix}$, and hence

$$\begin{aligned}\|(a_{ij}) + (b_{ij})\|_f &\le \|(C_1, C_1')\| \left\| \begin{pmatrix} C_\ell \\ C_\ell' \end{pmatrix} \right\| = \sqrt{\|C_1\|^2 + \|C_1'\|^2}\sqrt{\|C_\ell\|^2 + \|C_\ell'\|^2} \\ &\le \|(a_{ij})\|_f + \|(b_{ij})\|_f + 2\epsilon.\end{aligned}$$

Thus, the triangle inequality follows, and we have shown that $(\mathbb{C}[S], \|\cdot\|_f)$ satisfies all the axioms of an operator algebra seminorm. □

The above theorem implies some factorization results for polynomials and for analytic functions.

Corollary 18.2. *Let $(p_{ij}(z_1, \ldots, z_N))$ be a matrix of polynomials in N variables. Then $\|(p_{ij}(T_1, \ldots, T_N))\| < 1$ for all commuting N-tuples of contractions if and only if there exists an integer ℓ, scalar matrices C_i with $\|C_i\| < 1, 1 \le i \le \ell$, and diagonal matrices $D_i, 1 \le i \le \ell - 1$, of monomials, such that $(p_{ij}(z_1, \ldots, z_N)) = C_1 D_1 \cdots D_{\ell-1} C_\ell$.*

Proof. The first condition is equivalent to $\|(p_{ij})\|_u < 1$ in the operator algebra $(\mathcal{P}_N, \|\cdot\|_u)$. But $\mathcal{P}_N = \mathrm{OA}((\mathbb{Z}^+)^N)$, and the second condition comes from this identification and Theorem 18.1. □

Corollary 18.3. *Let (p_{ij}) be a matrix of polynomials in N variables, $N \le 2$. Then $\sup\{\|(p_{ij}(z))\| : z \in \mathbb{D}^N\} < 1$ if and only if there exists an integer ℓ, scalar matrices C_i with $\|C_i\| < 1, 1 \le i \le \ell$, and diagonal matrices $D_i, 1 \le i \le \ell - 1$, of monomials such that*

$$(p_{ij}) = C_1 D_1 \cdots D_{\ell-1} C_\ell.$$

Proof. Apply Corollary 18.2 and the fact that the universal norm and supremum norm are equal when $N \le 2$. □

Very little is known about constructing such factorizations. It is natural to wonder if there could exist an integer L such that every polynomial has a factorization as above with $\ell \le L$. The existence of a power-bounded operator that is not polynomially bounded (Theorem 10.9) implies that there cannot exist such an integer L. A proof of this fact is contained in Exercise 18.1. But we do not know a direct argument for the nonexistence of such an L.

The factorization formula in Corollary 18.2 gives an alternative approach to the questions of Chapter 5, namely, to determining if the operator algebras $(\mathcal{P}_N, \|\cdot\|_\infty)$ and $(\mathcal{P}_N, \|\cdot\|_u)$ are boundedly isomorphic. It is also unknown if these two operator algebras are completely boundedly isomorphic.

The above results allow us to easily obtain results of Nevanlinna and Agler characterizing polynomials p with $\|p\|_u < 1$.

To this end we need to introduce the concept of a positive definite function. Let X be a set, and let $\mathcal{H}$ be a Hilbert space. Then a function $K: X \times X \to B(\mathcal{H})$ is called *positive definite* provided that for every finite subset $\{x_1, \ldots, x_n\}$ of X, the operator matrix $(K(x_i, x_j))$ is positive definite. These functions should perhaps be called "completely" positive definite, but we stay with the classical definition.

The following elementary proposition makes it easy to produce examples of positive definite functions.

Proposition 18.4. *Let X be a set, and let $\mathcal{H}_1, \ldots, \mathcal{H}_m$ be Hilbert spaces.*

(i) If $F: X \to B(\mathcal{H}_1, \mathcal{H}_2)$ is any function, then $K: X \times X \to B(\mathcal{H}_1)$ given by $K(x, y) = F(x)F(y)^*$ is positive definite.

(ii) If $K_i: X \times X \to B(\mathcal{H}_i)$ are positive definite, then $K: X \times X \to B(\mathcal{H}_1 \oplus \cdots \oplus \mathcal{H}_n)$ given by $K(x, y) = K_1(x, y) \oplus \cdots \oplus K_n(x, y)$ is positive definite.

(iii) If $K_i: X \times X \to B(\mathcal{H}_1)$, $i = 1, \ldots, n$, are positive definite, then $K_1 + \cdots + K_n$ is positive definite.

(iv) If $F: X \to B(\mathcal{H}_2, \mathcal{H}_1)$ is any function and $K: X \times X \to B(\mathcal{H}_2)$ is positive definite, then $K_1: X \times X \to B(\mathcal{H}_1)$ defined by $K_1(x, y) = F(x)K(x, y)F(y)^*$ is positive definite.

Proof. Exercise 18.2. □

A factorization–representation theorem for positive definite functions implies that for every positive definite function $K: X \times X \to B(\mathcal{H}_1)$ there exists a Hilbert space $\mathcal{H}_2$ and $F: X \to B(\mathcal{H}_1, \mathcal{H}_2)$ such that $K(x, y) = F(x)F(y)^*$, but we shall not need that here.

The scalar-valued version of the following theorem is due to Nevanlinna [153].

Theorem 18.5 (Nevanlinna). *Let $F = (f_{ij})$ be an $m \times n$ matrix of analytic functions on $\mathbb{D}$. Then $\sup\{\|F(z)\|: |z| < 1\} \leq 1$ if and only if the function*

$K: \mathbb{D} \times \mathbb{D} \to M_m$ given by

$$K(z, w) = \frac{I - F(z)F(w)^*}{1 - z\bar{w}}$$

is positive definite.

Proof. If K is positive definite, then $0 \le K(z, z)$, which implies that $\|F(z)\| \le 1$. So assume that $\|F(z)\| \le 1$ for all z. It follows from standard techniques in complex analysis that such a function F is a pointwise limit of a sequence of matrices of polynomials $\{P_n\}$ with $\sup\{\|P_n(z)\|: |z| < 1\} < 1$. Assume that we can prove the theorem for such polynomials; then $K(z, w)$ will be the pointwise limit of the sequence of positive definite functions $K_n(z, w) = [I - P_n(z)P_n(w)^*]/(1 - z\bar{w})$ and hence will be positive definite.

Thus, it is enough to assume that $F = (p_{ij})$ is a matrix of polynomials with $\sup\{\|F(z)\|: |z| < 1\} < 1$. Hence by Corollary 18.3, we may factor $F(z) = C_1 D_1(z) \cdots D_{\ell-1}(z) C_\ell$, with $\|C_\ell\| < 1$.

First assume that $F(z)$ is the constant function C_1. Then $K(z, w) = [I - F(z)F(w)^*]/(1 - z\bar{w}) = (1 - z\bar{w})^{-1} P$, where $P = I - C_1 C_1^* \ge 0$. But $(1 - z\bar{w})^{-1} P = \sum_{k=0}^{\infty} z^k \bar{w}^k P$, and each term in this sum is a positive definite function of the form $G_k(z) G_k(w)^*$ with $G_k(z) = z^k P^{1/2}$.

Now assume that $F(z) = C_1 D_1(z)$. We have that

$$\begin{aligned}
&(1 - z\bar{w})^{-1}(I - F(z)F(w)^*) \\
&\quad = (1 - z\bar{w})^{-1}[I - C_1 C_1^* + C_1(I - D_1(z)D_1(w)^*)C_1^*] \\
&\quad = (1 - z\bar{w})^{-1}(I - C_1 C_1^*) + C_1 K(z, w) C_1^*
\end{aligned}$$

with $K(z, w) = (1 - z\bar{w})^{-1}(I - D_1(z)D_1(w)^*)$.

We have already seen that the first term in this sum is positive definite. Note that $K(z, w)$ is a diagonal matrix of functions of the form $(1 - z^n \bar{w}^n)/(1 - z\bar{w}) = 1 + z\bar{w} + z^2\bar{w}^2 + \cdots + z^{n-1}\bar{w}^{n-1}$ for various integers n. Each term in this sum is positive definite by Proposition 18.4(i). Further applications of Proposition 18.4 yield that $K(z, w)$ is positive definite and also that $C_1 K(z, w) C_1^*$ is positive definite. Thus, $(1 - z\bar{w})^{-1}(I - F(z)F(w)^*)$ is positive definite.

The proof now proceeds by induction on ℓ. By using an identity matrix if necessary we can always write $F(z) = C_1 D_1(z) \cdots C_\ell D_\ell(z)$. Now let $F(z) = C_1 D_1(z) G(z)$ where $G(z)$ is of length $\ell - 1$. Then,

$$\begin{aligned}
(1 - z\bar{w})^{-1}(I - F(z)F(w)^*) &= (1 - z\bar{w})^{-1}(I - C_1 D_1(z) D_1(w)^* C_1^*) \\
&\quad + C_1 D_1(z)[(1 - z\bar{w})^{-1}(I - G(z)G(w)^*)] D_1(w)^* C_1^*.
\end{aligned}$$

The first term in this sum is positive definite by the above calculation, and the factor $(1 - z\bar{w})^{-1}(I - G(z)G(w)^*)$ is positive definite by the inductive hypothesis. Hence, the second term in the sum and, consequently, the sum are positive definite by Proposition 18.4. Hence, the inductive step follows, and our proof is complete. □

Theorem 18.6 (Agler). *Let $F = (f_{ij})$ be an $m \times n$ matrix of analytic functions on $\mathbb{D}^2$. Then $\sup\{\|F(z_1, z_2)\|: |z_1| < 1, |z_2| < 1\} \le 1$ if and only if there exist positive definite functions $P_i: \mathbb{D}^2 \times \mathbb{D}^2 \to M_m$ such that*

$$\begin{aligned} I - F(z_1, z_2)F(w_1, w_2)^* &= (1 - z_1\bar{w}_1)P_1(z_1, z_2, w_1, w_2) \\ &\quad + (1 - z_2\bar{w}_2)P_2(z_1, z_2, w_1, w_2). \end{aligned}$$

Proof. If such a decomposition exists, then $I - F(z_1, z_2)F(z_1, z_2)^* \ge 0$ and so $\|F(z_1, z_2)\| \le 1$ for all (z_1, z_2).

To prove the converse, as in the proof of Theorem 18.5, it is enough to assume that F is a matrix of polynomials and that its supremum is strictly less than 1. Thus, F can be factored as in the statement of Corollary 18.3. The only difference between this factorization and the factorization appearing in the proof of Nevanlinna's theorem is that the diagonal factors are now monomials in z_1 and z_2.

However, note that

$$\begin{aligned} 1 - z_1^{n_1} z_2^{n_2} \bar{w}_2^{n_2} \bar{w}_1^{n_1} &= 1 - z_1^{n_1}\bar{w}_1^{n_1} + z_1^{n_1}\left(1 - z_2^{n_2}\bar{w}_2^{n_2}\right)\bar{w}_1^{n_1} \\ &= (1 - z_1\bar{w}_1)P_1(z_1, w_1) + (1 - z_2\bar{w}_2)\left[z_1^{n_1} P_2(z_2, w_2)\bar{w}_1^{n_1}\right], \end{aligned}$$

where $P_1(z_1, w_1) = 1 + (z_1\bar{w}_1) + \cdots + (z_1\bar{w}_1)^{n_1 - 1}$ and $P_2(z_2, w_2) = 1 + z_2\bar{w}_2 + \cdots + (z_2\bar{w}_2)^{n_2 - 1}$ are positive definite functions. From this it follows that for each diagonal factor we have that

$$\begin{aligned} I &- D(z_1, z_2)D(w_1, w_2)^* \\ &= (1 - z_1\bar{w}_1)Q_1(z_1, w_1) + (1 - z_2\bar{w}_2)Q_2(z_1, z_2, w_1, w_2) \end{aligned}$$

with Q_1 and Q_2 positive definite.

With these observations, the proof can now proceed by induction as in the proof of Theorem 18.5. □

For functions of N variables, $N > 2$, we no longer have equality of the supremum norm over $\mathbb{D}^N$ and the universal norm for N commuting contractions. For this reason Agler introduces the algebra $H_u^\infty(\mathbb{D}^N)$. A function is in the unit ball of $M_{m,n}(H_u^\infty(\mathbb{D}^N))$ if and only if it is the pointwise limit of a sequence of matrices of polynomials whose norms in

$$M_{m,n}((\mathcal{P}_N, \|\cdot\|_u))$$

are less than or equal to 1. Thus $A_u(\mathbb{D}^N) \subseteq H_u^\infty(\mathbb{D}^N)$ completely isometrically. We also have that $H_u^\infty(\mathbb{D}^N) \subseteq H^\infty(\mathbb{D}^N)$, but this containment is not isometric. However, it is unknown whether or not the containment is strict.

In fact it is easily seen (Exercise 18.2) that $H_u^\infty(\mathbb{D}^N) = H^\infty(\mathbb{D}^N)$ if and only if $A_u(\mathbb{D}^N) = A(\mathbb{D}^N)$ if and only if there exists a constant K_N such that $\|p\|_u \leq K_N\|p\|_\infty$ for all polynomials. This is just another formulation of the open problem raised in Chapter 5: whether or not the N-variable von Neumann inequality holds up to a constant.

Theorem 18.6 has been extended to $H_u^\infty(\mathbb{D}^N)$. However, the "easy" implication is no longer as obvious, and so we only record the result.

Theorem 18.7 (Agler). *Let $F = (f_{ij})$ be an $m \times n$ matrix of functions in $H_u^\infty(\mathbb{D}^N)$. Then $\|F\|_u \leq 1$ if and only if there exist positive definite functions $P_i \colon \mathbb{D}^N \times \mathbb{D}^N \to M_m$, $1 \leq i \leq N$, such that*

$$I - F(z)F(w)^* = (1 - z_1\bar{w}_1)P_1(z, w) + \cdots + (1 - z_N\bar{w}_N)P_N(z, w),$$

where $z = (z_1, \ldots, z_N)$ and $w = (w_1, \ldots, w_N)$.

The proof that $\|F\|_u \leq 1$ implies that F has the desired decomposition is similar to the proof of Theorem 18.6 and is left as Exercise 18.3.

There are currently few criteria known for when two elements of a semigroup can be separated by a contractive representation. Theorem 18.1 gives us at least one new criterion.

Theorem 18.8. *Let S be a semigroup and let s_1, s_2 be in S. Then $\pi(s_1) = \pi(s_2)$ for every contractive representation of S if and only if for every $\epsilon > 0$ there exists a factorization $s_1 - s_2 = C_1 D_1 \cdots D_{\ell-1} C_\ell$ such that $\|C_1\| \cdots \|C_\ell\| < \epsilon$ with C_i scalar matrices and D_i diagonal matrices of semigroup elements.*

Proof. We have that $\pi(s_1) = \pi(s_2)$ for every contractive representation if and only if $\|s_1 - s_2\|_{\mathrm{OA}(S)} = 0$. By Theorem 18.1, $\|s_1 - s_2\|_{\mathrm{OA}(S)} = 0$ is equivalent to the latter factorization condition. □

We now turn our attention to the analogous problem of determining when a normed algebra $\mathcal{B}$ with an identity of norm 1 can be represented isometrically as an algebra of operators on a Hilbert space.

To this end let $\mathrm{Rep}(\mathcal{B})$ denote the collection of unital contractive homomorphisms from $\mathcal{B}$ into $B(\mathcal{H})$ for some Hilbert space $\mathcal{H}$. We wish to use $\mathrm{Rep}(\mathcal{B})$

to define a "maximal" operator algebra structure on $\mathcal{B}$, which we shall denote MAXA($\mathcal{B}$), to distinguish it from the maximal operator space structure on the normed space $\mathcal{B}$. For (b_{ij}) in $M_{m,n}(\mathcal{B})$ we set

$$\|(b_{ij})\|_{\mathrm{MAXA}(\mathcal{B})} = \sup\{\|(\pi(b_{ij}))\|\colon \pi \in \mathrm{Rep}(\mathcal{B})\}$$

when Rep($\mathcal{B}$) is nonempty, and set it equal to 0 when Rep($\mathcal{B}$) is empty. Note that if $J = \{b\colon \pi(b) = 0 \text{ for all } \pi \in \mathrm{Rep}(\mathcal{B})\}$, then J is a two-sided ideal, with $J = \mathcal{B}$ when Rep($\mathcal{B}$) is empty, and that $\|\cdot\|_{\mathrm{MAXA}(\mathcal{B})}$ is only a seminorm on $\mathcal{B}$ but a norm on $\mathcal{B}/J$. Moreover, for b in $\mathcal{B}$, $\|b\|_{\mathrm{MAXA}(\mathcal{B})} \le \|b + J\|_{\mathcal{B}/J}$, where $\|\cdot\|_{\mathcal{B}/J}$ denotes the quotient norm.

Clearly, the vector space $\mathcal{B}/J$ equipped with this matrix norm satisfies the axioms to be an operator algebra. We call this the *maximal operator algebra structure on* $\mathcal{B}$ and denote this operator algebra by MAXA($\mathcal{B}$). Note that MAXA($\mathcal{B}$) = MAXA($\mathcal{B}/J$) completely isometrically for J as above.

An example is in order. Let $\mathcal{B}$ be the space of power series with summable coefficients,

$$\mathcal{B} = \left\{\sum_{n=0}^{\infty} a_n z^n\colon \sum_{n=0}^{\infty} |a_n| < +\infty\right\} \quad \text{with} \quad \left\|\sum_{n=0}^{\infty} a_n z^n\right\| = \sum_{n=0}^{\infty} |a_n|.$$

It is well known that $\mathcal{B}$, equipped with the usual Cauchy product of power series, is a Banach algebra. In fact, $\mathcal{B}$ can be identified with $\ell^1(\mathbb{Z}^+)$ equipped with the convolution product. It is easily seen that $\pi \in \mathrm{Rep}(\mathcal{B})$ if and only if $\pi(z)$ is a contraction. Hence, $J = (0)$ in this case, and by von Neumann's inequality,

$$\left\|\sum_{n=0}^{\infty} a_n z^n\right\|_{\mathrm{MAXA}(\mathcal{B})} = \left\|\sum_{n=0}^{\infty} a_n z^n\right\|_{\infty} \le \sum_{n=0}^{\infty} |a_n|,$$

where the middle member is the supremum over the disk. Thus, MAXA($\mathcal{B}$), in this case, can be identified, completely isometrically, with the norm-dense subalgebra of $A(\mathbb{D})$ consisting of power series with summable coefficients.

The above definition of MAXA($\mathcal{B}$) is extrinsic, and we wish to describe it intrinsically.

Theorem 18.9. *Let $\mathcal{B}$ be a normed algebra with an identity of norm 1, and let (b_{ij}) be in $M_{m,n}(\mathcal{B})$. Then*

$$\|(b_{ij})\|_{\mathrm{MAXA}(\mathcal{B})} = \inf\{\|C_1\| \cdots \|C_\ell\|\},$$

where the infimum is taken over all ways to factor

$$(b_{ij}) = C_1 D_1 \cdots C_{\ell-1} D_{\ell-1} C_\ell$$

as a product of arbitrarily many matrices over $\mathcal{B}$ of arbitrary size, with $C_1, \ldots, C_\ell$ scalar matrices and $D_1, \ldots, D_{\ell-1}$ diagonal matrices each of whose diagonal entries is an element of $\mathcal{B}$ of norm less than or equal to one.

Proof. The proof is similar to the proof of Theorem 18.1. If we denote the quantity on the right hand side of the above equation as $\|(b_{ij})\|_f$, then it suffices to prove that these satisfy the BRS conditions. We leave the details of the proof as Exercise 18.4. □

Corollary 18.10. *Let $\mathcal{B}$ be a normed algebra with an identity of norm 1. Then there exists a unital isometric homomorphism from $\mathcal{B}$ into $B(\mathcal{H})$ for some Hilbert space $\mathcal{H}$ if and only if for every b in $\mathcal{B}$, $\|b\|$ is given by the infimum appearing in Theorem 18.9.*

Corollary 18.11. *Let $\mathcal{A}$ be a unital operator algebra and let $K \geq 1$. Then the following are equivalent:*

(i) every unital contractive homomorphism ρ from $\mathcal{A}$ into $B(\mathcal{H})$ is completely bounded with $\|\rho\|_{\mathrm{cb}} \leq K$,

(ii) the identity map $\mathrm{id}\colon \mathcal{A} \to \mathrm{MAXA}(\mathcal{A})$ *is completely bounded with $\|\mathrm{id}\|_{\mathrm{cb}} \leq K$,*

(iii) for every m, n and (a_{ij}) in $M_{m,n}(\mathcal{A})$ with $\|(a_{ij})\| < 1$, for some ℓ there is a factorization $(a_{ij}) = C_1 D_1 \cdots C_{\ell-1} D_{\ell-1} C_\ell$ with C_i scalar, $\|C_1\| \cdots \|C_\ell\| < K$, and D_i diagonal with entries from $\mathcal{A}$ of norm less than or equal to 1.

Corollary 18.11 gives a new tool for answering questions about when contractive homomorphisms of operator algebras are completely contractive. Recall that when X is a compact subset of $\mathbb{C}$, and $\mathcal{R}(X)$ denotes the algebra of rational functions on X, then it is still unknown whether or not every unital contractive homomorphism of $\mathcal{R}(X)$ is completely contractive. Even for X a "nice" two-holed region this is unknown. By Corollary 18.11, this question is equivalent to deciding whether or not every matrix of functions from $\mathcal{R}(X)$ has a factorization as in Corollary 18.11 with $K = 1$.

Agler [1] has proven that when X is an annulus, then every contractive homomorphism is completely contractive. However, a direct proof of the annulus case, using Corollary 18.11, is still unavailable. Such a proof might shed new light on the unknown cases.

In Chapter 11, we proved that if X is a nice n-holed domain, then every contractive homomorphism ρ is completely bounded with $\|\rho\|_{\mathrm{cb}} \leq 2n + 1$, so applying Corollary 18.11 yields certain factorizations on these domains.

In Chapter 14, we saw that for G any absorbing, absolutely convex set in $\mathbb{C}^n, n \geq 5$, there exist unital contractive homomorphisms of $A(G)$ that are not completely contractive. Thus, there exist matrices of functions from $A(G)$ whose factorization norm is larger than their supremum norm. That is, $A(G) \neq \mathrm{MAXA}(A(G))$, completely isometrically.

Parrott's example shows that $\mathrm{MAXA}((\mathcal{P}_N, \|\cdot\|_\infty)) \neq (\mathcal{P}_N, \|\cdot\|_\infty)$, completely isometrically, for $N \geq 3$. In addition to it being unknown whether $(\mathcal{P}_N, \|\cdot\|_\infty)$ and $(\mathcal{P}_N, \|\cdot\|_u)$ are completely boundedly isomorphic, it is also unknown if $\mathrm{MAXA}((\mathcal{P}_N, \|\cdot\|_\infty))$ is completely boundedly isomorphic to either $(\mathcal{P}_N, \|\cdot\|_\infty)$ or $(\mathcal{P}_N, \|\cdot\|_u)$.

We close this chapter with a factorization theorem that applies to many situations. Suppose that we are given a unital algebra $\mathcal{A}$ together with unital operator algebras $\mathcal{A}_i$ and unital homomorphisms $\rho_i \colon \mathcal{A}_i \to \mathcal{A}$ such that the union of the images, $\bigcup_i \rho_i(\mathcal{A}_i)$, generates $\mathcal{A}$ as an algebra. For two examples of this situation, consider the case where $\mathcal{A} = \mathcal{A}_1 \otimes \mathcal{A}_2$ is the tensor product, with $\rho_1(a_1) = a_1 \otimes 1$ and $\rho_2(a_2) = 1 \otimes a_2$, and the case where $\mathcal{A} = \mathcal{A}_1 *_{\mathbb{C}} \mathcal{A}_2$ is the free product amalgamated over $\mathbb{C}$, with $\rho_1(a_1) = a_1 * 1$ and $\rho_2(a_2) = 1 * a_2$.

We wish to endow $\mathcal{A}$ with an operator algebra norm that respects the algebras $\mathcal{A}_i$. To this end we call a unital homomorphism $\pi \colon \mathcal{A} \to B(\mathcal{H})$ *admissible* provided that for every i, $\pi \circ \rho_i \colon \mathcal{A}_i \to B(\mathcal{H})$ is completely contractive.

We endow $\mathcal{A}$ with a matricial seminorm by defining the norm of a matrix of elements from $\mathcal{A}$ to be the supremum over all admissible representations into all operator algebras. The set of elements of norm 0 is a two-sided ideal $\mathcal{I}$ in $\mathcal{A}$, and the resulting matricial norm on the quotient $\mathcal{A}/\mathcal{I}$ makes it into an operator algebra. We shall call this operator algebra seminorm the *maximal operator algebra seminorm induced* by the given family of inclusions.

In the case where $\mathcal{A} = \mathcal{A}_1 \otimes \mathcal{A}_2$ is the tensor product of two operator algebras, the maximal operator algebra seminorm induced by the inclusion of $\mathcal{A}_1$ and $\mathcal{A}_2$ into $\mathcal{A}$ is the maximal tensor norm of [169] discussed in Chapter 12. When both algebras are C^*-algebras, this is, of course, the maximal C^*-tensor norm. In the case where $\mathcal{A} = \mathcal{A}_1 *_{\mathbb{C}} \mathcal{A}_2$, the maximal operator algebra seminorm induced by the inclusion of $\mathcal{A}_1$ and $\mathcal{A}_2$ into $\mathcal{A}$ is the full free product amalgamated over $\mathbb{C}$ introduced in Chapter 17.

The above definition of this norm is extrinsic, since it needs *all* representations into operator algebras. Using the ideas of this chapter, we would like to characterize this norm intrinsically.

Theorem 18.12. *Let $\mathcal{A}_j$, $j = 1, \ldots, m$, be unital operator algebras, let $\mathcal{A}$ be a unital algebra, and let $\pi_j \colon \mathcal{A}_j \to \mathcal{A}$ be unital homomorphisms such that the union of their images generates $\mathcal{A}$ algebraically. If A is in $M_n(\mathcal{A})$ for some n*

and if $\|A\|$ denotes the maximal operator algebra seminorm induced by these algebras, then

$$\|A\| = \inf\{\|A_1\| \cdots \|A_m\|: A = \pi_{i_1}(A_1) \cdots \pi_{i_m}(A_m)\},$$

where the infimum is taken over all ways to represent A as a product of images of matrices of elements from each $\mathcal{A}_j$ in any order and involving matrices of arbitrary sizes.

Proof. To prove this theorem one only needs to note that the inf on the right hand side is larger than the maximal operator algebra seminorm induced by these algebras, and then check that the right hand side defines a seminorm that satisfies the BRS axioms of an abstract operator algebra. The inclusion of each $\mathcal{A}_j$ into this latter operator algebra will yield an admissible representation of $\mathcal{A}$. We leave the details of this proof as Exercise 18.6. □

Notes

Theorems 18.1 and 18.8 and Corollaries 18.2 and 18.3 were proven in [29].

Theorems 18.6 and 18.7 were first proven in [2].

Theorem 18.12 first appeared in [166], where it was applied to some questions in interpolation theory. For further connections between interpolation and the abstract theory of operator algebras see [167].

Exercises

18.1 Let T be a power-bounded operator, with $\|T^n\| \leq C$ for all n. Show that if $p(z)$ has a factorization of the type given by Corollary 18.3, then $\|p(T)\| \leq C^\ell$. Deduce that if every polynomial had such a factorization with $\ell \leq L$, then T would be polynomially bounded with $\|p(T)\| \leq C^L \|p\|_\infty$ for all p. Apply Theorem 10.9 to deduce the nonexistence of such an integer L.

18.2 Prove Proposition 18.4.

18.3 Let $F \in M_{m,n}(H_u^\infty(\mathbb{D}^N))$. Prove that if $\|F\|_u \leq 1$ then $I - F(z)F(w)^*$ can be decomposed as claimed in Theorem 18.7.

18.4 Complete the proof of Theorem 18.9.

18.5 Let $\mathcal{A}$ denote the algebra of Laurent polynomials. Fix $R > 1$, and endow $\mathcal{A}$ with the extrinsic operator algebra norm obtained by taking the supremum over all representations π of the Laurent polynomials determined by $\pi(z) = T$, where T is an invertible operator satisfying $\|T\| \leq R$ and $\|T^{-1}\| \leq R$. Use factorization to give an intrinsic formula for this norm.

18.6 Supply the details of the proof of Theorem 18.12.

Chapter 19

Similarity and Factorization

In the last chapter we saw how the abstract characterization of operator algebras led to a number of factorization formulas for certain universal operator algebras. However, this theory was an isometric theory. In this chapter we focus on the isomorphic theory of operator algebras and applications to similarity questions.

We present Pisier's remarkable work on similarity degree and factorization degree, and Blecher's characterization of operator algebras up to cb isomorphism.

Pisier's work shows that for an operator algebra $\mathcal{B}$, every bounded homomorphism is completely bounded if and only if the type of factorization occurring in the study of MAXA($\mathcal{B}$) can be carried out with uniform control on the number of factors needed. The least such integer is the factorization degree of the algebra.

Pisier's work has a number of deep implications in the study of bounded representations of groups and in the study of Kadison's similarity conjecture. We focus primarily on Kadison's conjecture, that every bounded homomorphism of a C^*-algebra into $B(\mathcal{H})$ is similar to a $*$-homomorphism. Thus, we will show that Kadison's conjecture is equivalent to the existence of an integer d such that every C^*-algebra has factorization degree at most d.

A pivotal role in Pisier's work is played by the universal operator algebra of an operator space. So we begin this chapter by examining the construction of the universal algebra of a vector space in some detail.

Given a vector space V, let $V^{\otimes n}$ denote the tensor product of V with itself n times, and set $V^{\otimes 0} = \mathbb{C}$. Given an elementary tensor $x = v_1 \otimes \cdots \otimes v_n \in V^{\otimes n}, n \neq 0$, and $y = w_1 \otimes \cdots \otimes w_m \in V^{\otimes m}, m \neq 0$, we set $x \odot y = v_1 \otimes \cdots \otimes v_n \otimes w_1 \otimes \cdots \otimes w_m \in V^{\otimes(n+m)}$. If $n = 0$ and $x = \lambda$, we set $\lambda \odot y = (\lambda w_1) \otimes \cdots \otimes w_m$, and if $m = 0$ and $y = \mu$, we set $x \odot \mu = v_1 \otimes \cdots \otimes (\mu v_n)$. This operation extends to give a bilinear pairing from $V^{\otimes n} \times V^{\otimes m}$ into $V^{\otimes(n+m)}$, which we still denote by $\odot$.

We let $\mathcal{F}(V) = \sum_{n=0}^{\infty} \oplus V^{\otimes n}$ denote the set of finite direct sums of the tensor powers of V. For $x = \sum \oplus x_n$ in $\mathcal{F}(V)$, we call x_n the *homogeneous term of degree n* of x.

For $x = \sum \oplus x_n$ and $y = \sum \oplus y_n$ in $\mathcal{F}(V)$ we extend the definition of our bilinear pairing by setting $x \odot y = \sum \oplus z_n$, where $z_n = \sum_{k=0}^{n} x_k \odot y_{n-k}$. This formula now extends $\odot$ to a bilinear map, still denoted $\odot$, from $\mathcal{F}(V) \times \mathcal{F}(V)$ to $\mathcal{F}(V)$.

It is easily seen that $\mathcal{F}(V)$ equipped with its usual vector space structure and the product $\odot$ becomes a unital algebra. The algebra $\mathcal{F}(V)$ is called the *Fock algebra of* V. It has the following important universal property. If $\mathcal{A}$ is any unital algebra and $\varphi\colon V \to \mathcal{A}$ is any linear map, then there exists a unique extension of φ to a unital homomorphism $\pi_\varphi\colon \mathcal{F}(V) \to \mathcal{A}$. Clearly, for $x = v_1 \otimes \cdots \otimes v_n$ an elementary tensor, $\pi_\varphi(x) = \varphi(v_1)\cdots\varphi(v_n)$. Note that if $(\lambda\varphi)(v) = \lambda\varphi(v)$, then $\pi_{(\lambda\varphi)}(\sum \oplus x_n) = \sum \lambda^n \pi_\varphi(x_n)$.

Now if V is an operator space, we wish to endow $\mathcal{F}(V)$ with a family of operator algebra structures.

For each $c > 0$ and for $(x_{ij}) \in M_{m,n}(\mathcal{F}(V))$ we set

$$\|(x_{ij})\|_c = \sup\{\|(\pi_\varphi(x_{ij}))\|\colon \|\varphi\|_{\mathrm{cb}} \le c\},$$

where the supremum is taken over all Hilbert spaces $\mathcal{H}$ and all $\varphi\colon V \to B(\mathcal{H})$ satisfying $\|\varphi\|_{\mathrm{cb}} \le c$. We let $\mathrm{OA}_c(V)$ denote $\mathcal{F}(V)$ equipped with the above operator algebra norm $\|\cdot\|_c$.

Alternatively, we could let V_c denote the operator space V but equipped with the norm $\|(v_{ij})\|_c = c\|(v_{ij})\|$, so that $\varphi\colon V_c \to B(\mathcal{H})$ satisfies $\|\varphi\|_{\mathrm{cb}} \le 1$ if and only if $\varphi\colon V \to B(\mathcal{H})$ satisfies $\|\varphi\|_{\mathrm{cb}} \le c$. Then it is easily seen that the identity map $i\colon V \to V_c$ induces a completely isometric isomorphism $\pi_i\colon \mathrm{OA}_c(V) \to \mathrm{OA}_1(V_c)$.

The above definition of $\mathrm{OA}_c(V)$ is extrinsic, and, following the ideas from earlier chapters, we wish to give an intrinsic characterization of $\mathrm{OA}_c(V)$ via a factorization formula. By using V_c, as above, it will be enough to study $\mathrm{OA}_1(V)$.

To this end, we wish to consider all ways to factor elements of $M_{m,n}(\mathcal{F}(V))$ as products, using the operation $\odot$, of rectangular block-diagonal matrices where each block is either a scalar matrix or a matrix over V. For example,

$$\begin{aligned}&\lambda \oplus v_1 \oplus (v_2 \otimes v_3 + v_4 \otimes v_5)\\ &= (1,1) \odot \left(\begin{array}{c|ccc} \lambda & 0 & 0 & 0 \\ \hline 0 & v_1 & v_2 & v_4 \end{array}\right) \odot \left(\begin{array}{cc|c} 1 & 0 & 0 \\ 0 & 1 & 0 \\ \hline 0 & 0 & v_3 \\ 0 & 0 & v_5 \end{array}\right) \odot \begin{pmatrix} 1 \\ 1 \\ 1 \end{pmatrix}.\end{aligned}$$

In keeping with the L^∞ condition, we wish to define the norm of such a rectangular block-diagonal matrix as the maximum of the norm of the matrix over V and the norm of the scalar matrix.

Theorem 19.1. *Let $(x_{ij}) \in M_{m,n}(\mathrm{OA}_1(V))$. Then*

$$\|(x_{ij})\|_1 = \inf\{\|B_1\| \cdots \|B_\ell\|\},$$

where the infimum is over all ℓ and all ways to factor $(x_{ij}) = B_1 \odot \cdots \odot B_\ell$ as a product in $\mathcal{F}(V)$ of rectangular block-diagonal matrices where each block is either scalar or a matrix over V.

Proof. Let $\|(x_{ij})\|_f$ denote the infimum given on the right hand side. Given such a factorization and a completely contractive map $\varphi\colon V \to B(\mathcal{H})$, it is easy to see that $(\pi_\varphi(x_{ij})) = B_1^\varphi \cdots B_\ell^\varphi$, where B_i^φ is obtained by replacing the scalar part of B_i by scalar multiples of $I_\mathcal{H}$ and each element v of V by $\varphi(v)$. Since $\|B_i^\varphi\| \le \|B_i\|$, it follows that $\|(x_{ij})\|_1 \le \|(x_{ij})\|_f$.

To complete the proof it suffices to prove that $\mathcal{F}(V)$ equipped with the family of matrix norms $\|\cdot\|_f$ satisfies the BRS axioms. For then we can represent $(\mathcal{F}(V), \|\cdot\|_f)$ completely isometrically as an algebra of operators on $B(\mathcal{H})$ via $\rho\colon \mathcal{F}(V) \to B(\mathcal{H})$, and let $\varphi\colon V \to B(\mathcal{H})$ be defined by $\varphi = \rho|_V$. Then $\pi_\varphi = \rho$ and $\|(x_{ij})\|_f = \|(\pi_\varphi(x_{ij}))\| \le \|(x_{ij})\|_1$.

The proof that $(\mathcal{F}(V), \|\cdot\|_f)$ satisfies the BRS axioms is similar to the proofs in Chapter 18 and is left as an exercise (Exercise 19.1). □

For many applications of $\mathrm{OA}_c(V)$ it is useful to have further information about the norm structure. In particular, for $x = \sum \oplus x_n$ it is important to know the relationship between $\|x_n\|_c$ and $\|x\|_c$.

Proposition 19.2. *For $n \ge 0$, define $\Gamma_n\colon \mathrm{OA}_c(V) \to \mathrm{OA}_c(V)$ via*

$$\Gamma_n\left(\sum \oplus x_j\right) = x_n.$$

Then Γ_n is a complete contraction.

Proof. Given $\varphi\colon V \to B(\mathcal{H})$, $\|\varphi\|_{\mathrm{cb}} \le c$, let $\varphi_\theta(v) = e^{i\theta}\varphi(v)$. Then for $x = \sum \oplus x_j$ we have $\pi_\varphi(x_n) = \frac{1}{2\pi i}\int_0^{2\pi} e^{-in\theta}\pi_{\varphi_\theta}(x)\,d\theta$. Hence, $\|\pi_\varphi(x_n)\| \le \frac{1}{2\pi}\int_0^{2\pi} \|\pi_{\varphi_\theta}(x)\| d\theta \le \|x\|_c$, and it follows that Γ_n is a contraction by taking the supremum over all such φ. The proof that Γ_n is a complete contraction is similar. □

Proposition 19.3. *Let $n \geq 1$. If $V^{\otimes n}$ is endowed with the Haagerup tensor norm, then the inclusion of $V^{\otimes n}$ into $\mathrm{OA}_1(V)$ is a complete isometry.*

Proof. For $x \in V^{\otimes n}$ the Haagerup tensor norm of x is defined as the infimum of x over certain ways to represent x using the "faux" product $\odot$. Comparing these factorizations of x with Theorem 19.1, we see that the factorizations used to define the Haagerup tensor product are precisely the products of the form used in Theorem 19.1 when no scalar matrices are allowed. Thus, $\|x\|_1 \leq \|x\|_h$.

To prove the other inequality, pick unital operator algebras $\mathcal{A}_1, \ldots, \mathcal{A}_n$ and complete isometries $\gamma_i\colon V \to \mathcal{A}_i$. Let $\mathcal{A} = \mathcal{A}_i *_{\mathbb{C}} \cdots *_{\mathbb{C}} \mathcal{A}_n$; then by Theorem 17.6, the map $\gamma\colon V^{\otimes n} \to \mathcal{A}$ defined by $\gamma(v_1 \otimes \cdots \otimes v_n) = \gamma_1(v_1) \cdots \gamma_n(v_n)$ is a complete isometry from $V^{\otimes n}$ endowed with the Haagerup norm into $\mathcal{A}$.

Define $\varphi\colon V \to M_{n+1}(\mathcal{A})$ via

$$\varphi(v) = \begin{pmatrix} 0 & \gamma_1(v) & 0 & \cdots & 0 \\ \vdots & \ddots & \ddots & \ddots & \vdots \\ \vdots & & \ddots & \ddots & 0 \\ 0 & \cdots & \cdots & 0 & \gamma_n(v) \\ 0 & \cdots & \cdots & 0 & 0 \end{pmatrix}.$$

Then

$$\pi_\varphi(v_1 \otimes \cdots \otimes v_n) = \varphi(v_1) \cdots \varphi(v_n) = \begin{pmatrix} 0 & \cdots & 0 & \gamma_1(v_1) \cdots \gamma_n(v_n) \\ 0 & \cdots & 0 & 0 \\ \vdots & & \vdots & \vdots \\ 0 & \cdots & 0 & 0 \end{pmatrix},$$

and hence $\|x\|_h = \|\gamma(x)\|_h = \|\pi_\varphi(x)\| \leq \|x\|_1$ for x in $V^{\otimes n}$.

Thus, $\|x\|_h = \|x\|_1$ for $x \in V^{\otimes n}$. The proof for matrices over $V^{\otimes n}$ is identical. □

To obtain the analogous result for $\mathrm{OA}_c(V)$ one simply needs to use the identification of $\mathrm{OA}_c(V)$ with $\mathrm{OA}_1(V_c)$. Since the map $\varphi(x) = c^{-1}x$ is a complete isometry from V to V_c, we see that the inclusion of $V^{\otimes n}$ into $\mathrm{OA}_c(V)$ satisfies $\|x\|_c = c^n \|x\|_h$ for x in $V^{\otimes n}$.

The first application of the above result is to a generalization of the BRS theorem. We wish to extend the theorem to obtain a characterization of operator algebras up to complete isomorphism, characterizing algebras either lacking a unit or having a unit of norm larger than one or having a product that is only completely bounded.

We claim that all of these cases can be subsumed under a single case, namely, the case of an algebra with a unit e such that $\|e\| = 1$ but where we only have that $\|(a_{ij})(b_{ij})\| \leq M\|(a_{ij})\|\|(b_{ij})\|$, i.e., that the multiplication is completely bounded. For example, if $\|e\| = c^{-1}$ and $\|(a_{ij})(b_{ij})\| \leq M\|(a_{ij})\|\|(b_{ij})\|$, then setting $\|a\|_c = c\|a\|$ yields a completely isomorphic norm such that $\|e\|_c = 1$ but $\|(a_{ij})(b_{ij})\|_c \leq c^{-1}M\|(a_{ij})\|_c\|(b_{ij})\|_c$. For further details on the reduction to this case, see Exercises 19.2 and 19.3. Exercise 19.4 illustrates some of the further subtleties that arise when dealing with quotients of nonunital algebras.

Theorem 19.4. *Let $\mathcal{A}$ be an L^∞-matrix-normed algebra with unit e, $\|e\| = 1$, such that for all n and all (a_{ij}), (b_{ij}) in $M_n(\mathcal{A})$,*

$$\|(a_{ij})(b_{ij})\| \leq M\|(a_{ij})\|\|(b_{ij})\|.$$

Then there exists a unital operator algebra $\mathcal{B}$ and a unital completely bounded algebra isomorphism $\pi\colon \mathcal{B} \to \mathcal{A}$ such that $\|\pi\|_{\mathrm{cb}}\|\pi^{-1}\|_{\mathrm{cb}} \leq (\sqrt{M} + 1)^2$.

Proof. Let $V = \mathcal{A}$, fix $0 < c < 1$ to be determined later, and let $\varphi\colon V \to \mathcal{A}$ be given by $\varphi(v) = cv$. By the universal property of $\mathcal{F}(V)$ there exists a homomorphism $\pi_\varphi\colon \mathcal{F}(V) \to \mathcal{A}$ that is onto. We wish to prove that π_φ is completely bounded as a map from $\mathrm{OA}_1(V)$ to $\mathcal{A}$.

To this end, note that if $(x_{ij}) \in M_{p,q}(V^{\otimes n})$ with $\|(x_{ij})\|_1 = \|(x_{ij})\|_h < 1$, then (x_{ij}) factors as $(x_{ij}) = (v^1_{ij}) \odot (v^2_{ij}) \odot \cdots \odot (v^n_{ij})$ with $\|(v^\ell_{ij})\| < 1$ in $M_{p_\ell,q_\ell}(V)$ for some integers p_ℓ, q_ℓ. Consequently,

$$(\pi_\varphi(x_{ij})) = \big(\varphi(v^1_{ij})\big)\big(\varphi(v^2_{ij})\big)\cdots\big(\varphi(v^n_{ij})\big) = c^n\big(v^1_{ij}\big)\big(v^2_{ij}\big)\cdots\big(v^n_{ij}\big),$$

where the latter products are in $\mathcal{A}$. Hence,

$$\|(\pi_\varphi(x_{ij}))\| \leq c^nM^{n-1}\big\|\big(v^1_{ij}\big)\big\|\big\|\big(v^2_{ij}\big)\big\|\cdots\big\|\big(v^n_{ij}\big)\big\| < c^nM^{n-1},$$

and so $\|\pi_\varphi\|_{\mathrm{cb}} \leq c^nM^n$ when restricted to the subspace of homogeneous terms of degree $n \geq 1$. Note for $n = 0$, $\|\pi_\varphi\|_{\mathrm{cb}} = 1$. Now, given any (x_{ij}) in $M_{p,q}(\mathrm{OA}_1(V))$, we may decompose it as a sum of homogeneous terms of degree n, $(x_{ij}) = \sum_n (x^{(n)}_{ij})$, and by Proposition 19.2, $\|(x^{(n)}_{ij})\|_1 \leq \|(x_{ij})\|_1$. Thus,

$$\|(\pi_\varphi(x_{ij}))\| \leq \sum_{n=0}^{\infty}\big\|\big(\pi_\varphi\big(x^{(n)}_{ij}\big)\big)\big\| = \left(1 + \sum_{n=1}^{\infty} c^nM^{n-1}\right)\|(x_{ij})\|,$$

and so $\|\pi_\varphi\|_{\mathrm{cb}} \leq (1 - cM + c)/(1 - cM)$ as a map from $\mathrm{OA}_1(V)$ to $\mathcal{A}$. If we let $\mathcal{B} = \mathrm{OA}_1(V)/\ker(\pi_\varphi)$, then $\mathcal{B}$ is a unital operator algebra and the induced map $\pi = \dot{\pi}_\varphi$ is an algebraic isomorphism with $\|\pi\|_{\mathrm{cb}} = \|\pi_\varphi\|_{\mathrm{cb}}$.

Finally, given any $a \in \mathcal{A}$, we have $\pi_\varphi(c^{-1}a) = a$ and so $\|\pi^{-1}\|_{\mathrm{cb}} \leq c^{-1}$. Choosing $c = \frac{1}{M+\sqrt{M}}$ yields the desired result. □

It is not clear what the best constant for $\|\pi\|_{\text{cb}}\|\pi^{-1}\|_{\text{cb}}$ is in the above theorem. One can show that necessarily $\|\pi\|_{\text{cb}}\|\pi^{-1}\|_{\text{cb}} \geq \sqrt{M}$, where M is the least constant bounding the norm of a product, but we have been unable to achieve this bound.

Theorem 19.5 (Blecher). *Let $\mathcal{A}$ be an algebra that is also an L^{∞}-matrix-normed space. If the product on $\mathcal{A}$ is completely bounded, then for every $\epsilon > 0$ there exists a Hilbert space $\mathcal{H}$ and a one-to-one completely bounded homomorphism $\pi\colon \mathcal{A} \to B(\mathcal{H})$ such that $\pi^{-1}\colon \pi(\mathcal{A}) \to \mathcal{A}$ is also completely bounded with $\|\pi\|_{\text{cb}}\|\pi^{-1}\|_{\text{cb}} < 1 + \epsilon$.*

Proof. Let $V = \mathcal{A}$, fix $0 < c < 1$ to be determined later, and let $\varphi\colon V \to \mathcal{A}$ be given by $\varphi(v) = cv$ as in the proof of Theorem 19.4. Let $\mathcal{F}^0(V) \subset \mathcal{F}(V)$ denote the ideal obtained by deleting the $\mathbb{C}$-summand.

The canonical homomorphism π_{φ}, restricted to $\mathcal{F}^0(V)$, which we denote by π_{φ}^0, is still onto. We let $\text{OA}_1^0(V) \subseteq \text{OA}_1(V)$ denote $\mathcal{F}^0(V)$ equipped with the L^{∞}-matrix norm that it inherits as a subspace.

Computing $\|\pi_{\varphi}^0\|_{\text{cb}}$ as in the proof of Theorem 19.4, we have that $\|\pi_{\varphi}^0\|_{\text{cb}} \leq \sum_{n=1}^{\infty} c^n M^{n-1} = \frac{c}{1-cM}$, where M is the cb norm of the product map. If we let $J = \text{OA}_1^0(V)/\ker(\pi_{\varphi}^0)$, then the induced map $\pi = \dot{\pi}_{\varphi}^0$ is an algebraic isomorphism between $\mathcal{A}$ and J with $\|\pi\|_{\text{cb}} \leq \frac{c}{1-cM}$, while $\|\pi^{-1}\|_{\text{cb}} \leq c^{-1}$ as before.

Since $\ker(\pi_{\varphi}^0)$ is an ideal in $\text{OA}_1(V)$, we have that J is an ideal in the unital operator algebra $\mathcal{B} = \text{OA}_1(V)/\ker(\pi_{\varphi}^0)$. The result now follows on choosing c so that $\frac{1}{1-cM} < 1 + \epsilon$ and representing $\mathcal{B}$ completely isometrically as an algebra of operators on a Hilbert space. □

In situations where both Theorem 19.4 and Theorem 19.5 apply, the principal difference is that Theorem 19.5 allows for representations of $\mathcal{A}$ such that $\pi(\mathcal{A})$ is a subalgebra of $B(\mathcal{H})$ whose unit is an idempotent of norm greater than one. These more general representations allow for the lack of dependence on M in $\|\pi\|_{\text{cb}}\|\pi^{-1}\|_{\text{cb}}$.

This lack of dependence on M implies that $\pi(\mathcal{A})$ must have a product whose cb norm is strictly less than 1. To illustrate how this can occur, consider $\mathcal{B} \subseteq B(\mathcal{H})$ any unital subalgebra. Then for any $t > 0$,

$$\mathcal{B}_t = \left\{ \begin{pmatrix} b & tb \\ 0 & 0 \end{pmatrix} : b \in \mathcal{B} \right\} \subseteq B(\mathcal{H} \oplus \mathcal{H})$$

is an algebra with unit $\left(\begin{smallmatrix} I & tI \\ 0 & 0 \end{smallmatrix}\right)$ of norm $\sqrt{1+t^2}$. This algebra has a "super" contractive multiplication, since

$$\left\| \begin{pmatrix} b_1 & tb_1 \\ 0 & 0 \end{pmatrix} \begin{pmatrix} b_2 & tb_2 \\ 0 & 0 \end{pmatrix} \right\| \leq \frac{1}{\sqrt{1+t^2}} \left\| \begin{pmatrix} b_1 & tb_1 \\ 0 & 0 \end{pmatrix} \right\| \left\| \begin{pmatrix} b_2 & tb_2 \\ 0 & 0 \end{pmatrix} \right\|.$$

We now wish to turn our attention to Pisier's theory of factorization and similarity degree. This theory is a remarkable refinement of Corollary 18.11 to the study of when bounded homomorphisms are completely bounded.

We say that a unital operator algebra $\mathcal{A}$ has *factorization pair* (d, K) provided that for all m, n every $(a_{ij}) \in M_{m,n}(\mathcal{A})$ with $\|(a_{ij})\| < 1$ has a factorization

$$(a_{ij}) = C_1 D_1 C_2 \cdots C_d D_d C_{d+1}$$

where $C_1, \ldots, C_{d+1}$ are scalar matrices with $\|C_1\| \cdots \|C_{d+1}\| < K$ and $D_1, \ldots, D_d$ are diagonal matrices with entries from the unit ball of $\mathcal{A}$. We say that $\mathcal{A}$ has *factorization degree d* provided that d is the least integer such that $\mathcal{A}$ has factorization pair (d, K) for some constant K.

We say that $\mathcal{A}$ has a *similarity pair* (d, K) provided that every unital bounded homomorphism $\rho \colon \mathcal{A} \to B(\mathcal{H})$ is completely bounded with $\|\rho\|_{\text{cb}} \leq K\|\rho\|^d$. We say that $\mathcal{A}$ has *similarity degree d* provided that d is the least integer such that $\mathcal{A}$ has a similarity pair (d, K) for some constant K.

One connection between factorization pairs and similarity pairs is immediate.

Proposition 19.6. *Let $\mathcal{A}$ be a unital operator algebra. If $\mathcal{A}$ has factorization pair (d, K), then $\mathcal{A}$ has similarity pair (d, K).*

Proof. Let $\|(a_{ij})\| < 1$. Then the factorization of (a_{ij}) into $d+1$ scalar matrices and d diagonal matrices of norm less than or equal to 1 yields a factorization of $(\rho(a_{ij}))$ into the same scalar matrices and d diagonal matrices each of norm less than or equal to $\|\rho\|$. Hence $\|(\rho(a_{ij}))\| \leq K\|\rho\|^d$. □

We shall prove, conversely, that if $\mathcal{A}$ has similarity degree d, then $\mathcal{A}$ has factorization degree d.

The first step in this process is to realize the connection between the Haagerup tensor product of the operator space $\text{MAX}(\mathcal{A})$ with itself and the factorization degree. To this end, let $V_N = \text{MAX}(\mathcal{A}) \otimes_h \cdots \otimes_h \text{MAX}(\mathcal{A})$ denote the tensor product of N copies of $\text{MAX}(\mathcal{A})$ endowed with the Haagerup tensor norm. Also, let $\gamma_N \colon V_N \to \mathcal{A}$ be the completely contractive map induced by the product, so that $\gamma_N(a_1 \otimes \cdots \otimes a_N) = a_1 \cdots a_N$.

If K_N denotes the kernel of γ_N, then, since γ_N is onto, V_N/K_N is linearly isomorphic to $\mathcal{A}$. We let $\|\cdot\|_{(N)}$ denote the operator space structure induced on $\mathcal{A}$ by this identification.

Thus, for $(a_{ij}) \in M_{m,n}(\mathcal{A})$ we have that $\|(a_{ij})\|_{(1)} = \|(a_{ij})\|_{\mathrm{MAX}(\mathcal{A})}$ and $\|(a_{ij})\|_{(N)} < 1$ if and only if $(a_{ij}) = (\gamma_N(u_{ij}))$ for some $(u_{ij}) \in M_{m,n}(V_N)$ with $\|(u_{ij})\|_{V_N} < 1$. The following summarizes the properties of these norms.

Proposition 19.7. *Let $\mathcal{A}$ be a unital operator algebra, and let*

$$(a_{ij}) \in M_{m,n}(\mathcal{A}).$$

Then

(i) $\|(a_{ij})\|_{(N)} < 1$ *if and only if* (a_{ij}) *can be factored as*

$$(a_{ij}) = C_1 D_1 C_2 D_2 \cdots C_N D_N C_{N+1}$$

where $C_1, \ldots, C_{N+1}$ *are scalar matrices with* $\|C_i\| < 1$ *and* $D_1, \ldots, D_N$ *are diagonal matrices with entries from the unit ball of* $\mathcal{A}$,

(ii) *for all* N, $\|(a_{ij})\|_{(N+1)} \leq \|(a_{ij})\|_{(N)}$,

(iii) $\|(a_{ij})\|_{\mathrm{MAXA}(\mathcal{A})} = \lim_{N \to \infty} \|(a_{ij})\|_{(N)}$.

Proof. To prove (i), write $(a_{ij}) = (\gamma_N(u_{ij}))$ with $\|(u_{ij})\|_{V_N} < 1$. By the definition of the Haagerup tensor norm,

$$(u_{ij}) = \left(x_{ij}^1\right) \odot \cdots \odot \left(x_{ij}^N\right),$$

where $\|(x_{ij}^k)\|_{\mathrm{MAX}(\mathcal{A})} < 1$.

By Theorem 14.2, $(x_{ij}^k) = R_k D_k S_k$, where R_k and S_k are scalar matrices with $\|R_k\| < 1$, $\|S_k\| < 1$, and D_k is a diagonal matrix whose entries are from the unit ball of $\mathcal{A}$.

Since γ_N is the map that replaces the faux product $\odot$ by the actual product in $\mathcal{A}$, we have that

$$(a_{ij}) = R_1 D_1 S_1 R_2 D_2 S_2 \cdots R_N D_N S_N,$$

and the proof of (i) is completed by taking $C_1 = R_1, C_2 = S_1 R_2, \ldots, C_N = S_{N-1} R_N, C_{N+1} = S_N$.

Statement (ii) follows by noting that $\gamma_N(u_{ij}) = \gamma_{N+1}(u_{ij} \otimes e)$, where e denotes the identity of $\mathcal{A}$.

Statement (iii) follows from (i) and Theorem 18.9. □

The following is an immediate application of Proposition 19.7 and illustrates the importance of these norms.

Proposition 19.8. *Let $\mathcal{A}$ be a unital operator algebra. Then $\mathcal{A}$ has factorization pair (d, K) if and only if $\|(a_{ij})\|_{(d)} \leq K\|(a_{ij})\|$ for all m, n, and $(a_{ij}) \in M_{m,n}(\mathcal{A})$.*

We are now prepared to examine algebras for which every bounded homomorphism is completely bounded. Note that if this is the case, then for every $c > 1$ there must exist a constant $M = M_c$ such that if $\|\rho\| \le c$ then $\|\rho\|_{\text{cb}} \le M$. For if this were not true, then one could form the direct sum of a sequence of homomorphisms ρ_n with $\|\rho_n\| \le c$ and $\|\rho_n\|_{\text{cb}} \ge n$ to obtain a single homomorphism with $\|\rho\| \le c$ but $\|\rho\|_{\text{cb}} = +\infty$.

Proposition 19.9. *Let $\mathcal{A}$ be a unital operator algebra, let $c > 1$, and assume that for every unital homomorphism $\rho\colon \mathcal{A} \to B(\mathcal{H})$ with $\|\rho\| \le c$ we have $\|\rho\|_{\text{cb}} \le M$. Then for every m, n and every $A = (a_{ij}) \in M_{m,n}(\mathcal{A})$ with $\|A\| < \frac{1}{M}$, there exists an integer ℓ and elements $Y_k \in M_{m,n}(\mathcal{A})$, $0 \le k \le \ell$, such that*

(i) $A = Y_0 + Y_1 + \cdots + Y_\ell$,
(ii) Y_0 *is a scalar matrix times the identity of* $\mathcal{A}$ *with* $\|Y_0\| < 1$,
(iii) $\|Y_k\|_{(k)} < c^{-k}$ *for* $1 \le k \le \ell$.

Proof. Let $\text{MAX}(\mathcal{A})_c$ denote the operator space $\text{MAX}(\mathcal{A})$ but with

$$\|(x_{ij})\|_{\text{MAX}(\mathcal{A})_c} = c\|(x_{ij})\|_{\text{MAX}(\mathcal{A})}.$$

The identity map on $\mathcal{A}$ induces a completely contractive homomorphism $\pi\colon \text{OA}_1(\text{MAX}(\mathcal{A})_c) \to \mathcal{A}$. Let J denote the kernel of this homomorphism, and let $\text{OA}_1(\text{MAX}(\mathcal{A})_c)/J$ be the quotient operator algebra, so that

$$\dot{\pi}\colon \text{OA}_1(\text{MAX}(\mathcal{A})_c)/J \to \mathcal{A}$$

is one-to-one and onto. Let

$$\rho = \dot{\pi}^{-1}\colon \mathcal{A} \to \text{OA}_1(\text{MAX}(\mathcal{A})_c)/J$$

be the inverse of this map. Since $\|\rho(a)\| = \|a + J\| \le \|a\|_c = c\|a\|$, we have that $\|\rho\| = c$.

Thus, since $\text{OA}_1(\text{MAX}(\mathcal{A})_c)/J$ is an abstract unital operator algebra, it must be the case that $\|\rho\|_{\text{cb}} \le M$. Consequently, if $\|(a_{ij})\| < \frac{1}{M}$ in $M_{m,n}(\mathcal{A})$, then there exists an element $X = (x_{ij})$ in the open unit ball of $\text{OA}_1(\text{MAX}(\mathcal{A})_c)$ with $\pi(x_{ij}) = a_{ij}$. Decomposing $X = X_0 + \cdots + X_\ell$ as a sum of homogeneous terms, $X_k \in M_{m,n}(\text{MAX}(\mathcal{A})_c^{\otimes k})$, we have by Proposition 19.2 that $\|X_k\|_{\text{OA}_1(\text{MAX}(\mathcal{A})_c)} \le \|X\|_{\text{OA}_1(\text{MAX}(\mathcal{A})_c)} < 1$.

Applying Proposition 19.3, we have that the norm of the element X_k in OA_1 $(\text{MAX}(\mathcal{A})_c)$ is the same as the norm of X_k in the k-fold Haagerup tensor product of $\text{MAX}(\mathcal{A})_c$ with itself. Since each factor is multiplied by c we see that $\|X_k\|_{V_k} = c^{-k}\|X_k\|_{\text{OA}_1(\text{MAX}(\mathcal{A})_c)} < c^{-k}$. Thus, if we let Y_k be the image in $M_{m,n}(\mathcal{A})$ of X_k under the product map, then by definition $\|Y_k\|_{(k)} < c^{-k}$ for $1 \le k \le \ell$, $A = Y_0 + Y_1 + \cdots + Y_\ell$, and Y_0 is as claimed. □

We are now in a position to prove Pisier's main factorization theorem. The following elementary lemma is useful.

Lemma 19.10. *Let V be a vector space with norms $\|\cdot\|$ and $|||\cdot|||$. If there exists $r, 0 < r < 1$, such that every $v \in V$ decomposes as $v = v_1 + v_2$ with $|||v_1||| \leq a\|v\|$, $\|v_2\| \leq r\|v\|$, then $|||v||| \leq \frac{a}{1-r}\|v\|$.*

Proof. Exercise 19.5. □

Theorem 19.11 (Pisier). *Let $\mathcal{A}$ be a unital operator algebra and let $c > 1$. If every unital homomorphism $\rho\colon \mathcal{A} \to B(\mathcal{H})$ with $\|\rho\| \leq c$ satisfies $\|\rho\|_{\text{cb}} \leq M$, then $\mathcal{A}$ has factorization pair*

$$\left(d, \frac{M(c^{d+1} - 1)}{c^{d+1} - c^d - M}\right)$$

for any d satisfying $M < c^d(c - 1)$.

Proof. Let $A \in M_{m,n}(\mathcal{A})$ with $\|A\| < 1/M$. Write $A = Y_0 + Y_1 + \cdots + Y_\ell$ as in Proposition 19.9. Then

$$\begin{aligned}\|Y_0 + \cdots + Y_d\|_{(d)} &\leq \|Y_0\|_{(d)} + \|Y_1\|_{(d)} + \cdots + \|Y_d\|_{(d)} \\ &\leq \|Y_0\| + \|Y_1\|_{(1)} + \cdots + \|Y_d\|_{(d)} \\ &\leq 1 + c^{-1} + \cdots + c^{-d},\end{aligned}$$

while

$$\|Y_{d+1} + \cdots + Y_\ell\| \leq c^{-(d+1)} + \cdots + c^{-\ell} \leq \frac{c^{-(d+1)}}{1 - c^{-1}} = \frac{c^{-d}}{c - 1}$$

by Proposition 19.7(ii).

Thus the hypotheses of Lemma 19.10 are met for $|||\cdot||| = \|\cdot\|_{(d)}$ with $a = M(1 + c^{-1} + \cdots + c^{-d})$, $r = Mc^{-d}/(c - 1)$. Hence, $\|A\|_{(d)} \leq K\|A\|$ with $K = \frac{a}{1-r}$, for every m, n, and $A \in M_{m,n}(\mathcal{A})$. Thus, by Proposition 19.8, $\mathcal{A}$ has factorization pair (d, K). □

Corollary 19.12 (Pisier). *Let $\mathcal{A}$ be a unital operator algebra. Then $\mathcal{A}$ has factorization degree d if and only if $\mathcal{A}$ has similarity degree d.*

Proof. We have already seen that if $\mathcal{A}$ has factorization pair (d, K) then $\mathcal{A}$ has similarity pair (d, K).

So assume that $\mathcal{A}$ has similarity pair (d, K). Then for any $c > 1$, $\|\rho\| \leq c$ implies $\|\rho\|_{\text{cb}} \leq Kc^d = M$. To verify that $\mathcal{A}$ has factorization degree d_1 we need $Kc^d < c^{d_1}(c-1)$; but c is arbitrary, and taking $c > K+1$ allows us to choose $d_1 = d$. □

It is interesting to note that given a similarity pair (d, K), the value of the constant K_1 that one obtains for the factorization pair (d, K_1) using the above results is $K_1 = K(c^{d+1}-1)/[c-(K+1)]$, which is always larger than K. It is not known if similarity pair (d, K) implies factorization pair (d, K).

We now return to the topic of Kadison's conjecture, mentioned in Chapter 9. Kadison conjectured that every bounded homomorphism of a C^*-algebra into the algebra of bounded operators on a Hilbert space is similar to a $*$-homomorphism. By the results in Chapter 9 we see that this is equivalent to requiring that every bounded homomorphism of a C^*-algebra be completely bounded.

The above results lead to the following conclusions.

Corollary 19.13 (Pisier). *The following are equivalent:*

(i) *the answer to Kadison's conjecture is affirmative,*
(ii) *every bounded homomorphism of a C^*-algebra into an operator algebra is completely bounded,*
(iii) *for each $c > 1$ there exists $M > 1$ such that if π is a bounded homomorphism from a C^*-algebra into an operator algebra with $\|\pi\| \leq c$, then $\|\pi\|_{cb} \leq M$,*
(iv) *there exists $c_1 > 1$ and $M_1 > 1$ such that if π is a bounded homomorphism from a C^*-algebra into an operator algebra with $\|\pi\| \leq c_1$, then $\|\pi\|_{cb} \leq M_1$,*
(v) *there exists an integer d and constant K such that (d, K) is a similarity pair for every C^*-algebra,*
(vi) *there exists an integer d such that every C^*-algebra has similarity degree less than or equal to d,*
(vii) *there exists an integer d and constant K_1 such that (d, K_1) is a factorization pair for every C^*-algebra,*
(viii) *there exists an integer d such that every C^*-algebra has factorization degree less than or equal to d.*

Proof. We leave the proof as Exercise 19.6. □

Notes

Our presentation of similarity degree and factorization degree follows Pisier's ([184], [188], and [193]) fairly closely. Equivalent versions of many of our propositions can be found there. Although the concepts of similarity pairs and factorization pairs are not mentioned explicitly, they play a role in many of his proofs.

Theorem 19.5 is due to Blecher [21], but the proof that we present here is closer to Pisier's [193]. Blecher's original proof introduced and used the concept of a *quantum variable.* Theorem 19.4 seems to be new, although its proof is essentially the same as that of Theorem 19.5.

The factorization appearing in Theorem 19.1 does not seem to have been noticed before.

Further results on the theory of factorization degree and similarity degree for operator algebras and groups can be found in [188], [189], [190], and [193].

Exercises

19.1 Prove that $(\mathcal{F}(V), \|\cdot\|_f)$ of Theorem 19.1 satisfies the BRS axioms.

19.2 Let $\mathcal{A}$ be an L^∞-matrix-normed algebra such that $\|(a_{ij})(b_{ij})\| \leq M\|(a_{ij})\|\|(b_{ij})\|$. Let $\mathcal{A}_1 = \mathcal{A} \oplus \mathbb{C}$ denote the algebra obtained by adjoining a unit e to $\mathcal{A}$. Show that $\mathcal{A}_1$ can be given an L^∞-matrix-norm structure such that $\|e\| = 1$ and for $(a_{ij}), (b_{ij})$ in $M_n(\mathcal{A}_1)$, $\|(a_{ij})(b_{ij})\| \leq (M+2)\|(a_{ij})\|\|(b_{ij})\|$.

19.3 Let $\mathcal{A}$ be an L^∞-matrix-normed algebra, let $\mathcal{B}$ be an operator algebra, and let $\varphi\colon \mathcal{A} \to \mathcal{B}$ be a completely bounded algebra isomorphism with completely bounded inverse. Prove that for $(a_{ij}), (b_{ij})$ in $M_n(\mathcal{A})$, one has $\|(a_{ij})(b_{ij})\| \leq \|\varphi^{-1}\|_{\mathrm{cb}}\|\varphi\|_{\mathrm{cb}}^2\|(a_{ij})\|\|(b_{ij})\|$.

19.4 Let $I_0 = \{f \in A(\mathbb{D})\colon f(0) = 0\}$, let $0 < r < 1$, and let $I_{0,r} = \{f \in A(\mathbb{D})\colon f(0) = f(r) = 0\}$. Prove that $I_0/I_{0,r}$ is an L^∞-matrix-normed algebra with a completely contractive multiplication and an identity $e = r^{-1}z + I_{0,r}$. Show that $\|e\| = r^{-1}$ and that $\dim(I_0/I_{0,r}) = 1$. (Thus, $I_0/I_{0,r}$ is a one-dimensional algebra only completely isomorphic to $\mathbb{C}$.)

19.5 Prove Lemma 19.10.

19.6 Prove Corollary 19.13.

Bibliography

[1] J. Agler, Rational dilation on an annulus, *Ann. of Math.* **121** (1985), 537–564.

[2] J. Agler, On the representation of certain holomorphic functions defined on a polydisc, in *Topics in Operator Theory: Ernst D. Hellinger Memorial Volume*, edited by L. de Branges, I. Gohberg, and J. Rovnyak, Operator Theory and Applications, Vol. 48, Birkhäuser-Verlag, Basel, 1990, 47–66.

[3] J. Agler, Some interpolation theorems of Nevanlinna–Pick type, *J. Operator Theory*, to appear.

[4] A.B. Aleksandrov and V. Peller, Hankel operators and similarity to a contraction, *Internat. Math. Res. Notices* (1996), 263–275.

[5] T. Ando, On a pair of commutative contractions, *Acta Sci. Math.* **24** (1963), 88–90.

[6] W.B. Arveson, Subalgebras of C^*-algebras, *Acta Math.* **123** (1969), 141–224.

[7] W.B. Arveson, Subalgebras of C^*-algebras II, *Acta Math.* **128** (1972), 271–308.

[8] W.B. Arveson, *An Introduction to C^*-Algebras*, Springer-Verlag, New York, 1976.

[9] W.B. Arveson, Notes on extensions of C^*-algebras, *Duke Math. J.* **44** (1977), 329–355.

[10] D. Avitzour, Free products of C^*-algebras, *Trans. Amer. Math. Soc.* **271** (1982), 423–436.

[11] C. Badea and V. Paulsen, Schur multipliers and operator-valued Foguel–Hankel operators, *Indiana Univ. Math. J.* **50** (2001), 1509–1522.

[12] B.A. Barnes, The similarity problem for representations of a B^*-algebra, *Michigan Math. J.* **22** (1975), 25–32.

[13] B.A. Barnes, When is a representation of a Banach algebra Naimark related to a $*$-representation?, *Pacific J. Math.* **72** (1977), 5–25.

[14] G. Bennett, Schur multipliers, *Duke Math. J.* **44** (1977), 603–639.

[15] C.A. Berger, A strange dilation theorem, Abstract 625–152, *Notices Amer. Math. Soc.* **12** (1965), 590.

[16] C.A. Berger and J.G. Stampfli, Norm relations and skew dilations, *Acta Sci. Math.* **28** (1967), 191–195.

[17] C.A. Berger and J.G. Stampfli, Mapping theorems for the numerical range, *Amer. J. Math.* **89** (1967), 1047–1055.

[18] D.P. Blecher, Tensor products of operator spaces II, *Canad. J. Math.* **44** (1992), 75–90.

[19] D.P. Blecher, The standard dual of an operator space, *Pacific J. Math.* **153** (1992), 15–30.

[20] D.P. Blecher, Generalizing Grothendieck's program, in *Function Spaces*, edited by K. Jarosz, Lecture Notes in Pure and Applied Math., Vol. 136, Marcel Dekker, 1992.
[21] D.P. Blecher, A completely bounded characterization of operator algebras, *Math. Ann.* **303** (1995), 227–240.
[22] D.P. Blecher, Modules over operator algebras, and maximal C^*-dilation, *J. Funct. Anal.* **169** (1999), 251–288.
[23] D.P. Blecher, The Shilov boundary of an operator space and the characterization theorems, *J. Funct. Anal.* **182** (2001), 280–343.
[24] D.P. Blecher, Multipliers and dual operator algebras, *J. Funct. Anal.* **183** (2001), 498–525.
[25] D.P. Blecher, E.G. Effros, and V. Zarikian, One-sided M-ideals and multipliers in operator spaces, I, *Pacific J. Math.*, to appear.
[26] D.P. Blecher and C. Le Merdy, On quotients of function algebras, and operator algebra structures on ℓ_p, *J. Operator Theory* **34** (1995), 315–346.
[27] D.P. Blecher, P.S. Muhly, and V.I. Paulsen, Categories of operator modules–Morita equivalence and projective modules, *Mem. Amer. Math. Soc.* **143**, No. 681 (2000).
[28] D.P. Blecher and V.I. Paulsen, Tensor products of operator spaces, *J. Funct. Anal.* **99** (1991), 262–292.
[29] D.P. Blecher and V.I. Paulsen, Explicit constructions of universal operator algebras and applications to polynomial factorization, *Proc. Amer. Math. Soc.* **112** (1991), 839–850.
[30] D.P. Blecher and V.I. Paulsen, Multipliers of operator spaces and the injective envelope, *Pacific J. Math.* **200** (2001), 1–17.
[31] D.P. Blecher, Z. J. Ruan, and A.M. Sinclair, A characterization of operator algebras, *J. Funct. Anal.* **89** (1990), 188–201.
[32] D.P. Blecher and R.R. Smith, The dual of the Haagerup tensor product, *J. London Math. Soc.* **45** (1992), 126–144.
[33] J. Bourgain, On the similarity problem for polynomially bounded operators on Hilbert space, *Israel J. Math.* **54** (1986), 227–241.
[34] A. Brown and C. Pearcy, *Introduction to Operator Theory I: Elements of Functional Analysis*, Springer-Verlag, New York, 1977.
[35] J.W. Bunce, Representations of strongly amenable C^*-algebras, *Proc. Amer. Math. Soc.* **32** (1972), 241–246.
[36] J.W. Bunce, Approximating maps and a Stone–Weierstrass theorem for C^*-algebras, *Proc. Amer. Math. Soc.* **79** (1980), 559–563.
[37] J.W. Bunce, The similarity problem for representations of C^*-algebras, *Proc. Amer. Math. Soc.* **81** (1981), 409–413.
[38] J.F. Carlson, and D.N. Clark, Projectivity and extensions of Hilbert modules over $A(\mathbb{D}^N)$, *Michigan Math. J.* **44** (1997), 365–373.
[39] J.F. Carlson and D.N. Clark, Cohomology and extensions of Hilbert modules, *J. Funct. Anal.* **128** (1995), 278–306.
[40] J. Carlson, D.N. Clark, C. Foias, and J. Williams, Projective Hilbert $A(\mathbb{D})$-modules, *New York J. Math.* **1** (1994), 26–38, electronic.
[41] M.D. Choi, Positive linear maps on C^*-algebras, *Canad. J. Math.* **24** (1972), 520–529.
[42] M.D. Choi, A Schwarz inequality for positive linear maps on C^*-algebras, *Illinois J. Math.* **18** (1974), 565–574.
[43] M.D. Choi, Completely positive linear maps on complex matrices, *Lin. Alg. and Appl.* **10** (1975), 285–290.

[44] M.D. Choi, A simple C*-algebra generated by two finite order unitaries, *Canad. J. Math.* **31** (1979), 887–890.

[45] M.D. Choi, Some assorted inequalities for positive linear maps on C^*-algebras, *J. Operator Theory* **4** (1980), 271–285.

[46] M.D. Choi, Positive linear maps, in *Operator Algebras and Applications*, edited by R.V. Kadison, *Proceedings of Symposia in Pure Mathematics*, Vol. 38, AMS, Providence, 1982.

[47] M.D. Choi and E.G. Effros, The completely positive lifting problem for C^*-algebras, *Ann. of Math.* **104** (1976), 585–609.

[48] M.D. Choi and E.G. Effros, Separable nuclear C^*-algebras and injectivity, *Duke Math J.* **43** (1976), 309–322.

[49] M.D. Choi and E.G. Effros, Injectivity and operator spaces, *J. Functional Anal.* **24** (1977), 156–209.

[50] M.D. Choi and E.G. Effros, Nuclear C^*-algebras and injectivity. The general case, *Indiana Univ. Math. J.* **26** (1977), 443–446.

[51] M.D. Choi and E.G. Effros, Nuclear C^*-algebras and the approximation property, *Amer. J. Math.* **100** (1978), 61–79.

[52] E. Christensen, Perturbation of operator algebras II, *Indiana Univ. Math. J.* **26** (1977), 891–904.

[53] E. Christensen, Extensions of derivations, *J. Funct. Anal.* **27** (1978), 234–247.

[54] E. Christensen, Extensions of derivations II, *Math. Scand.* **50** (1982), 111–122.

[55] E. Christensen, On non-selfadjoint representations of operator algebras, *Amer. J. Math.* **103** (1981), 817–834.

[56] E. Christensen, Similarities of II_1 factors with property Γ, *J. Operator Theory* **15** (1986), 281–288.

[57] E. Christensen, E.G. Effros, and A.M. Sinclair, Completely bounded multilinear maps and C^*-algebraic cohomology, *Invent. Math.* **90** (1987), 279–296.

[58] E. Christensen and A.M. Sinclair, Representations of completely bounded multilinear operators, *J. Funct. Anal.* **72** (1987), 151–181.

[59] E. Christensen and A.M. Sinclair, A survey of completely bounded operators, *Bull. London Math. Soc.* **21** (1989), 417–448.

[60] E. Christensen and A.M. Sinclair, On von Neumann algebras which are complemented subspaces of $B(H)$, *J. Funct. Anal.* **122** (1994), 91–102.

[61] E. Christensen and A.M. Sinclair, Module mappings into von Neumann algebras and injectivity, *Proc. London Math. Soc. (3)* **71** (1995), 618–640.

[62] E. Christensen and A.M. Sinclair, Completely bounded isomorphisms of injective von Neumann algebras, *Proc. Edinburgh Math. Soc.* **32** (1989), 317–327.

[63] P. Chu, Finite dimensional representations of function algebras, Ph.D. Thesis, University of Houston, 1992.

[64] D.N. Clark and S. Ferguson, Submodules of L^2 of the N-torus, *Oper. Theory Adv. Appl.* **115** (2000), 113–122.

[65] W. Cohn, S. Ferguson, and R. Rochberg, Boundedness of higher Hankel forms, factorization in potential spaces and derivations, *Proc. London Math. Soc. (3)* **82** (2001), 110–130.

[66] B. Cole and J. Wermer, Ando's theorem and sums of squares, *Indiana Univ. Math. J.* **48** (1999), 767–791.

[67] A. Connes, Classification of injective factors, *Ann. of Math.* **104** (1976), 585–609.

[68] J.B. Conway, *Subnormal Operators*, Pitman, Boston, 1981.

[69] M.J. Crabb and A.M. Davie, von Neumann's inequality for Hilbert space operators, *Bull. London Math. Soc.* **7** (1975), 49–50.

[70] A.T. Dash, Joint spectral sets, *Rev. Roumaine Math. Pures et Appl.* **16** (1971), 13–26.

[71] K.R. Davidson and V.I. Paulsen, On polynomially bounded operators, *J. Reine u. Angew. Math.* **487** (1997), 153–170.

[72] J. Dixmier, Les moyennes invariante dans les semi-groupes et leurs applications, *Acta Sci. Math. Szeged* **12** (1950), 213–227.

[73] J. Dixmier, *C^*-Algebras*, North-Holland, New York, 1977.

[74] P.G. Dixon, Banach algebras satisfying the non-unital von Neumann inequality, *Bull. London Math. Soc.* **27** (1995), 359–362.

[75] P.G. Dixon, The von Neumann inequality for polynomials of degree greater than two, *J. London Math. Soc. (2)* **14** (1976), 369–375.

[76] R.G. Douglas, *Banach Algebra Techniques in Operator Theory*, Academic Press, New York, 1972.

[77] R.G. Douglas and V.I. Paulsen, Completely bounded maps and hypo-Dirichlet algebras, *Acta Sci. Math.* **50** (1986), 143–157.

[78] N. Dunford and J.T. Schwartz, *Linear Operators I; General Theory*, Interscience, New York, 1958.

[79] E.G. Effros, Aspects of non-commutative order, Notes for a lecture given at the Second US–Japan Seminar on C^*-Algebras and Applications to Physics (April 1977).

[80] E.G. Effros, Advances in quantized functional analysis, *Proceedings of the International Congress of Mathematicians*, Berkeley, 1986, 906–916.

[81] E.G. Effros and A. Kishimoto, Module maps and Hochschild–Johnson cohomology, *Indiana Univ. Math. J.* **36** (1987), 257–276.

[82] E.G. Effros and Z.J. Ruan, On matricially normed spaces, *Pacific J. Math.* **132** (1988), 243–264.

[83] E.G. Effros and Z.J. Ruan, A new approach to operator spaces, *Canad. Math. Bull.* **34** (1991), 329–337.

[84] E.G. Effros and Z.J. Ruan, On the abstract characterization of operator spaces, *Proc. Amer. Math. Soc.* **119** (1993), 579–584.

[85] E.G. Effros and Z.J. Ruan, Self duality for the Haagerup tensor product and Hilbert space factorization, *J. Funct. Anal.* **100** (1991), 257–284.

[86] E.G. Effros and Z.J. Ruan, Recent development in operator spaces, in *Current Topics in Operator Algebras, Proceedings of the ICM-90 Satellite Conference Held in Nara (August 1990)*, World Scientific, River Edge, N.J., 1991, 146–164.

[87] E.G. Effros and Z.J. Ruan, Representations of operator bimodules and their applications, *J. Operator Theory* **19** (1988), 137–157.

[88] E.G. Effros and Z.J. Ruan, *Operator Spaces*, Oxford Univ. Press, Oxford, 2000.

[89] L. Fejer, Über trigonometrische Polynome, *J. für Math.* **146** (1915), 53–82.

[90] S. Ferguson, Polynomially bounded operators and Ext groups, *Proc. Amer. Math. Soc.* **124** (1996), No. 9, 2779–2785.

[91] S. Ferguson, Backward shift invariant operator ranges, *J. Funct. Anal.* **150** (1997), 526–543.

[92] S. Ferguson, The Nehari problem for the Hardy space of the torus, *J. Operator Theory* **40** (1998), 309–321.

[93] S. Ferguson, Higher-order Hankel forms and cohomology groups, *Oper. Theory Adv. Appl.*, to appear.

[94] S. Ferguson and M. Lacey, Weak factorization for H^1 of the bidisk, preprint.
[95] S. Ferguson and R. Rochberg, Higher-order Hilbert–Schmidt Hankel forms on the ball, preprint.
[96] S. Ferguson and C. Sadosky, Characterizations of bounded mean oscillation on the polydisk in terms of Hankel operators and Carleson measures (with C. Sadosky), *J. Anal. Math.* **81** (2000), 239–267.
[97] P.A. Fillmore, *Notes on Operator Theory*, Van Nostrand-Reinhold, New York, 1970.
[98] S.R. Foguel, A counterexample to a problem of Sz.-Nagy, *Proc. Amer. Math. Soc.* **15** (1964), 788–790.
[99] C. Foias, Sur certains théorèmes de J. von Neumann, concernant les ensembles spectreaux, *Acta Sci. Math.* **18** (1957), 15–20.
[100] C. Foias and A.E. Frazho, *The Commutant Lifting Approach to Interpolation Problems*, Operator Theory: Advances and Applications, Vol. 44, Birkhäuser, Basel, 1990.
[101] C. Foias and J.P. Williams, On a class of polynomially bounded operators, unpublished preprint.
[102] M. Frank and V.I. Paulsen, Injective envelopes of C^*-algebras as operator modules, *Pacific J. Math.*, to appear.
[103] T. Gamelin, *Uniform Algebras*, Prentice-Hall, Englewood Cliffs, N.J., 1969.
[104] D. Gaspar and A. Racz, An extension of a theorem of T. Ando, *Michigan Math. J.* **16** (1969), 377–380.
[105] I.C. Gohberg and M.G. Krein, *Introduction to the Theory of Linear Nonselfadjoint Operators*, Translations of Mathematical Monographs, Vol. 18, AMS, Providence, 1969.
[106] U. Haagerup, Solution of the similarity problem for cyclic representations of C^*-algebras, *Ann. of Math.* **118** (1983), 215–240.
[107] U. Haagerup, Decomposition of completely bounded maps on operator algebras, unpublished preprint.
[108] U. Haagerup, Injectivity and decomposition of completely bounded maps, in *Operator Algebras and Their Connection with Topology and Ergodic Theory*, Springer Lecture Notes in Mathematics, Vol. 1132, 1985, 170–222.
[109] D.W. Hadwin, Dilations and Hahn decompositions for linear maps, *Canad. J. Math.* **33** (1981), 826–839.
[110] P.R. Halmos, On Foguel's answer to Nagy's question, *Proc. Amer. Math. Soc.* **15** (1964), 791–793.
[111] P.R. Halmos, Ten problems in Hilbert space, *Bull. Amer. Math. Soc.* **76** (1970), 887–933.
[112] M. Hamana, Injective envelopes of operator systems, *Publ. RIMS Kyoto Univ.* **15** (1979), 773–785.
[113] M. Hamana, Injective envelopes of dynamical systems, in *Operator Algebras and Operator Theory*, Pitman Research Notes 271, Longman, 69–77.
[114] E. Heinz, Ein v. Neumannscher Satz über beschrankte Operatoren im Hilbertschcn Raum, *Göttinger Nachr.* (1952), 5–6.
[115] D.A. Herrero, A Rota universal model for operators with multiply connected spectrum, *Rev. Roum. Math. Pures et Appl.* **21** (1976), 15–23.
[116] D.A. Herrero, *Approximation of Hilbert Space Operators*, Pitman, Boston, 1982.
[117] J.A.R. Holbrook, On the power bounded operators of Sz.-Nagy and Foias, *Acta Sci. Math.* **29** (1968), 299–310.
[118] J.A.R. Holbrook, Multiplicative properties of the numerical radius in operator theory, *J. Reine Angew. Math.* **237** (1969), 166–174.

[119] J.A.R. Holbrook, Inequalities governing the operator radii associated with unitary p-dilations, *Michigan Math. J.* **18** (1971), 149–159.
[120] J.A.R. Holbrook, Spectral dilations and polynomially bounded operators, *Indiana Univ. Math. J.* **20** (1971), 1027–1034.
[121] J.A.R. Holbrook, Distortion coefficients for crypto-contractions, *Lin. Alg. and Appl.* **18** (1977), 229–256.
[122] J.A.R. Holbrook, Distortion coefficients for crypto-unitary operators, *Lin. Alg. and Appl.* **19** (1978), 189–205.
[123] T. Huruya, On compact completely bounded maps of C^*-algebras, *Michigan Math. J.* **30** (1983), 213–220.
[124] T. Huruya, Linear maps between certain non-separable C^*-algebras, *Proc. Amer. Math. Soc.* **92** (1984), 193–197.
[125] T. Huruya, Decompositions of linear maps into non-separable C^*-algebras, *Publ. RIMS, Kyoto Univ.* **21** (1985), 645–655.
[126] T. Huruya and J. Tomiyama, Completely bounded maps of C^*-algebras, *J. Operator Theory* **10** (1983), 141–152.
[127] M. Junge and G. Pisier, Bilinear forms on exact operator spaces and $B(H) \otimes B(H)$, *GAFA* **5** (1995), 329–363.
[128] R.V. Kadison, A generalized Schwarz inequality and algebraic invariants for C^*-algebras, *Ann. Math.* **56** (1952), 494–503.
[129] R.V. Kadison, On the orthogonalization of operator representations, *Amer. J. Math.* **77** (1955), 600–620.
[130] T. Kato, Some mapping theorems for the numerical range, *Proc. Japan Acad.* **41** (1965), 652–655.
[131] E. Kirchberg, C^*-nuclearity implies CPAP, *Math. Nachr.* **76** (1977), 203–212.
[132] E. Kirchberg, The derivation and the similarity problem are equivalent, *J. Operator Theory* **36** (1996), 59–62.
[133] R.A. Kunze and E.M. Stein, Uniformly bounded representations and harmonic analysis of the 2×2 real unimodular group, *Amer. J. Math.* **82** (1960), 1–62.
[134] C. Lance, On nuclear C^*-algebras, *J. Funct. Anal.* **12** (1973), 157–176.
[135] C. Lance, Tensor products and nuclear C^*-algebras, in *Operator Algebras and Applications*, edited by R.V. Kadison, Proceedings of Symposia in Pure Mathematics, Vol. 38, AMS, Providence, 1982.
[136] A. Lebow, On von Neumann's theory of spectral sets, *J. Math. Anal. and Appl.* **7** (1963), 64–90.
[137] C. LeMerdy, Representations of a quotient of a subalgebra of $B(X)$, *Math. Proc. Cambridge Phil. Soc.* **119** (1996), 83–90.
[138] R.I. Loebl, Contractive linear maps on C^*-algebras, *Michigan Math. J.* **22** (1975), 361–366.
[139] R.I. Loebl, A Hahn decomposition for linear maps, *Pacific J. Math.* **65** (1976), 119–133.
[140] B. Mathes, A completely bounded view of Hilbert–Schmidt operators, *Houston J. Math.* **17** (1991), 404–418.
[141] B. Mathes, Characterizations of row and column Hilbert space, *J. London Math. Soc. (2)* **50** (1994), 199–208.
[142] B. Mathes and V.I. Paulsen, Operator ideals and operator spaces, *Proc. Amer. Math. Soc.* **123** (1995), 1763–1772.
[143] M.J. McAsey and P.S. Muhly, Representations of non-self-adjoint crossed products, *Proc. London Math. Soc. (3)* **47** (1983), 128–144.
[144] W. Mlak, Unitary dilations of contraction operators, *Rozprawy Mat.* **46** (1965), 1–88.

[145] W. Mlak, Unitary dilations in case of ordered groups, *Ann. Polon. Math.* **17** (1966), 321–328.

[146] W. Mlak, Positive definite contraction valued functions, *Bull. Acad. Polon. Sci. Ser. Sci. Math. Astronom. Phys.* **15** (1967), 509–512.

[147] W. Mlak, Absolutely continuous operator valued representations of function algebras, *Bull. Acad. Polon. Sci. Ser. Sci. Math. Astronom. Phys.* **17** (1969), 547–550.

[148] W. Mlak, Decompositions and extensions of operator valued representations of function algebras, *Acta. Sci. Math. (Szeged)* **30** (1969), 181–193.

[149] W. Mlak, Decompositions of operator-valued representations of function algebras, *Studia Math.* **36** (1970), 111–123.

[150] M.A. Naimark, On a representation of additive operator set functions, *C. R. (Dokl.) Acad. Sci. URSS* **41** (1943), 359–361.

[151] M.A. Naimark, Positive definite operator functions on a commutative group, *Bull. (Izv.) Acad. Sci. URSS (Ser. Math.)* **7** (1943), 237–244.

[152] J. von Neumann, Eine Spektraltheorie für allgemeine Operatoren eines unitären Raumes, *Math. Nachr.* **4** (1951), 258–281.

[153] R. Nevanlinna, Ueber beschrankte Funktionen, die in gegebene Punkten vorgeschriebene Werte annehmen, *Ann. Acad. Sci. Fenn.*, No. 732 (1929).

[154] T. Oikhberg, Hilbertian operator spaces with few completely bounded maps, preprint.

[155] L. Page, Bounded and compact vectorial Hankel operators, *Trans. Amer. Math. Soc.* **150** (1970), 529–540.

[156] S.K. Parrott, Unitary dilations for commuting contractions, *Pacific J. Math.* **34** (1970), 481–490.

[157] S.K. Parrott, On a quotient norm and the Sz.-Nagy–Foias lifting theorem, *J. Funct. Anal.* **30** (1978), 311–328.

[158] V.I. Paulsen, Completely bounded maps on C^*-algebras and invariant operator ranges, *Proc. Amer. Math. Soc.* **86** (1982), 91–96.

[159] V.I. Paulsen, Every completely polynomially bounded operator is similar to a contraction, *J. Funct. Anal.* **55** (1984), 1–17.

[160] V.I. Paulsen, Completely bounded homomorphisms of operator algebras, *Proc. Amer. Math. Soc.* **92** (1984), 225–228.

[161] V.I. Paulsen, *Completely Bounded Maps and Dilations*, Pitman Research Notes in Mathematics Series, Vol. 146, Pitman Longman (Wiley), 1986.

[162] V.I. Paulsen, Three tensor norms for operator spaces, in *Mappings of Operator Algebras*, edited by H. Araki and R.V. Kadison, Proceedings of the Japan–US Joint Seminar in Operators, Univ. of Pennsylvania, 1988, Progress in Mathematics, Vol. 84, Birkhäuser, Boston, 1990, 251–264.

[163] V.I. Paulsen, Representations of function algebras, abstract operator spaces and Banach space geometry, *J. Funct. Anal.* **109** (1992), 113–129.

[164] V.I. Paulsen, The maximal operator space of a normed space, *Proc. Edinburgh Math. Soc.* **39** (1996), 309–323.

[165] V.I. Paulsen, Resolutions of Hilbert modules, *Rocky Mountain J. Math.* **27** (1997), 271–297.

[166] V.I. Paulsen, Operator algebras of idempotents, *J. Funct. Anal.* **181** (2001), 209–226.

[167] V.I. Paulsen, Matrix-valued interpolation and hyperconvex sets, *Integral Eqns. Operation Theory* **41** (2001), 38–62.

[168] V.I. Paulsen, G. Popescu, and D. Singh, On Bohr's inequality, *J. London Math. Soc.*, to appear.

[169] V.I. Paulsen and S.C. Power, Tensor products of non-self-adjoint operator algebras, *Rocky Mountain J. Math.* **20** (1990), 331–350.
[170] V.I. Paulsen and R.R. Smith, Multilinear maps and tensor norms on operator systems, *J. Funct. Anal.* **73** (1987), 258–276.
[171] V.I. Paulsen and C.Y. Suen, Commutant representations of completely bounded maps, *J. Operator Theory* **13** (1985), 87–101.
[172] C. Pearcy, An elementary proof of the power inequality for the numerical radius, *Michigan Math. J.* **13** (1966), 289–291.
[173] G.K. Pedersen, *C^*-Algebras and Their Automorphism Groups*, Academic Press, London, 1979.
[174] V. Peller, Estimates of functions of power bounded operators on Hilbert space, *J. Operator Theory* **7** (1982), 341–372.
[175] V. Peller, Estimates of functions of Hilbert space operators, similarity to a contraction and related function algebras, in *Linear and Complex Analysis Problem Book*, edited by V.P. Havin, S.V. Hruscev, and N.K. Nikolskii, Springer Lecture Notes 1043, 1994, 199–204.
[176] G. Pisier, The operator Hilbert space OH, complex interpolation and tensor norms, *Memoirs Amer. Math. Soc.* **122**, 585 (1996).
[177] G. Pisier, Non-commutative vector valued L_p-spaces and completely p-summing maps, *Astérisque* **247** (1998), 1–131.
[178] G. Pisier, Sur les opérateurs factorisables par OH, *Comptes Rendus Acad. Sci. Paris Sér. I* **316** (1993), 165–170.
[179] G. Pisier, *Factorization of Linear Operators and the Geometry of Banach Spaces*, CBMS (Regional Conferences of the AMS), Vol. 60, 1986; reprinted with corrections, 1987.
[180] G. Pisier, Completely bounded maps between sets of Banach space operators, *Indiana Univ. Math. J.* **39** (1990), 251–277.
[181] G. Pisier, Exact operator spaces, presented at Colloque sur les Algèbres d'Opérateurs, in *Recent Advances in Operator Algebras (Orleans, 1992)*, Soc. Math. France, *Astérisque* **232** (1995), 159–186.
[182] G. Pisier, Projections from a von Neumann algebra onto a subalgebra, *Bull. Soc. Math. France* **123** (1995), 139–153.
[183] G. Pisier, Dvoretzky's theorem for operator spaces and applications, *Houston J. Math.* **22** (1996), 399–416.
[184] G. Pisier, *Similarity Problems and Completely Bounded Maps*, 2nd, expanded edition, Springer-Verlag Lecture Notes, Vol. 1618, 2001.
[185] G. Pisier, A simple proof of a theorem of Kirchberg and related results on C^*-norms, *J. Operator Theory* **35** (1996), 317–335.
[186] G. Pisier, Remarks on complemented subspaces of von Neumann algebras, *Proc. Roy. Soc. Edinburgh* **121A** (1992), 1–4.
[187] G. Pisier, Quadratic forms in unitary operators, *Lin. Alg. and Appl.* **267** (1997), 125–137.
[188] G. Pisier, The similarity degree of an operator algebra, *St. Petersburg Math. J.* **10** (1999), 103–146.
[189] G. Pisier, The similarity degree of an operator algebra II, *Math. Z.* **234** (2000), 53–81.
[190] G. Pisier, Similarity problems and length, *Taiwanese J. Math.* **5** (2001), 1–17.
[191] G. Pisier, A polynomially bounded operator on Hilbert space which is not similar to a contraction, *J. Amer. Math. Soc.* **10** (1997), 351–369.
[192] G. Pisier, A simple proof of a theorem of Jean Bourgain, *Michigan Math. J.* **39** (1992), 475–484.

[193] G. Pisier, *An Introduction to the Theory of Operator Spaces*, Cambridge Univ. Press, to appear.
[194] G. Pisier, Multipliers of the Hardy space H^1 and power bounded operators, *Colloq. Math.* **88** (2001), 57–73.
[195] F. Pop, A.M. Sinclair, and R.R. Smith, Norming C^*-algebras by C^*-subalgebras, *J. Funct. Anal.* **175** (2000), 168–196.
[196] S. Pott, Standard models under polynomial positivity conditions, *J. Operator Theory* **41** (1999), 365–389.
[197] E. Ricard, Decompositions de H^1, multiplicateurs de Schur et espaces d'opérateurs, Ph.D. Thesis, Université Pierre et Marie Carie (Paris VI) 2001.
[198] F. Riesz and B. Sz.-Nagy, *Functional Analysis*, 2nd edition, New York, 1955.
[199] J.R. Ringrose, Automatic continuity of derivations of operator algebras, *J. London Math. Soc.* **5** (1972), 432–438.
[200] A.R. Robertson, Injective matricial Hilbert spaces, *Math. Proc. Cambridge Philos. Soc.* **110** (1991), 183–190.
[201] A.R. Robertson and S. Wassermann, Completely bounded isomorphisms of injective operator systems, *Bull. London Math. Soc.* **21** (1989), 285–290.
[202] G.C. Rota, On models for linear operators, *Comm. Pure Appl. Math.* **13** (1960), 468–472.
[203] Z.J. Ruan, Subspaces of C*-algebras, *J. Funct. Anal.* **76** (1988), 217–230.
[204] Z.J. Ruan, Injectivity and operator spaces, *Trans. Amer. Math.* **315** (1989), 89–104.
[205] W. Rudin, *Function Theory on Polydisks*, Benjamin, New York, 1969.
[206] B. Russo and H.A. Dye, A note on unitary operators in C^*-algebras, *Duke Math. J.* **33** (1966), 413–416.
[207] S. Sakai, *C^*-Algebras and W^*-Algebras*, Springer-Verlag, New York, 1971.
[208] D. Sarason, On spectral sets having connected complement, *Acta Sci. Math.* **26** (1965), 289–299.
[209] D. Sarason, Generalized interpolation in H^∞, *Trans. Amer. Math. Soc.* **127** (1969), 179–203.
[210] J.J. Schaffer, On unitary dilations of contractions, *Proc. Amer. Math. Soc.* **6** (1955), 322.
[211] A. Shields, Weighted shift-operators and analytic function theory, in *Topics in Operator Theory*, edited by C. Pearcy, AMS, Providence, 1974.
[212] A.M. Sinclair and R.R. Smith, *Hochschild Cohomology of von Neumann Algebras*, LMS Lecture Note Series, Cambridge Univ. Press, Cambridge, 1995.
[213] A.M. Sinclair and R.R. Smith, Hochschild cohomology for von Neumann algebras with Cartan subalgebras, *Amer. J. Math* **120** (1998), 1043–1057.
[214] R.R. Smith, Completely bounded maps between C^*-algebras, *J. London Math. Soc.* **27** (1983), 157–166.
[215] R.R. Smith, Private communication.
[216] R.R. Smith, Completely contractive factorization of C^*-algebras, *J. Funct. Anal.* **64** (1985), 330–337.
[217] R.R. Smith, Completely bounded module maps and the Haagerup tensor product, *J. Funct. Anal.* **102** (1991), 156–175.
[218] R.R. Smith and J.D. Ward, Matrix ranges for Hilbert space operators, *Amer. J. Math.* **102** (1980), 1041–1081.
[219] R.R. Smith and J.D. Ward, Locally isometric liftings from quotient C^*-algebras, *Duke Math. J.* **47** (1980), 621–631.
[220] R.R. Smith and D. Williams, The decomposition property for C^*-algebras, *J. Operator Theory* **16** (1986), 51–74.

[221] W.F. Stinespring, Positive functions on C^*-algebras, *Proc. Amer. Math. Soc.* **6** (1955), 211–216.
[222] E. Stormer, *Positive Linear Maps of C^*-Algebras*, Lecture Notes in Physics, Vol. 29, Springer-Verlag, Berlin, 1974, 85–106.
[223] E. Stormer, Extension of positive maps into $B(H)$, *J. Funct. Anal.* **66** (1986), 235–254.
[224] I. Suciu, Unitary dilations in case of partially ordered groups, *Bull. Acad. Polon. Sci. Ser. Math. Astr. Phys.* **15** (1967), 271–275.
[225] C.-Y. Suen, Completely bounded maps on C^*-algebras, *Proc. Amer. Math. Soc.* **93** (1985), 81–87.
[226] C.-Y. Suen, The unique representation of a self-adjoint bounded linear functional, *Proc. Amer. Math. Soc.* **95** (1985), 58–62.
[227] B. Sz.-Nagy, On uniformly bounded linear transformations in Hilbert space, *Acta Sci. Math. Szeged* **11** (1947), 152–157.
[228] B. Sz.-Nagy, Sur les contractions de l'espace de Hilbert, *Acta Sci. Math.* **15** (1953), 87–92.
[229] B. Sz.-Nagy, Completely continuous operators with uniformly bounded iterates, *Magyar Tud. Akad. Mat. Kutato Int. Kaze* **4** (1959), 89–93.
[230] B. Sz-Nagy, Products of operators of class C_ρ, *Rev. Roumaine Math. Pures Appl.* **13** (1968), 897–899.
[231] B. Sz.-Nagy and C. Foias, *Harmonic Analysis of Operators on Hilbert Space*, American Elsevier, New York, 1970.
[232] T. Takasaki and J. Tomiyama, On the geometry of positive maps in matrix algebras, *Math. Z.* **184** (1983), 101–108.
[233] M. Takesaki, *Theory of Operator Algebras I*, Springer-Verlag, Berlin, 1979.
[234] J. Tomiyama, On the transpose map of matrix algebras, *Proc. Amer. Math. Soc.* **88** (1983), 635–638.
[235] J. Tomiyama, Recent development of the theory of completely bounded maps between C^*-algebras, *Publ. Res. Inst. Math. Sci.* **19** (1983), 1283–1303.
[236] N.Th. Varopoulos, On an inequality of von Neumann and an application of the metric theory of tensor products to operator theory, *J. Funct. Anal.* **16** (1974), 83–100.
[237] N.Th. Varopoulos, A theorem on operator algebras, *Math. Scand.* **37** (1975), 173–182.
[238] D. Voiculescu, Norm-limits of algebraic operators, *Rev. Roumaine Math. Pures et Appl.* **19** (1974), 371–378.
[239] J.L. Walsh, The approximation of harmonic functions by harmonic polynomials and by harmonic rational functions, *Bull. Amer. Math. Soc.* **35** (1929), 499–544.
[240] M.E. Walter, Algebraic structures determined by 3 by 3 matrix geometry, preprint.
[241] J.P. Williams, Schwarz norms for operators, *Pacific J. Math* **24** (1968), 181–188.
[242] G. Wittstock, Ein operatorwertiger Hahn–Banach Satz, *J. Funct. Anal.* **40** (1981), 127–150.
[243] G. Wittstock, On matrix order and convexity, in *Functional Analysis: Surveys and Recent Results*, Math. Studies, Vol. 90, North-Holland, Amsterdam, 1984, 175–188.
[244] S.L. Woronowicz, Nonextendible positive maps, *Comm. Math. Phys.* **51** (1976), 243–282.
[245] S.L. Woronowicz, Positive maps of low dimensional matrix algebras, *Rep. Math. Phys.* **10** (1976), 165–183.

[246] M. Youngson, Completely contractive projections on C^*-algebras, *Quart. J. Math. Oxford* **34** (1983), 507–511.
[247] C. Zhang, Representations of operator spaces, *J. Operator Theory* **33** (1995), 327–351.
[248] C. Zhang, Completely bounded Banach–Mazur distance, *Proc. Edinburgh Math. Soc. (2)* **40** (1997), 247–260.
[249] C. Zhang, Representation and geometry of operator spaces, Ph.D. Thesis, Univ. of Houston, 1995.
[250] W. Werner, Small K-groups for operator systems, preprint.

Index